From reviews of *How to Make War*

"A remarkable book . . . which may play an important role in the debate that is just beginning over U.S. defense policy"   —*Forbes*

"Everything needed to make intelligent decisions about modern warfare, including how best to prepare for it"   —*The Detroit Free Press*

"Documented in detail as copious as the American public's fascination with war is fervent . . . A program for the coming events"
        —*The Washington Post*

"Some real thinking about what makes armed forces work or fail to work, and what all the data will mean in battle"—*Philadelphia Inquirer*

"A substantial tool for drawing independent evaluations about a nation's military might"   —*Marine Corps Gazette*

"Comprehensive and occasionally terrifying . . . Statistics and logistics, projections and prophecies, are all a part of Dunnigan's stock-in-trade."
        —*Booklist*

"His approach is neither pro-war nor anti-war—it is a balancing of facts and figures, precedents and trends, the human element and the mechanics."   —*Grand Rapids Press*

"Startling"   —*The Los Angeles Times*

HOW TO MAKE WAR

A Comprehensive Guide to Modern Warfare

JAMES F. DUNNIGAN

QUILL

NEW YORK 1983

Library of Congress Cataloging in Publication Data

Dunnigan, James F.
  How to make war.

  Bibliography: p.
  Includes index.
  1. Military art and science.   2. Arms and armor.
3. War.   I. Title.
U102.D835   1983   355   82-23065
ISBN 0-688-00780-5
ISBN 0-688-01975-7 (pbk.)

Printed in the United States of America

First Quill Edition

1 2 3 4 5 6 7 8 9 10

BOOK DESIGN BY BERNARD SCHLEIFER

To my parents, whose love of learning and hatred of weapons gave me the proper perspective. To all the professional military men I have known, for sharing their secrets with me and demonstrating how they can be pacifists, warriors and patriots without contradiction.

ACKNOWLEDGMENTS

THE FOLLOWING PEOPLE were invaluable in their assistance in reading and criticizing the manuscript: Ray Macedonia, Austin Bay, Doug MacCaskill, Jim Simon, Steve Patrick, Frank Dunnigan, Sterling Hart, Kathy Bay, Jay Jacobson, Dave Robertson, Ed Seeley, and numerous others who preferred to contribute anonymously or whose names were never made known to me.

Some of these people are military professionals. Some are civilian analysts of military affairs, while others are simply interested citizens. These groups represent a cross section of people who can use this book. Their comments, often quite extensive, led to many beneficial alterations in the manuscript.

CONTENTS

PART 4: THE HUMAN FACTORS

PART 5: SPECIAL WEAPONS

PART 6: WARFARE BY THE NUMBERS

PART 7: MOVING THE GOODS

PART 8: TOOLS OF THE TRADE

1

HOW TO BECOME AN EFFECTIVE ARMCHAIR GENERAL

WARFARE IS AN ever-present terror in our lives. Still clouded by obscurity and confused by myths, the process of warfare is unknown to most of us. The mass media help create and perpetuate these myths. And often the appointed experts are equally ill-informed.

When a war breaks out, these myths become apparent as distortions. There is no question about it, warfare is something to be avoided and prevented whenever possible. Operating from misunderstandings, leaders and citizens are much more likely to get involved in war. One of the constants of history is that a nation rarely goes to war until it has convinced itself that victory is attainable and worth the cost. In reality, warfare is never worth the cost. Those who start wars generally regret it. Those who avoid war often feel they have missed a golden opportunity to right some wrong. Real warfare is ugly, destructive and remembered fondly only by those who survived it without getting too close.

This book removes some of the obscurity and destroys a few of the myths.

The Principles of War

To understand how the military mind operates one should know the central "truths" most military commanders are taught. These principles of war have been distilled from man's long history of warfare. They reflect reality. Were they followed to the letter, there would probably be no war. The principles of war preach, above all, that you must know what you are doing and that you must not overextend yourself. These principles vary somewhat from nation to nation, but the following describes the more common and important ones.

Maintenance of the objective. This means choosing an objective and sticking with it. In warfare the commander often operates with very little information about what is going on. As the situation develops, there is a temptation to change objectives. This wastes time and energy. History has shown that the army that consistently pursues its original goal is most likely to succeed. An example is found in the Arab-Israeli wars. The Israelis ruthlessly maintained their objectives, ignoring temptations to surround bypassed Arab formations. This straightforward attitude always resulted in the destruction of far larger Arab forces. By contrast, the Egyptians, in 1973, changed their plan after crossing the Suez Canal. Instead of digging in to receive the Israeli counterattack, they launched attacks of their own. This resulted in heavy Egyptian losses, which set the stage for a successful Israeli crossing of the canal.

Economy of force. This is otherwise known as not overdoing it. No one ever has sufficient forces to accomplish everything. Economy of force dictates carefully parceling out forces for each phase of the operation. Most important is the maintenance of a large reserve. If nothing else, once all your committed forces get hopelessly tangled up, you will still have control of the reserve. Invariably the reserve snatches survival from the jaws of disaster. During World War II, the German Army maintained a reserve no matter how desperate the situation at the front. This habit alone may have prolonged the war by at least a year.

Security. It's not sheer bloody-mindedness that causes captured spies to be shot in wartime. Information can usually be calculated in lives saved or lost. If you know what the enemy is up to while concealing your own plans, your chances of success increase immensely. The crucial Battle of Midway in 1942 was won largely because the United States knew of Japanese plans, from having broken their codes, while the Japanese knew little of the U.S. forces' deployment.

Flexibility. This may seem a contradiction of the maintenance of the objective principle, but it isn't. Flexibility in planning, thought and action is otherwise known as common sense. Maintenance of the objective does not imply ignoring the obvious. If your orders are to take a ridge line, and you determine that the easiest way to do this would be to go around it and surround the enemy position, that's being flexible. If while moving around the ridge you discover that the enemy is abandoning his positions, you should send a force to occupy the ridge. That's being flexible and maintaining the objective at the same time.

Entropy. Entropy is not usually considered one of the principles of war, but it has been a constant throughout military history. In practice, entropy means that after an initial shock, the war will settle down to a steady grind. Once a war gets started, casualty and movement rates become predictable. In combat, personnel losses average a few percent a day per division. Against enemy opposition, even mechanized forces do not move farther than some 20 km a day. There are exceptions, and the exceptions may win battles. Over the course of an entire war, however, entropy takes over. Don't let flashy press reports fool you; exceptions tend to get published far more than day-to-day averages.

How to Find the Right Questions

Warfare, to put it bluntly, is just a job. There are techniques the successful practitioners must learn and tools they must master. As in any other profession, conditions change constantly. Practitioners must adapt to these changes by correctly answering the questions raised by changed conditions. But warfare, like testing flashbulbs, cannot be practiced. This makes it difficult to determine the important questions.

Below are some of the ones that are raised in this book.

What does war cost? Appalled by the size of this year's defense budget? With worldwide arms spending exceeding $700 billion, you have plenty of company. Moreover, as the chapter on the cost of war points out, while peacetime defense is expensive and detrimental to the economy, a war would bankrupt the United States. That chapter gives details on the American and Russian defense budgets. The chapters on combat operations rate the relative worth of the various weapons bought. The chapter on logistics gives more details on the matériel needed to carry on a war. Using the chapters on the cost of war, logistics and attrition, you can do your own calculations on the cost of a current or a future war. Although the cost of war is not often mentioned in the press, most governments are well aware of it. This cost is often a major element in the decision to wage war or to seek a less expensive means of achieving national goals. These chapters explain why most modern wars either are short or eventually bankrupt the participants. The Iran-Iraq War is a good example of a "war of bankruptcy."

Is the Russian fleet a growing menace? The section on naval operations explains that establishing naval supremacy isn't as easy as it looks in the newspapers, especially from the Russian perspective. These chapters put into perspective potential naval operations in any part of the world, under any circumstances. They demonstrate that as powerful as the Russian fleet appears, it is vastly outmatched by Western naval forces.

Is the threat of nuclear war increasing? The chapter on strategic nuclear weapons reveals a few surprises about what might happen: for example, ICBM's with nerve-gas warheads. What probably won't happen is the end of the world. The reasons? Primarily fear of massive use of the weapons and the unlikelihood that the weapons will actually work. Yet anything is possible. Read and study the details and decide for yourself. The chapter on tactical nuclear weapons points out a number of factors influencing weapons reliability and effects that are not normally published in the open press. Nuclear weapons may well be used in the future. Who will use them and how will they be used? Read these chapters and draw your own conclusions.

Who's on first in Afghanistan, the Middle East, Africa, Asia? The chapter on the armed forces of the world puts this topic into perspective. The information on each nation's armed forces indicates the potential resolution of such conflicts. Other chapters can be consulted to gain a more complete understanding of the possible outcomes. All countries have armed forces, but not all have an effective military organization. Except for the top ten military nations in the world, effective offensive warfare is not a realistic possibility. The most pressing danger is that more militarily competent countries will be drawn into a local squabble. With the information contained in this chapter you can quickly assess who might do what to whom.

Who gets hurt? In modern warfare few combatants are exposed to enemy fire and fewer still actually fight. They rarely even see an enemy soldier, except as a corpse or a prisoner. The sections on ground, naval and air combat demonstrate this in detail. These sections also add accurate detail to frequently misleading news accounts of combat.

How are current wars being fought? The chapters on various aspects of military operations give details not usually found in other sources. The chapters on the human factors are also crucial, as they are often ignored or misinterpreted.

What is all this talk about electronic warfare? A separate chapter lays it all out, step by step. The widespread introduction of electronics has profoundly changed the ways in which wars are fought. This electronic equipment has led to overconfidence, overspending and sometimes increased military effectiveness. More than anything else, electronics have led to uncertainty, as there is no practical experience with what these devices will actually do in a major war. It is important to understand the potential, limitations and current status of electronic weapons and equipment.

A Few Notes on Approach

A half-serious maxim among military historians contends that you can determine which army is more effective by looking at their uniforms. The best-dressed army is generally the least effective. A fresh coat of paint makes any weapon appear awesome. How, then, do we determine which weapon is better than the other? My solution is to combine historical trends and aggregation to produce a numerical evaluation of weapons and the units that use them. This approach works especially well if combined with a study of the trends in leadership and manpower quality. You may not agree with some of my evaluations, but at least you'll have a point from which to start your argument. This book is for those people who ask questions rather than simply accept the most obvious answer.

Because much of this book's subject matter is normally classified secret, or worse, information had to be obtained from whatever sources were open. Because of my long experience with this type of information gathering, I am confident that this is as accurate a picture of modern warfare as you are going to get. If you have access to classified information, you're probably not sure you can trust it anyway. But that's another story (see the chapter on intelligence). Any errors in fact or interpretation are my own.

The metric system has been used for the most part. Units of

distance are measured in kilometers. To convert to statute (British-American) miles, multiply kilometers (or km) by 1.6. To convert to nautical miles, multiply by 1.8. Weights are in metric tons (2240 pounds). Often nonmetric measures are used to enable British or American readers to grasp scale better.

Things Russian are called Russian throughout. Purists will insist that the proper term is "Soviet." The purists are correct. I thought it easier to call Russians Russians, for they have been around far longer than the Soviet Union. Russians consider themselves Russian. Russia's neighbors see no difference between czar and commissar. The Soviet Union may not last, but Russia will endure.

There are two major military systems in the world today, the "Western" and the "Russian." The United States and Western Europe produce the vast majority of the equipment used by Western nations. Those same nations tend also to use Western unit organizations, tactics and doctrine. The same applies to countries that obtain most of their equipment from Russia. Therefore, much reference is made to Western and Russian nations. Since over half the military might in the world is found in just two nations—the United States and Russia—this is a reasonable, albeit somewhat simplistic, practice.

Part 1
GROUND COMBAT

2

THE POOR
BLOODY INFANTRY

IMAGINE YOURSELF IN a suburban town that has been abandoned by most of its inhabitants. Those few that remain are likely to take a shot at you, but the real danger is from explosive shells that fall from the sky at seemingly random intervals. Your only protection is to seek shelter in ruined buildings or to dig a hole in the rain-sodden ground. You have not had a hot meal or a bath for five weeks, and are subsisting on cold food out of a can or pouch. Your small group of ragged companions waits for instructions to come over a radio. You will be told to move either in one direction where there are fewer explosions and people shooting at you, or in another direction where there is more mayhem. Your only escape from this nightmare is to receive an injury or be killed.

A piece of science fiction? No, just the life of the average combat infantryman. Often we have heard unbelievable stories of suffering inflicted upon the infantry. The usual reaction is: "How do they [the soldiers] stand it?" The answer is: only for so long. Studies during World War II indicated that after as few as 100 to 200 days of combat stress, the average infantryman is a mental and physical wreck, incapable of further performance. Most infantry-men don't survive that long. With a minimum daily casualty rate of 2 percent, the chances of keeping body and soul together for 150 days are slim.

The infantry, by definition, takes the brunt of the fighting. In ancient armies infantrymen were the poorest soldiers with the weakest weapons, spears and shields. They were put in front to absorb the enemy's arrows and spears, and to keep the other side's infantry occupied. Meanwhile, the wealthier, and usually noble, horsemen in the rear waited for the right moment to charge to victory or retreat from defeat. Win or lose, the infantry took most of the casualties.

Four thousand years have not changed substantially the role of the infantry. Today far more firepower is tossed about. Battles are often fought along continuous fronts thousands of kilometers long instead of a few hundred meters of battle line on some dusty field.

Candidates for the Infantry

Studies during World War II discovered that the most effective infantryman was a fellow of average intelligence, or above, with good mechanical skills, just the sort of person you would want to have serve in the artillery or run a tank. Because armies have become so much more technical, the more capable manpower tends to be sent everywhere but the infantry.

It is impossible, however, to have good infantry if you get only the most stupid and inept recruits. The leaders of infantry units must be of good quality, otherwise the infantry is completely lost, and these leaders, particularly the sergeants, come from the ranks. The net result is that the infantry takes the recruits it can get and makes the best of it.

The Infantry Unit

Every infantryman depends on his unit for support: physical, moral, mental, medical. The primary unit is the squad of nine to fifteen men. It is usually commanded by a sergeant and consists of two or more fire teams. Each team is responsible for some sort of heavy weapon (machine gun, automatic rifle, antitank weapon). Since World War I, it has been learned that the firepower of the infantry is in its crew-served heavy weapons, not the rifles of the individual infantryman. The interesting reason is that when the shooting starts, a single infantryman seeks cover and does not use his weapon. If he does use his rifle, he will most likely just fire blindly into the air. A crew-served weapon, such as a machine gun with one man firing and another feeding in the ammunition, tends to be used because the men on the crew reassure each other.

Loneliness and the presence of death are a devastating combination that plunges most soldiers into a frozen panic. It is not unusual for units new to combat to hit the ground en masse at the sound of a few rifle shots. Once on the ground, out of sight of its leaders, a battalion could take hours to get up and start moving again. Knowledgeable leadership, a spirit of cooperation and mutual support can turn an infantry unit from a panic-stricken mob into a cohesive, effective combat unit.

The infantryman works with the other members of his squad and knows many of the other members of the platoon. When not in the field the platoon often lives in the same large room or building. When out in the field the infantrymen live wherever they can. A hole in the ground is a popular place because of the protection from enemy fire. These holes are usually two- or three-man affairs. Some get rather elaborate, with logs or other materials reinforcing the basic hole in the ground.

Infantry Organization

Squads. Nine to 15 men comprise a squad. Armed with rifles, machine guns and antitank weapons. Commanded by a corporal or a sergeant. Platoons usually advance on a frontage of 50 to 100 meters. Defending, they are responsible for *twice* that area (200 meters wide and deep).

Platoons. Three or 4 squads are a platoon (30 to 50 men). Commanded by a sergeant or a lieutenant. In the attack, platoons advance on a front of 100 to 150 meters, with 2 platoons up front and one behind. In defense, two to three times that area is covered (up to 500 meters' frontage) usually by concentrating several squads for all-around defense.

Company. Three or 4 platoons form a company (100 to 250 men). Additional weapons are mortars and heavier machine guns as well as special sensor equipment. Commanded by a lieutenant, captain or major. Advances in the attack on a frontage of 500 to 1000 meters. In defense, holds twice that much.

Battalion. Three to 5 companies are a battalion (400 to 1500 men). This unit might have tanks. The battalion could also have antiaircraft weapons and special sensor equipment. Commanded by a major or a lieutenant-colonel. In the attack, advances on a frontage of 1 to 3 km. Defends twice that much.

Larger units. Larger units such as regiments, divisions and armies usually contain smaller and smaller proportions of infantry. By definition, they are no longer referred to as infantry units.

Combat engineers. Called "pioneers" by most armies, combat engineers are specialists in exotic weapons and engineering equipment. They are also expected to be infantry trained and in emergencies are used just like infantry. Their weapons include mines, special explosives, flamethrowers and any new device that seems to fall within their area of expertise. Their specialty is setting up particularly elaborate defenses quickly. Engineers are also expert at demolishing enemy defenses quickly, which qualifies them as combat troops. They are highly respected by the infantry. This is primarily because of their activities in front of the infantry when assisting an assault on particularly dense enemy defenses, especially concrete-reinforced "fixed" defenses. Because of their specialists' training, combat engineers are used as regular infantry only in emergencies. They are not as easily replaced as regular infantry.

The Infantryman's Combat Activities

Reserve. Not in contact with the enemy, but still available for combat. This is an opportunity to rest units that have been in heavy combat. The troops catch up on sleep, bathe, receive mail and write letters, eat a hot meal, receive replacements for lost men and equipment, train. Usually, but not always, all this occurs in an area that is not under enemy fire. Often the accommodations aren't much better than those at the front.

Movement. In the old days most of the infantrymen walked. Some still do. The majority of infantry units today are mechanized. They ride in armored personnel carriers (APC's), trucks or helicopters. Aside from the risk of ambush from the air or ground, movement affords a good opportunity to sleep. Veterans learn this quickly and practice it diligently. The vehicle drivers are only human, and if great care is not taken, they also will doze off. A characteristic of mechanized units moving around a lot are frequent accidents as driver fatigue increases.

Meeting engagement. In the early stages of a war a lot of "meeting engagements" are anticipated between advancing friendly and enemy units. They are expected to be spirited and hectic. The Russians believe this will be the most common form of combat in the next war. In fact, the Russians have come to designate any initial contact with the enemy as a meeting engagement. In any case, the infantryman will find himself either ambushed or, after a warning, sent off in some direction as if on patrol to ambush someone else. Only the most confident and experienced troops look forward to movement-type fighting.

Construction. Most of the time, no one makes a big fuss. Once a soldier gets a taste of infantry combat, he quickly develops pacifist attitudes. But even in an inactive situation, the infantry still has casualties every day to stray enemy (and friendly) fire. Thus, there is a ready-made incentive to prepare for an increased level of enemy activity. Even with the widespread use of APC's, troops still dig. The APC's themselves are dug in if time and resources (a bulldozer blade on a tank or an APC) permit. The objectives are to get "wired in" and "dug in." Digging comes first. The wiring consists of adding the frills, putting out the sensors, laying and burying telephone wire. If available, barbed wire and barricades are erected. Security is also improved by laying minefields and positioning weapons so that they can be fired "blind," through smoke, fog or darkness, at likely enemy avenues of approach. The artillery observers register their guns and plot the best locations for instant barrages, so that in the heat of combat the order for a particular pattern of shells on a certain area of terrain can be given as tersely as possible. This is especially important because the enemy often uses electronic warfare to blot out radio communications, and artillery fire cuts telephone wires. Defenders are often reduced to using colored flares to call for artillery barrages (as in 1918).

All troops in a position must be drilled on where everything is, including supplies. Routes protected from enemy fire are established for getting around, and out of, the position. Finally, they must make provision for things going from bad to worse. The routes for withdrawal to another defensive position, and the new position itself, must be identified. The troops must be drilled on the sequence of a withdrawal; otherwise it can easily turn into a rout and a great slaughter.

All this preparation takes some skill, not much time (twenty-four hours will do wonders) and a lot of effort. Throughout it all the infantry must beware of occasional artillery and snipers.

Patrol. This is the most dangerous and tension-producing activity for an infantryman. Squads or platoons go into enemy territory to collect information on enemy activities. If the friendly and enemy forces are separated by a few kilometers of no-man's-land, then this area will have a lot of patrol activity. The main goals are to prevent the enemy from sneaking up on you and to prepare to surprise him yourself. The biggest danger in a patrol is an ambush. Modern electronic sensors have helped somewhat—microphones or, more recently, devices that recognize the sound or heat signature of men and equipment. Battlefield radars are also used if the area is open enough. But it's still dangerous to go into a wooded or built-up area on patrol. If you don't run into an enemy patrol that sees you before you see them, you may run into an enemy ambush, unless you can set up yours first. All this activity takes place day and night. Daytime ambushes are set up during the night. Also set up are booby traps and land mines. At night there is also activity to remove the other fellow's mines and booby traps. Just sitting in your hole can be dangerous enough; getting out and patrolling is very unhealthy.

Defense. The other fellow attacks. You might get a warning from your patrols. When it starts, you get hit with a lot of artillery fire, and perhaps poison gas and nuclear weapons. The survivors must get out of the bottom of holes or dugouts, quickly get antitank weapons into action and provide some direction for their own artillery. Since you are still under cover and the advancing enemy is not, there is a chance of surviving. Remember that modern warfare is very spread out. As part of a platoon strongpoint, you occupy one of 10 or so holes in the ground in a circular area 100 to 200 meters in diameter. The advancing enemy might not even be coming your way (they know all about using artillery to neutralize defending units they must bypass). If you do have the misfortune to be in the way, you can only hope all the defending firepower will stop the enemy. You may survive being overrun, given most armies' goal of getting through front-line positions and into the rear area so as to shoot up supply and artillery units.

If push comes to shove, there's always surrender, assuming the battered attacker is in a compassionate mood. However, given the fact that an average of 70 percent of the passengers of destroyed APC's are killed or injured, the enemy might be bloody minded. Although it is not much written about, prisoners are often not taken during opposed attacks, especially if individuals or small groups are trying to give up. The attacker doesn't want to spare any troops to guard prisoners, particularly since he needs all the help he can get to complete the attack successfully. And then there are all those troops that are wounded and are more in need of attention than enemy prisoners. And that is why defeated defenders attempt to hide or sneak away rather than test the tender mercies of the attacker through surrender. Veteran troops know this; otherwise they wouldn't be veterans.

On the other hand, the defender has a number of advantages. First, he is under cover, usually dug in. He is difficult to see. The attacker is nervously aware of this invisibility, and this often leads to an attack breakdown. The attacker often observes other attacking troops being hit. The defender also sees this and is encouraged. Because defending casualties are usually not seen by other defenders, there is no similar discouragement for the defender. The defender also has the advantage of knowing that relative safety is as close as the bottom of his hole in the ground. When attacking troops attempt to seek such safety, the assault breaks down. You can't attack when you are flat on your face. Therefore, attacks usually succeed only if the troops are well trained and led (a rare combination), and/or the defender has been all but obliterated by artillery and other firepower. If the defender cannot maintain a continuous front of fire, the attacker will be stopped only in front of the surviving positions. These positions will then be hit in the flank or rear and eliminated. The continuous front needn't contain a lot of troops; one functioning machine gun can hold up infantry on a front of 200 meters or more.

In the defense you never win. There are only various degrees of defeat. If you stop the enemy attack once, there will most likely be another one. If you do manage to clobber the attacker, you are often rewarded with an order to attack. You can only hope that the enemy will flee and you will pursue. Even then, the enemy will eventually stop and resist. If he stops and surrenders, the war is over. But that only happens once per war.

An astute commander attempts to have defending troops fall back before the next attack hits, particularly if he calculates that this next attack will overwhelm his defense. Even if he feels he can hold, a successful and well-timed fallback will force the enemy to waste a lot of firepower, fuel and energy attacking an empty position, thus weakening the enemy for the next attack.

Attacking. In some cases this is not much different than going on patrol, particularly if it is a night attack. But it is the most dreaded of all infantry operations. No matter how well planned, an attack means that you must get up and expose yourself to enemy fire. Ideally, the artillery smashes the defender to the point where the infantry simply walks in, collects a few prisoners and keeps on going. Attacks rarely happen that way.

If patrolling and intelligence gathering are first-rate, you will know the position of most but rarely all of the enemy positions. If the artillery fire is plentiful (it never is), you can destroy most of these positions. If the leaders planning the attack are skillful enough, you can knock out key positions, which then allows you to get around the others. This way you can destroy them with minimal loss. If the attacking troops are adept enough, they work as a deadly team, avoiding enemy fire and eliminating defending positions systematically. If, if, if . . .

The keys are skill, preparation and, above all, information. History proves that a successful attack is won before it begins. The norm is not enough time, not enough resources, not enough skill, not enough information. Even against an unskilled defender, everything must go right to achieve minimal losses. Remember, a defender can fire off a shot in relative safety and not even be detected. The attackers may eventually get him but not before they've had a few more casualties. In an attack the quality of the troops is the critical factor. Poorly trained, poorly led troops often do not even press home an attack against any opposition, and when they do, they take heavy casualties. Some things never change and this is one of them. Reckless bravery does not help much, as it just gets more attackers killed. The Arab-Israeli wars have shown this. The Japanese *banzai* attacks in World War II did the same.

Pursuit. Once the enemy is on the run, you must chase down the defeated remnants before they can reform and defend again. Pursuit is deceptively dangerous. You never know when the enemy will stop and ambush you, and with what. The resistance might be just a few diehards unable, or unwilling, to retreat any farther. It might also be fresh enemy units, strong enough to stop an attacker cold. The watchword of pursuit is speed. Go so fast that you overtake the fleeing defender, along a parallel route, of course. The ideal situation is to set up your own ambush, then collect more prisoners or kill the enemy off in comparative safety. Again, the deciding factors are skill in patrolling and intelligence gathering, as well as the ability to deploy rapidly against any resistance despite fatigue.

The Standard of Living at the Front

It is very low. The overriding goal is not to get hit by flying objects. This requires being inconspicuous, as what the enemy can't see he can't shoot at, at least not deliberately. Shellfire is less deliberate and more difficult to hide from. For this reason most infantrymen become like hobbits—they live underground. At that, it's an uneasy life. There's much work to be done. Positions must be prepared and maintained. Equipment must also be looked after. Enemy fire and Mother Nature conspire to keep everything dirty, damaged and generally on the verge of breakdown.

Security is the major consideration. From 20 percent to 50 percent of the troops are on guard at all times, doing little more than manning their weapons and waiting for the enemy to do something. Patrols must be sent out. All this is immediate and a matter of life and death.

A routine is established and the work divided so that troops have some time for sleeping and eating and not much else. There are always emergencies and distractions to disrupt those not "at work"—from an enemy attack to random shellfire.

The uneasy nature of life under enemy fire is not conducive to rest and relaxation. About 10 percent of all casualties are attributed to combat fatigue, the cumulative effect of little sleep, poor food (usually cold and consumed in an unappetizing atmosphere), dreary living conditions, and the constant threat of random death or mutilation. If it rains, you usually get wet. If it's cold, you bundle up as best you can. If it stays damp, you are in constant danger of all those afflictions that arise from constant exposure, like trench foot (your toes literally rot). If it's a tropical climate, you can rot all over, plus contract numerous tropical diseases. All these afflictions can be avoided only by constantly striving to keep dry and medicated, which requires discipline and the availability of medicines and dry clothing. Staying clean is nearly impossible, as you are living in the dirt.

Normally the only solution to the constant threat of wastage from these living conditions is rotating the troops out of the front line periodically. Two weeks in, one week out is ideal. Since there are rarely enough troops to go around, four weeks in and a few days out is more usual.

Even without the immediate presence of the enemy, life in the field in wartime is an ultimately degenerating experience. Most infantry units have to maintain enormous quantities of equipment. A 900-man U.S. Mech Infantry battalion has nearly 100 vehicles, mostly armored personnel carriers, to maintain. At least 4 man-hours a day per tracked vehicle are required to keep the vehicles running. Add to this 60 machine guns, 60 antitank guided-missile systems, over 100 pieces of communications equipment (primarily radios and electronic sensors), plus generators, stoves, mainte-nance supplies, personal weapons and other gear. Living in barracks, the troops can easily spend over 20 hours a week per man just keeping the equipment in shape. When in the field, time is also spent moving or in contact with the enemy. The troops are capable of only so much. Eventually they and their equipment begin to waste away. It's an eternal truth of warfare that when campaigning, even without deadly contact with the enemy, an army will even-tually wear itself out.

Tools of the Trade

LIGHT INFANTRY WEAPONS

First, the infantryman has his personal weapons. These are pistols (generally useless), bayonets (useful for domestic chores), grenades and rifles. The rifle is the most effective weapon, particularly the modern assault rifle. Increasingly it is of small caliber—5.56mm, the same as the American .22. Unlike machine guns, which are fed bullets from 100+ bullet belts, rifles have built-in magazines holding 20 to 40 rounds. On full automatic they fire off 30 rounds in less than 3 seconds. Also unlike machine guns, the rifles have lighter barrels, which will overheat after 100 rounds fired in less than a few minutes. An average infantryman usually will not carry more than 10 or 20 magazines. The real killing is done with the heavy weapons.

The characteristics of grenades, the infantryman's "personal artillery," have been vastly distorted by the media. Fragmentation grenades weigh about a pound, can be thrown a maximum of 40 meters and injure about 50 percent of those within 6 meters of the explosion. Less than 10 percent of the wounded will die, thus making grenades one of the least lethal weapons in the infantry-man's arsenal. In theory, grenades have a 3- to 5-second fuse.

Quality control being what it is, these fuses are sometimes a little longer or shorter. Grenades are favored for fighting at night or in built-up areas. Experienced infantrymen handle them with great care. Unlike as shown in the movies, fragmentation grenades create a small, smokeless explosion, and no one can pull the pin that arms the grenade with his teeth.

Not all grenades are the usual fragmentation type. Other important types are smoke (for concealment), high explosive (for close-in work; mostly blast and few fragments, to stun the guy 5 meters in front of you without worrying about fragments), thermal (for burning holes in metal), illumination (for turning night into day) and marker (to let off colored smoke to show someone upstairs where to drop something).

HEAVY INFANTRY WEAPONS

The crew-served, heavy infantry weapons consist of machine guns, grenade launchers, explosive devices, antitank weapons and mortars. Theoretically, all infantrymen should be adept at all these weapons. In practice, everyone specializes. In particular, the mortars and antitank guided missiles (ATGM's) tend to be manned only by specialists. All infantrymen are generally trained in the use of machine guns, grenade launchers and antitank grenade rocket launchers.

SPECIAL EQUIPMENT

Some types of equipment are handled by most infantrymen under close supervision. Minefields and booby traps can be emplaced by an average infantryman if carefully supervised, but these highly explosive devices are very dangerous if mishandled. The troops know it and work a lot more efficiently with them if there is an expert around. Other devices are dangerous only if they don't work, particularly sensors and other communications equipment like radios, telephones and signal flares. This equipment is particularly essential for effective defense. Sensors let you know that the enemy is coming and from which direction. Commo (communications) equipment lets you know where you are and is the vital link to artillery, air power and resupply support.

PERSONAL EQUIPMENT

There are more mundane but essential tools: the vital folding shovel entrenching tool, a bayonet (used for everything *but* stabbing someone), a gas mask and other chemical protectors like ointments, antidotes, special clothing. The most respected articles of protective clothing are the flak vest and helmet. Only some Western armies use flak vests. The latest versions of these vests, which actually extend below the waist, weigh under ten pounds, are plastic, and will stop all shell fragments and most bullets. They are uncomfortable in warm weather, but combat veterans swear by them. In most armies helmets are still made of steel. The newer plastic ones are lighter and offer better protection. And then there is the first-aid kit, a mini-drugstore with pep pills, aspirin, antibiotics, bandages, pain-killers and other controlled substances. Not everything that goes into the first-aid pouch is official issue, or even legal. Infantrymen learn to equip themselves as best they can. Speaking of unofficial equipment, veteran infantrymen often obtain a pistol or even a shotgun. Transistor radios continue to be present. They provide one way of finding out what's happening, assuming that stations are still broadcasting. For personal comfort there are sleeping bags, spare clothing (often socks, if nothing else), sunglasses (handy for surviving nuclear explosions unblinded), canteens, messkits and food (candy is particularly favored). As most troops now move around in APC's, the amount of equipment carried has increased. Losing the APC, therefore, has a great adverse effect on morale.

COMBAT VALUES

Taking into account firepower, mobility and protection, weapons can be assigned an estimated numerical combat value.

	W Ger	US	Russia
Transport	20	20	10
Men	1	1	1
Tank	1200	1100	1000
APC	320	120	200
SP How.	900	900	500
ATGM	300	300	200
MG	3	3	2
Med Mort	220	220	150
Hvy Mort	360	360	240
AA Gun	500	200	400
SP SAM	400	400	300
Lt SAM	75	75	45
ATRL	80	30	80

Combat Values

This chart assigns a numerical combat value to each weapon, which is a combination of the weapon's destructive power, ease of use, mobility and reliability. All things being equal, the attacker needs at least a 3-to-1 advantage in combat power to have a reasonable chance of success. Six to one or better is preferred. See the chapter on attrition for more details.

Values are given for three nations' armies: W GER (West Germany), US (United States) and USSR (Russia). The values show the national differences in weapon design.

TRANSPORT is assigned a value, per vehicle, which represents the ability of the unit's unarmed vehicles to support the weapons. Fuel, ammunition and often the weapons themselves must be carried. The quantity and quality of the unit's transport determines how effective the weapons will be.

MEN. Each man, and his personal weapons, is assigned a base reference value of 1. In addition, each man is responsible for some other job, either operating a larger crew-served weapon or providing some form of support. By itself, armed only with personal weapons, the infantry is relatively helpless against most other modern weapons.

TANK is a heavily armored, tracked vehicle carrying a large gun in a turret.

APC is an armored personnel carrier.

SP HOW is a self-propelled howitzer (artillery).

ATGM is an antitank guided-missile launcher.

MG is a machine gun, not including the infantry's automatic rifles. Each Russian infantry battalion also contains 4 35mm automatic grenade launchers with an effective range of 800 meters. Machine guns mounted on tanks are not included.

MED MORT are medium mortars (81mm to 82mm); light artillery, often carried in an APC.

HVY MORT are heavy mortars (120mm) that are mechanically similar to medium mortars and often mounted in APC's.

AA GUN is a self-propelled, antiaircraft 20 to 30mm machine gun.

SP SAM (self-propelled, surface-to-air-missile launcher) is a vehicle carrying missiles and associated radar and control equipment.

LT (light) SAM is a one-man, shoulder-fired antiaircraft missile.

ATRL (antitank rocket launcher) is portable, carried by individual soldiers.

U.S. INFANTRY UNITS

U.S. infantry units tend to distribute their firepower more evenly among smaller elements than those of other nations.

Number of Units>	Battalion	Combat Power %	Company	Combat Power %	Platoon	Combat Power %
	1		3		9	
Men	901	2	170	2	38	2
Tank	0	0	0	0	0	0
APC	88	29	18	28	4	26
SP How.	0	0	0	0	0	0
ATGM	62	51	14	54	4	65
MG	61	1	16	1	3	0
Med Mort	9	5	3	8	0	0
Hvy Mort	4	4	0	0	0	0
AA Gun	0	0	0	0	0	0
SP SAM	0	0	0	0	0	0
Lt SAM	5	1	0	0	0	0
ATRL	72	6	18	7	4	6
Combat Power		36		7.78		1.85
% Combat Power		100		64		46
% Manpower		100		57		38

RUSSIAN INFANTRY UNITS

Russian infantry units tend to be smaller than those in other armies.

Number of Units>	Regiment	Combat Power %	Battalion	Combat Power %	Company	Combat Power %	Platoon	Combat Power %
	1		3		9		27	
Men	2443	2	409	2	108	2	33	2
Tank	40	33	0	0	0	0	0	0
APC	130	21	34	37	10	41	3	40
SP How.	18	7	0	0	0	0	0	0
ATGM	147	24	34	37	10	41	3	40
MG	224	0	67	1	19	1	6	1
Med Mort	0	0	0	0	0	0	0	0
Hvy Mort	18	4	6	8	0	0	0	0
AA Gun	4	1	0	0	0	0	0	0
SP SAM	4	1	0	0	0	0	0	0
Lt SAM	36	1	9	2	0	0	0	0
ATRL	81	5	27	12	9	15	3	16
Combat Power		123		18.15		4.87		1.49
% Combat Power		100		44		36		33
% Manpower		100		50		40		36

Infantry Battalions, Companies and Platoons

This chart shows the organization, weapons and combat power of U.S. and Russian infantry units. The organization and weaponry of the Russians are used by the vast majority of Communist countries as well as by most nations that accept military aid from Russia. The U.S. system, a variation of the Western European system, uses larger, more heavily armed units. The Germans (in 1945 and the 1970's), the Israelis and other significant Western military powers have recognized that the leaner organization of the Russians is superior in combat. Variants of the Russian system are thus replacing the United States system in many countries.

NUMBER OF UNITS is the number of each unit in the superior unit to the left side of the chart (i.e., 3 companies in a battalion, 9 platoons in a battalion). See combat values chart notes (2-1) for weapons explanations.

COMBAT POWER is the total combat power rating of the unit divided by 1000. Based on individual weapon ratings and their quantity in the unit.

% COMBAT POWER, % MANPOWER shows the percentage smaller units have of the next largest unit's combat power or manpower. For example, the 9 infantry platoons of the U.S. infantry battalion possess 46% of the battalion's combat power. The remaining combat power is in the battalion and company support units. One immediately obvious fact is that while Russians and Americans put about the same percentage of their manpower in the infantry platoons, Americans place more of their combat power directly in these platoons. The 9 infantry platoons of a U.S. infantry battalion contain 38 percent of the battalion's manpower but 46 percent of its combat power. The Russians, on the other hand, also have 36 percent of their manpower in the 27 infantry platoons of their infantry regiment, but only 33 percent of the combat power.

The difference stems from varying doctrines. The Russian system keeps a larger portion of the superior unit's combat power under central control. This combat power is applied, usually in large doses, through one of the subordinate units. The advantage of this central control is primarily just that, control. The Russian system does not rely on communications. If the regimental commander wants to get his reserve—over half his regiment's combat power—into action, he sends a messenger down the road to deliver the message. Or leads it himself. This central control is also practical. It eases supply and maintenance, given the lower levels of support in these areas. On the minus side, units think, train and operate best as units. A lot of last-minute shuffling lessens their combat power. Even Western units tend to "cross-attach," i.e., an infantry battalion exchanges one company with a tank battalion. The Russians distribute tank platoons and other reserve units to the infantry companies. An infantry platoon or squad is attached to a pure tank unit.

U.S. INFANTRY UNITS

BATTALION (901 men). The mechanized infantry battalion contains 3 infantry companies, 1 combat-support company (224 men, 21 M113's, 16 TOW ATGM's, 4

DRAGON ATGM's, 4 107mm mortars, 5 Redeye SAM's) and 1 headquarters company (167 men, 13 M113's).

COMPANY (170 men). The mechanized infantry company contains 3 rifle platoons, 1 headquarters section (10 men, 2 M113's), 1 maintenance section (11 men) and 1 weapons platoon (35 men, 4 M113's, 3 81mm mortars, 2 TOW ATGM's).

PLATOON (38 men). The mechanized rifle platoon contains a platoon headquarters (5 men, 1 M113, 4 DRAGON ATGM's), 3 rifle squads (each with 11 men, 1 M113 or IFV, 1 machine gun).

The U.S. Army issues ATRL's as rounds of ammunition (the tube is not reloadable). Machine guns are distributed, for all practical purposes, on a basis of 2 per 25 men. Each APC usually has a 12.7mm machine gun mounted on it. Experienced troops tend to "acquire" many more machine guns. Most Western armies follow the same general organization.

Practically all armored, and most nonarmored, vehicles have radios. There are approximately 71 trucks, including a few armored support vehicles, in an infantry battalion. They carry other personnel, maintenance tools, sensor equipment, supplies, and other weapons, like portable surface-to-air rockets and portable flamethrowers. Supplies include up to three or more days' worth of food, fuel and ammunition as well as spare parts for everything.

RUSSIAN INFANTRY UNITS

REGIMENT (2443 men). In the Russian Army the regiment serves the same function as the battalion in Western armies. That is, the regiment is capable of supporting itself in the field. To do this it must have various support capabilities (signal, maintenance, supply, specialist combat units). The Russian motor rifle regiment contains 3 motor rifle battalions, a regimental headquarters (42 men, 2 APC's), a tank battalion (241 men, 40 tanks, 3 APC's), a howitzer battalion (321 men, 18 152mm howitzers, usually self-propelled, an air-defense battery (64 men, 4 ZSU 23mm automatic cannon and 4 SA-9 missile systems), a reconnaissance company (57 men, 12 APC's), an antitank battery (57 men, 9 APC's, mounting a total of 45 AT-4 ATGM tubes), an engineer company (70 men with 2 self-propelled bridges), a maintenance company (66 men), a medical company (37 men), a motor-transport company (54 men with 40 4.5-ton trucks, each with a 10-ton trailer), a chemical-defense company (27 men and 3 decontamination rigs on trucks), a traffic-control platoon (20 military policemen), a supply-and-service platoon (12 men).

BATTALION (409 men). Contains 3 motor rifle companies, a mortar battery (48 men, 6 120mm mortars), a signal platoon (14 men), a medical section (9 men), a supply-and-maintenance platoon (12 men).

COMPANY (109 men). Contains 3 rifle platoons and a headquarters section (9 men, 1 APC).

PLATOON (33 men). Contains 3 rifle squads, each with 11 men, 1 APC, 2 machine guns and 1 ATRL.

DIVISIONAL EQUIPMENT AND COMBAT POWER

Russia has more divisions than any other army and divides them into categories of considerably different combat ability.

	Russia Motor Rifle	Combat Power %	Russia Tank	Combat Power %	US Mech Inf	Combat Power %	US Armored	Combat Power %	West German Mech Inf	Combat Power %	West German Armored	Combat Power %	Chinese Inf	Combat Power %
Trans-port	1500	2	1500	2	3500	11	3500	10	3200	10	3200	10	370	2
Men	13498	2	12380	2	16923	2	17110	2	17000	3	17500	3	12706	6
Tank	266	39	344	51	248	44	356	55	250	47	300	54	45	22
APC	673	20	627	19	498	10	515	9	337	17	300	14	0	0
Arty	148	11	94	7	75	11	75	9	92	13	92	12	48	12
ATGM	486	14	302	9	380	18	334	14	220	10	179	8	0	0
MG	920	0	550	0	1480	1	1360	1	980	0	940	0	168	0
Med Mort	0	0	0	0	54	2	36	1	0	0	0	0	189	14
Hvy Mort	54	2	36	1	53	3	53	3	30	2	24	1	12	1
AA Gun	110	6	110	6	24	1	24	1	36	3	36	3	30	6
SP SAM	46	2	46	2	24	2	24	1	0	0	0	0	0	0
Lt SAM	112	1	112	1	80	1	80	1	80	1	80	1	0	0
ATRL	260	3	192	2	1000	5	900	4	300	4	250	3	957	38
Combat Power	684		677		617				717		635		665	
# of units	15		15		6				4		4		6	
Combat Power	581		576		524				609				202	
# of units	31		9		2				2					
Combat Power	410		406										118	
# of units	73		21											

The other equipment is similar to that carried by the U.S. battalion. The regiment has 149 trucks.

Note that most Western armies have field kitchens at the company level. The Russians have them only at the regimental level. The U.S. company has 8 trucks for support, the Russian battalion has 7, the Russian company none. In general, the quality of life for the Russian soldier is lower. This puts the Russian soldier on a thinner edge of survival. He has less food and less medical, shelter and equipment-maintenance support. Compared with his counterpart in Western armies, the average Russian soldier is less experienced and less capable, at least in the peacetime conscript army. In wartime the Russian soldier becomes as experienced as his Western counterpart. But this method is wasteful of human life and equipment.

Divisional Equipment and Combat Power

See Combat Values chart notes for weapons explanations.

The basic unit of ground forces is the division, a force of 10,000 to 18,000 men, in most cases. There are over 1000 in the world today. The units shown in the chart represent about 85 percent of the world's divisions. The Russian organization alone represents over half. Chinese-style infantry divisions represent another quarter.

The chart shows weapons and their combat power. What it does not show is the sometimes considerable qualitative differences resulting from:

1. *Quality and quantity of support equipment.* Obvious items enhance combat ability, like engineer, signal and transportation support, plus more exotic things, such as electronic devices for electronic warfare, data processing, fire control. Also important are procedures for the effective movement of supply and maintenance and repair of equipment before, during and after combat.

2. *Training and doctrinal differences.* Different nations can use identical organization and equipment, but because of different approaches to the selection and training of troops, or the application of the doctrine, there will be substantial qualitative differences. Take, for example, Bulgaria and East Germany. Both use Russian organization and equipment. All things being equal, the East German unit will be substantially superior to the Bulgarian.

3. *Differences within a national army.* Nowhere is this more a consideration than in the Russian Army. See below.

RUSSIAN-STYLE DIVISIONS

Although all Russian divisions use the same organization, there are three different grades. The number and total combat value of each grade is shown at the bottom of the chart. The highest grade is the "groups of forces" in Eastern Europe. These comprise 30 divisions (15 tank and 15 infantry). The Russian divisions in the chart are based on these group of forces quality divisions. These units are always kept at full strength and are the first to receive new equipment, aside from units in western Russia that try out new weapons and equipment prior to large-scale distribution.

The next grade are the categories 1 and 2 divisions within Russia. These comprise 43 divisions (11 tank, 32 infantry). They are next in line for new equipment after the groups of forces divisions. Generally they have about 1500 fewer men, 50 to 100 fewer tanks, 40 to 60 fewer artillery pieces and generally lower equipment levels. Peacetime manning is only 50 percent to 75 percent in most cases, although 4 to 6 divisions are at full strength. Local reserves, men released from service in the past three years, can bring these units up to strength in a few days and be combat-ready in about a month. Overall combat value of these units would be 10 percent to 20 percent below the group of forces divisions.

Last, there are the category 3 divisions (20 tank, 72 infantry). Their equipment levels are similar to the category 2 divisions. Equipment is not only the oldest in the Russian inventory, but also the most poorly maintained. Manning levels on these divisions range from 10 percent to 30 percent, with a full complement of combat vehicles and weapons, most in storage. But transport has to be taken from the civilian economy. Much specialist equipment is either obsolete or not present. It would take up to three months to get these units ready for combat. Even then, their combat power would be 30 percent to 50 percent below that of the group of forces divisions.

In a special category are 80,000 airborne and marine troops. These are organized into 7 full-strength airborne divisions and 5 marine regiments. Two thirds are light infantry, without APC's or tanks.

Non-Russian divisions using Russian organization would be equivalent to category 2 or 3. In Europe these comprise 45 category 1 and category 2 divisions (14 tank, 31 infantry) and 12 category 3 divisions (3 tank, 9 infantry). Many other nations throughout the world use Russian-style organization, especially in the Middle East. Although their manning levels are close to 100 percent, they usually have less than half the combat power of a top-rated Russian division.

RUSSIAN MOTOR RIFLE DIVISION. A *division headquarters* (301 men), 3 *motor rifle regiments* (2443 men each), 1 *tank regiment* (1101 men), an *artillery group* (1643 men, 18 100mm guns, 18 122mm rocket launchers, 4 FROG launchers, 36 122mm howitzers, 18 152mm howitzers), an *air-defense regiment* (302 men, 24 SA-6, 6 SA-11), an *independent tank battalion* (241 men, 51 tanks), a *reconnaissance battalion* (300 men, 8 APC's, 6 tanks, motorcycles), an *engineer battalion* (380 men), a *signal battalion* (280 men), a *chemical-defense battalion* (150 men); *support troops* (729 men), consisting of a *maintenance battalion* (294 men), a *medical battalion* (158 men), a *motor-transport battalion* (217 men), a *traffic-control company* (60 military policemen).

RUSSIAN TANK DIVISION. Organized identically to the motor rifle division except for the following changes: 3 *tank regiments* (1510 men, 95 tanks, 23 APC's, 4 ZSU-23's, 4 SA-9's, 6 self-propelled bridges, 1 *motor rifle battalion)*, 1 *motor rifle regiment,* no *antitank battalion,* no *independent tank battalion.*

U.S. AND WEST GERMAN DIVISIONS

Both nations have similar divisions of total men and equipment. While there are a number of differences in how they use their divisions, one is major. The

Germans use their brigades more as independent units. In effect, the brigades become little divisions. This has a major advantage in that the battalions in the brigade are a permanent part of the brigade and therefore train together regularly.

The U.S. brigades in practice (in peacetime) keep the same battalions, although in theory the battalions are to be shifted around to meet different situations.

Another significant difference is the Germans' less lavish use of APC's and ATGM's. The rationale is that, with limited funds, it is better to put the money into fewer and better weapons. Thus, the German APC (the Marder) is far superior to the U.S. M-113. The Germans are more lavish with artillery and ammunition.

Overall the similarities are far more prominent than the differences. Like the Russians, the Americans and the West Germans utilize a "base" system for their divisions. That is, the combat support units are common for all divisions. Only the mix of combat units differentiates them. U.S. active divisions have lower equipment and manning levels for some of their divisions. These weaker units are shown at the bottom of the chart.

U.S. DIVISIONS. A *division headquarters* (190 men), 3 *brigade headquarters* (120 men each), a variable number of *tank* and *mechanized infantry battalions* (6 tank and 5 infantry produce an armored division; 4 tank and 6 infantry produce a mechanized infantry division), a *divisional artillery* (2560 men, 54 155mm howitzers, 12 203mm howitzers) an *air-defense battalion* (582 men, 24 Vulcan cannon and 24 Chaparral missiles), an *engineer battalion* (991 men, 40 APC's, 8 combat engineer tanks), a *reconnaissance battalion* (876 men, 27 tanks, 32 APC's, 38 ATGM's, 27 helicopters, 9 107mm mortars), a *signal battalion* (582 men), an *aviation company* (96 men, 10 aircraft), a *military police company* (201 men), a *chemical-defense company* (185 men), a *combat electronic-warfare intelligence battalion* (465 men), *support troops* (2253 men). Both infantry and armored divisions have 9341 men. Tank and infantry battalions for an armored division would be 7769 men, for an infantry division, 7582 men.

WEST GERMAN DIVISIONS. Each has 3 brigades. An *armored* (Panzer) *division* has 2 *armored brigades* and 1 *infantry brigade*. A *mechanized infantry* (Panzergrenadier) *division* has 1 *armored* and 2 *mechanized infantry brigades*. Each brigade has 4500 men. The division "base" is 3500 men consisting of an *artillery regiment* (12 175mm guns, 6 203mm howitzers, 16 110mm rocket launchers), a *reconnaissance battalion*, an *engineer battalion*, a *security battalion*, an *aviation company*, an *air-defense battalion* (36 armored, self-propelled 30mm guns) and *service* units. Additional *signal, engineer, artillery* and other *support* troops are distributed among the brigades.

A West German *tank brigade* contains a *brigade headquarters*, an *armored antitank company* (13 self-propelled ATGM's), an *engineer company* (5 self-propelled bridges), 1 *infantry battalion* (71 APC's, 6 120mm mortars), 2 *tank battalions* (54 tanks each), an *artillery battalion* (18 self-propelled 155mm howitzers), a *supply-and-maintenance battalion*, a *replacement battalion*.

A West German *mechanized infantry brigade* contains a *brigade headquarters*, an *armored antitank company* (16 self-propelled 90mm guns, 8 self-propelled

ATGM's), 2 *infantry battalions* and 1 *tank battalion,* plus other units found in the tank brigade.

The West Germans are in the middle of a reorganization that will give them tank and infantry battalions similar in size to Russian battalions. Instead of 3 of these battalions per brigade there will be 5 (2 33-tank battalions in the infantry brigade, or 3 in the tank brigade). The U.S. Army has been planning a similar reorganization and will probably eventually implement it.

The British and the French are also implementing a "brigade-oriented" organization. They have, in effect, 4000- to 8000-man divisions that are simply large brigades.

As the chart shows, the Americans and the West Germans have very few reserve and mobilization divisions. They do have primarily lighter, non-mechanized units suitable for rapid air movement, "intervention" in faraway parts of the world and rugged terrain.

Both nations have reserve units. The U.S. Army reserves are identical to the active units. The West German reserves are similar but not identical. Any mobilization program depends on trained manpower and usable equipment. The Russians maintain both in a very formal sense. The Western armies have less equipment on hand but a greater capacity to produce it. Therefore, depending on the length of a war and on the destruction of manufacturing facilities, Western nations like Germany will raise new units similar, if not identical, to those already in existence. During World War II, Germany raised the equivalent of 50 mechanized and armored divisions, plus more than 200 Chinese-style infantry divisions (see below). This was done with a similar population and much smaller industrial capacity than Germany possesses today.

CHINA

The Chinese divisions are actually closely related to the pre-1950 Russian divisions, before the Russians mechanized all their units. The Chinese system has several advantages. Most of the troops are long-term volunteers. Promotion is from the ranks. Troops generally remain in their original regiment for their entire military career. Marriage is allowed only for officers and senior noncommissioned officers. There are 30 pay grades from recruit to the highest officer. There are no ranks in the Western sense, only positions of increasing responsibility.

The Chinese disadvantages are obvious. Only heavy equipment moves by truck, the infantry walks. There is a pervasive lack of equipment. What there is is often obsolete by Western standards. The Chinese combat power is enhanced by the training, quality and superior morale of the troops, perhaps an increase of 10 percent to 30 percent, depending on the unit. Because of the lack of strategic mobility, the Chinese Army is primarily a defensive force.

CHINESE INFANTRY DIVISIONS. A *divisional headquarters* (696 men), 3 *infantry regiments* (each with 2817 men, 63 mortars, 42 machine guns, 303 ATRL's), a *tank regiment* (596 men, 34 tanks, 10 turret-less tanks known as assault guns), an *artillery regiment* (1135 men, 18 76/85/100mm guns, 12 122mm howitzers, 12 120/160mm mortars), a *signal battalion* (318 men), an *engineer battalion* (471 men), an *air-defense battalion* (515 men, 30 14.5/37/57mm guns), a *reconnaissance company* (135 men), an *antitank company* (100 men, 18 107mm recoil-less rifles), a *flamethrower company* (84 men, 27 flamethrowers), a *chemical-defense company* (100 men), a *military police company* (105 men). *Supply, maintenance and transport support* is controlled and provided by army-level units.

DENSITY OF INFANTRY WEAPONS

A small number of weapons is needed to defend every 200 meters of front. Armies tend to have far more weapons available than they need to stop an attacker.

Major Weapons per 200 Meters of Front
US Mech Infantry Battalion

Battalion Frontage>	Kilometers				
	1	2	3	4	5
MG	12	6	4	3	2
ATGM	12	6	4	3	2
APC	13	6	4	3	3
ATRL	14	7	5	4	3
AR	54	27	18	14	11

Major Weapons per 200 Meters of Front
Russian Motorized Rifle Regiment

Battalion Frontage>	Kilometers				
	2	4	6	8	10
MG	21	11	7	5	4
ATGM	11	5	4	3	2
APC	11	5	4	3	2
ATRL	8	4	3	2	2
AR	54	27	18	14	11
MBT	4	2	1	1	1

Density of Infantry Weapons

This chart shows the density of infantry weapons per 200 meters of unit frontage. For U.S. units the frontage is for an infantry battalion. For Russian units the frontage is for an infantry regiment.

BATTALION FRONTAGE is the size of the unit's frontage (1 to 5 km for the U.S. battalion, 2 to 10 km for the Russian regiment).

The weapons listed are: MG's (machine guns), ATGM's (antitank guided-missile launchers), APC's (armored personnel carriers), ATRL's (antitank rocket launchers), AR's (automatic rifles), MBT's (main battle tanks). Whether attacking or defending, this is the average density of weapons per 200 meters. Normally only 65 percent to 90 percent of this strength is actually up front. The remainder is held back as a reserve.

An attacking unit generally takes up only half the frontage it would use when defending. For example, a battalion that defends on a 4-km front would attack on a 2-km front. Russian units tend toward extremes, using very wide defending frontages (up to 10 km) and very narrow attack frontages (2 km). As you can see, a Russian regiment attacking a U.S. battalion defending on a 4-km front outnumbers the U.S. troops in the attack sector by 21 to 4 in machine guns, 11 to 3 in ATGM's, 11 to 3 in APC's, 8 to 5 in ATRL's and 4 to 0 in tanks.

This calculation of who is stronger is not difficult to make. A U.S. battalion is defending a 2-km front. It is being attacked by a Russian regiment. Look under the 2-km columns for both units. The American weapons are obviously outnumbered. But consider: For each 200 meters of front, the attacking APC's and tanks number 15. The defending antitank guided-missile systems number 6 plus 7 short-range antitank rocket launchers. The enemy vehicles are out in the open. It all comes down to the "exchange ratio"—how many attackers will be lost for each defender. In World War II, the ratio in antitank warfare was as high as 6 attackers for 1 defender (with heavy 76.2mm antitank guns). What the ratio is today is less certain because the ATGM does not operate like an antitank gun. Assume that the ratio is in the defender's favor, say 4 to 1. This would increase because most of the targets are lightly armored APC's, which can be shot up by 12.7mm machine guns. This pumps the ratio back up to 6 to 1. Therefore, the defender needs at least 3 ATGM's per 200 meters of front to destroy the Russians. To stop them requires less because the attacker usually falters before he is completely destroyed. About 50 percent of the attacker's losses generally convinces the attacker that he should pull back and reconsider before trying again.

The ATRL's are less effective and serve mainly as a margin of safety. You might lose 1 or 2 ATRL's for each enemy armored vehicle destroyed. This means there is still a chance of holding with only 1 ATGM per 200 meters. Once you have stopped the armored vehicles, the infantry is a much easier matter. Two or 3 machine guns or automatic rifles per 200 meters will discourage most infantry attacks. The defending commander knows what he has. He has an idea of what the Russian regiment has. Chances are neither side is up to strength, and this must be taken into account, especially by the defender.

Also to be taken into account is the terrain. If part of the defender's front is difficult terrain (marsh, thick woods, steep grades that tanks can't climb) his chances improve.

Then there are the effects of enemy fire and friendly artillery. If the Russians can muster their famous "200 guns per kilometer of front" and fire at you for 16 minutes, and if you are only in "hasty defense" (you've been preparing your position for less than a day), you will lose at least half your force. If you can also call in artillery, this will cause more casualties among the enemy infantry and APC's, but mainly it will disrupt the enemy attack, causing it to falter and probably fall back and try again.

If the enemy uses gas, this will increase the effectiveness of its artillery by three or more times. If you use mines, particularly the little two-pound "track buster" antivehicle mines, you have a good chance of stopping enemy tanks and APC's easily.

And so it goes, not quite as scientifically as it first appears. The defending commander must use a lot of judgment and guesswork in addition to calculation.

INFANTRY WEAPONS

There is little practical difference between most of the world's infantry weapons.

Weapon	Primary User	Caliber	Weapon Weight (lbs)	Ammo Weight (lbs/100)	Ammo in Weapon	Pract Rate of Fire/RPM	Eff Range (m)	Used For
FN/G3	W Ger	7.62	10.7	5.5	25	75	800	Standard infantry rifle
AK-47/M	Russia	7.62	8.8	6.2	30	90	400	Standard infantry rifle
M-16A1	US	5.56	8.1	3.5	20	80	600	Standard infantry rifle
AK-74	Russia	5.45	8.8	4.7	30	100	500	Standard infantry rifle
MG3	W Ger	7.62	39.6	6.1	200	200	1200	Standard LMG (new)
M-60 LMG	US	7.62	28.9	6.1	100	200	1200	Standard LMG (new)
PKM	Russia	7.62	18.4	6.1	250	200	1200	Standard LMG (new)
RPK	Russia	7.62	13.5	6.1	40	120	800	Standard squad LMG
RPK-74	Russia	5.45	13.2	6.1	40	120	600	Standard squad LMG
SAW	US	5.56	19.4	5.6	200	200	800	Standard squad LMG
SVD	Russia	7.62	10.1	7.3	10	20	800	Standard sniper rifle

Infantry Weapons

This chart shows the most common small arms used worldwide. The various other weapons utilized by the infantry will also be described in this section.

WEAPON is the weapon's official designation.

PRIMARY USER is the nation that is the principal user of the weapon as well as its designer and major manufacturer.

CALIBER is the diameter of the weapon's bullet, in millimeters (1 inch = 25.4mm).

WEAPON WEIGHT (LBS) is the weapon's loaded weight in pounds. This is more meaningful than the empty weight, as the weapon can only be used when loaded with ammunition.

AMMO WEIGHT LBS/100 is the weight of 100 rounds of ammunition. This includes magazine or metal link weight. Rifles usually have ammunition in magazines; machine guns are linked together with metal fasteners. Without magazines or links, ammunition has the following weights (rounds per pound): Russian, 5.45mm–43; U.S., 5.56mm–40; Western, 7.62mm–19; Russian, 7.62mm (short)–42; 7.62mm (long, used in the PKM and SVD)–32.

AMMO IN WEAPON is the number of rounds normally loaded in the weapon. Machine guns fire ammo from theoretically endless belts of linked rounds. As a practical matter, the belt is long enough to carry in a box hanging from the weapon. This allows the machine gun to be portable and handled by one man.

PRACT. RATE OF FIRE/RPM is the practical rate of fire in rounds per minute. Most of these weapons have theoretical rates of fire between 600 and 1300 rounds per minute. A number of factors prevent these high rates from being reached in practice. *Impaired accuracy at high rates of fire* is the major practical limitation. As an automatic weapon is fired, it recoils. Although modern weapons have reduced this recoil considerably, it still exists and throws off the aim. Bursts of five to ten rounds are generally more effective than steady streams of fire (don't believe what you see in the movies). A machine gun with a high rate of fire, used with small bursts of fire, becomes a long-range shotgun. This has proved to be the most effective way of killing people with machine guns. *Barrel overheating* is one of the more serious problems. Depending on the weather (heat and exposure to the sun are the worst), a machine-gun barrel quickly overheats if the practical rate of fire is exceeded for more than a few minutes. The results are that rounds fire without your pulling the trigger ("cooking off"), and there will be eventual failure of the barrel. For this reason, water-cooled machine guns were used up through World War II. It was found, however, that just as much damage could be avoided by keeping the rate of fire down. Another solution, which is still used on many machine guns, is an easily changed spare barrel. The Germans did this extensively and successfully during World War II and still do it today. When you fire from a fixed position and need rapid fire in a short period of time, a few spare barrels solve the overheating problem. *Jamming,* especially on belts, is another problem. Too much ammunition through the weapon at one time produces a mechanical jam. This takes at least a few seconds to clear and requires reloading the belt. It can also happen, but less frequently, in magazine-fed weapons. *Ammunition supply*—under the best of conditions, it takes five or more seconds to change magazines. More time is required to load a new belt.

EFF RANGE METERS is the average effective range of the weapon in meters. These are difficult standards to attain. With any of the weapons, a superb marksman can obtain hits at twice the stated ranges. As a practical matter, there aren't many marksmen in the ranks. Even if there were, the conditions of infantry combat rarely allow the time to get off a highly accurate shot. Most firing is done in bursts at fleeting targets. Those who have the misfortune to be targets have no incentive to make it any easier to get hit. On the battlefield you keep your head down and move quickly. If you don't, you get killed.

Other major factors in the effective range are the design of the weapon and its ammunition and troop training. A weapon that is designed to remain steady when fired will produce greater accuracy. This was one of the major reasons for the move to the smaller but faster 5.56mm round.

Ammunition design can also produce greater accuracy and more lethal results. A new Belgium 5.56mm round (the SS-109), for example, outperforms the existing 5.56mm *and* 7.62mm rounds. In any event, most 5.56mm rounds can go through 15mm aluminum APC armor at 100 meters. A 7.62mm round can do this at 400 meters. Against unprotected human targets, both sizes are effective to the limits of aimed fire (800 meters). Beyond 800 meters machine guns deliver indirect fire, which serves primarily to keep enemy heads down. An unprotected man being hit by a 7.62mm round can still be killed up to 2000 meters away, while the same is true at only 1000 meters for 5.56mm rounds. New 5.56mm rounds are being introduced that make this caliber lethal to 1500 meters.

Accuracy is about the same for the two calibers, with the 5.56mm round having a notable advantage at ranges over 400 meters. For example, 5.56mm and 7.62mm weapons (with bipods) firing 2- to 3-round bursts at a 6-foot-diameter target will, after firing 30 rounds, put the following percentages of their bullets in the target at the ranges indicated: (5.56mm/7.62mm): 300 meters–81/81, 400 meters–73/77, 600 meters–55/41, 700 meters (9-foot-diameter target)–67/37; 1000 meters (9 foot)–52/30. Contrary to popular myth, 5.56mm rounds are not deflected by underbrush any more than are 7.62mm rounds.

USED FOR is the primary use of the weapon.

Standard infantry rifle is the weapon most commonly used in the army. The FN/G3 is one of the many variants of the original Belgium FN rifle. It uses a full-sized, 7.62mm round and, although automatic, is basically an updated World War II weapon. The AK-47 is a copy of the German SG-44 assault rifle. The German weapon saw extensive use toward the end of World War II, and the Russians wisely adopted it. It uses a shorter 7.62mm round. The M-16 is a high-velocity, 22-caliber weapon first proposed in the late 1930's. World War II intervened, and it took nearly thirty years for the idea finally to be accepted. The AK-74 is a Russian copy of the M-16 and is just entering service. For a number of years the AK-47 (and its upgrade, the AKM) will continue in use.

Standard LMG (light machine gun) is the most widely used machine-gun type in any army, utilized by the infantry as well as being mounted in vehicles. Until recently it was felt that this weapon had to use the heavier "standard" 7.62mm round. The reasoning was that longer-range fire could be obtained only in this way. Recent advances in weapon and ammunition design have largely eliminated this problem. However, it may be possible to design an effective armor-piercing round for the 7.62mm machine guns that could penetrate the light armor on APC's. Meanwhile, many infantry units are stuck with supplying two sizes of ammunition. The MG3 is an upgrade of the very successful German MG 42 used in World War II. The M-60 and PKM follow the same pattern, although not as successfully.

Standard squad LMG (light machine gun) is a Russian innovation. Here they solve the ammunition supply problem somewhat by giving the troops a heavy-duty version of the standard rifle. This weapon can take belts, and with its heavier barrel, it provides more firepower. The United States is finally adopting such a weapon, while the Russians are introducing one to complement their new 5.45mm rifle.

Standard sniper rifle is commonly used in a specialized form of warfare. In most armies, 2 percent to 3 percent of the troops are trained and equipped as snipers. Their weapon generally is a nonautomatic rifle using a full size (7.62mm) round.

Other Infantry Weapons

HEAVY MACHINE GUNS

These are not exactly *infantry* weapons, as they are not portable and are usually mounted on a vehicle. However, except for their larger caliber (12.7mm to 40mm), they operate much like infantry machine guns. For the 12.7mm and 14.5mm calibers, they work in identical fashion, being fed by linked belts of ammunition. These weapons have longer ranges, although this advantage is somewhat limited by the ability of the human eye to see the target. Their most common use is against aircraft or armored vehicles. In a pinch, an infantryman will not hesitate to turn such awesome firepower against two-legged targets.

PISTOLS

These are rather useless on the battlefield. Generally carried by officers, they are more a badge of office than a practical weapon. Any rational commander needing an effective personal weapon grabs an automatic rifle. The low accuracy of pistols is the chief problem; it is difficult to hit a man-sized target at ranges over 25 meters. Even if you hit someone that close, a pistol bullet generally won't have as much stopping power as a rifle.

One solution to its low effectiveness was the *machine pistol*. This is a magazine-fed automatic weapon that fires pistol ammunition. Because of its high rate of fire, it is accurate to 100 meters. However, these weapons weigh almost as much as automatic rifles. With the introduction of the 5.56mm round, machine pistols lost whatever advantage they had in lightness and high firepower. At present, machine pistols and pistols in general are best suited for police work in built-up areas.

GRENADES

Grenades are small bombs weighing about a pound and, upon detonation, capable of wounding exposed personnel out to a radius of 15 meters. The ideal grenade injures anyone within 2 meters of the explosion, 75 percent of those who are 4 meters away, 50 percent of those 6 meters away, 25 percent of those 10 meters away, 5 percent to 10 percent of those 15 meters away, and less than 1 percent of those 20 meters away. Grenades as effective as this are a recent development. Most of the World War II vintage grenades sent a large part of their fragments into the ground or straight up.

Grenades were first developed to solve the problem of shooting around corners. Fighting in built-up areas would be much more costly for the attacker were it not for grenades. The simplest way to clear out a roomful of enemy troops is to heave in a grenade.

Grenades can be thrown no more than 40 meters (20 to 30 meters is more likely). Most are fragmentation devices containing 4 or 5 ounces of explosive. They have a 3- to 5-second fuse. Some grenades produce smoke or fire to provide cover from enemy observation.

Some grenades can be thrown farther by using a special fitting on the barrel of a rifle that allows a blank round to propel them 100 meters or more. More popular is a special grenade launcher, which resembles a shotgun. These launchers fire a smaller grenade (40mm, with half the effect of the larger ones) accurately to a range of about 100 meters.

MORTARS

Mortars are the infantry's personal artillery. They must be kept light, so the infantry can keep them handy. Although they are capable of high rates of fire (up to 30 rounds a minute), not much ammunition can be carried. After all, the infantry is under fire most of the time and moves a lot.

With all these restrictions, the chief virtue of the mortar is its ability to respond rapidly and accurately to the infantry's need for firepower. An additional advantage is the mortar's ability to hit targets behind obstacles. It fires its shells in a high trajectory.

The most common mortars are 81mm. They weigh about 100 pounds and fire an under-15-pound shell 3000 to 4000 meters. The heavier 120mm (U.S. 107mm) mortar weighs 700 pounds and fires a 33-pound shell 5000 to 6000 meters.

Like grenades, mortars fire incendiary and smoke shells. Another important type of ammunition is the illumination shell. Both smoke and illumination shells must usually be delivered on short notice. Here, mortars are often more dependable than artillery. In addition, not as many shells will be needed for smoke or illumination missions. One 120mm smoke round, for example, can deliver a smoke screen that persists for over five minutes.

Some nations still use smaller mortars (50mm to 60mm). These, however, are little more than grenade throwers.

MINES AND OTHER SURPRISES

Land mines and their cousins, booby traps, are classic infantry weapons. These are defensive and enable the infantry to defeat larger forces and armored vehicles. The chief limitations of mines are their weight and the time required to plant them. As a rule of thumb, it takes at least 1 ton of mines to cover every 100 meters. It takes at least 10 man-hours per ton of mines to place them.

Mines are surprise weapons; they are usually laid in areas covered by the fire of other infantry weapons to prevent the enemy from discovering their presence until it is too late. Similar to mines are booby traps, which are grenades, mines, or other explosives rigged with trip wires or other devices. A land mine planted under an inch or less of earth is just another form of booby trap.

Some mines are meant to be dug up and reused if the enemy never encounters them. This takes three or four times as long as it took to lay them.

ELECTRONIC AIDS AND OTHER GADGETS

The revolution in electronics has not left the infantry unaffected. Most useful are observation devices. The infantry now has its own radar and, even more useful, passive night-vision equipment. The latter are often called "starlight scopes"—vision devices which electronically magnify available light so that night no longer covers enemy movement. These are often attached to weapons or are simply used to detect troops that can be attacked with mortars or artillery.

Other sensors are covered in more detail in the chapter on electronic warfare.

3

TANKS: THE ARM OF DECISION

TANKS ACCOUNT FOR about a third of a mechanized army's firepower and 20 percent of its equipment cost, yet their crews account for less than 2 percent of its manpower. Tanks spend most of their time hiding or looking for a place to hide. They must do this in order to use most effectively their considerable firepower.

The "arm of decision" hasn't always operated this way. Traditionally there have been three distinct combat forces for land warfare. First, there was the infantry, which took a lot of abuse and was absolutely necessary. Then, there were the missile troops—spear-throwers, slingers, archers, artillery—who were protected by the infantry because the missile troops were better at killing the enemy at a distance than they were at defending themselves. Finally, there was the cavalry: infantry or missile troops on horses. Better armed, trained and motivated than their footbound associates, the cavalry were the shock troops. No country could afford the expense of an entire army of cavalry. Normally the cavalry was reserved either to turn a stalemate into a victory or mitigate a defeat.

When horse cavalry became obsolete in the early part of this century, its traditions were often transferred to the tank troops, with strange results. Initially, particularly during the early stages of World War II, many armor units attempted to storm their way through all opposition. They soon learned that the opposition could shoot back with deadly effect. Sitting behind all that armor, some tank crews feel invulnerable. Experienced tankers know better.

They also know that if they are careful, they can avoid getting hurt. The importance of being careful has become more recognized by tankers over the years.

What Tanks Cannot Do

World War II destroyed at great cost the various myths about what tanks could do. Each one of the tank "no-no's" in turn define what a tanker's life is all about:

Tanks cannot advance on the enemy without thorough and continuous reconnaissance. Tanks are large (25 feet long, 12 feet wide and 8 to 10 feet high), noisy like bulldozers and confining for the crew. When the tank is "buttoned up" with all hatches closed, the crew can see out only through slits and periscopes. This makes tanks vulnerable to infantry, especially in close terrain or in a built-up area. Tanks may look dangerous, but since they can only shoot at what they can see, a nimble infantryman can easily stay out of harm's way. The infantry knows this and tends to defend in broken terrain.

One successful counter-technique is the U.S. Army's "over-watch." Half a tank unit (a platoon of two to five tanks) gets into a position from which it can observe the route of the other half. The moving half of the unit will then move into positions overlooking the route for the next bound.

Tanks cannot operate by themselves. Ideally tanks support the infantry. The infantry advances on foot just behind the tanks. When opposition is encountered, the tanks clear it, then the advance continues. The infantry protects the tanks by acting as their eyes and ears; the tanks support the infantry with their firepower. Against opposition, tanks cannot efficiently operate by themselves. As the U.S. Marine Corps puts it, "Hunting tanks is fun and easy," particularly if the tanks have charged ahead of their infantry. Well-trained units do not do this.

In any case, the infantry will be killed off more easily than the tanks. Even with its own armored personnel carriers, the infantry is more vulnerable. To be effective the infantry must dismount and expose itself to enemy fire. If infantry and tanks operate together most efficiently, the tanks will still be around after most of the infantry has been lost. At that point the tanks advance only at great risk

Tanks cannot operate in massed formations. A massed formation (tanks operating closer than a hundred meters from each other) only attracts enemy fire, particularly antitank weapons and artillery. Artillery will not destroy a tank, but it can put it out of action by damaging the engine or the tracks and associated running gear. Missile-armed helicopters are at their best in a target-rich environment. Many new air- and artillery-delivered antitank weapons are basically area weapons that will destroy whatever is in a particular area. So keep the tanks spread out.

Tanks cannot use untrained crews. A good tank crew operates as a team. A team is created through the crew operating together for at least six months. This is not always accomplished, even in peacetime. When war comes the side with the best-trained crews prevails, particularly in the case of the defender. Attacking requires much more skill and teamwork. But in the defense a tank is the ideal antitank weapon. The primary requirement is choosing a good defensive position with a safe retreat route to the next defensive position. The Israelis used this technique with deadly effect on the Golan Heights in 1973. The Russian-trained Syrian tanks just kept on coming. In less than twenty-four hours hundreds of Syrian tanks were disabled with less than one Israeli tank lost for every ten Syrian vehicles. The Syrians violated a number of other rules, but mainly their crew skill was too low to prevent a massacre. More skillful crews would have chosen better-covered routes of advance and would have been able to return fire more effectively at the outnumbered Israelis.

Tanks cannot move long distances (more than a few hundred kilometers) without losing most vehicles to mechanical breakdown. Tanks are not built for moving long distances. Weighing 40 to 60 tons and moving on tracks, they are designed for slow cross-country movement (less than 30 km per hour) rather than for high-speed road marches (50 to 60 km per hour). Russian tanks break down, on the average, every 250 km, NATO vehicles, every 300 km. With proper maintenance support, these breakdowns can usually be repaired in minutes to a few hours. Even so, a division of 300 tanks moving 100 km (three hours' marching) will have a hundred or more breakdowns. Depending on the tanks' condition, the crew's maintenance training and the efficiency of the tank-maintenance units, a division will lose 2 percent to 20 percent of its vehicles per hour of marching. Most will eventually catch up. But the effect will be disorganized units, rundown vehicles and crews, and a generally less effective combat division. The preferred method is to make long moves by train or tank transporter truck. If tanks have to move long distances, it should be only against the enemy. Even then, it's not unusual for there to be more noncombat than combat losses, and a tank with a burned-out motor or a broken track is just as ineffective as one disabled by enemy fire.

Tanks cannot neglect routine maintenance. Tracked vehicles require a lot of maintenance. Eight man-hours a day is a good average for top condition. Not a bad idea considering lives may depend on it.

Tanks cannot operate without recovery and repair units (unless you want to abandon combat and noncombat losses). Packed into one 45-ton vehicle are no less than 6 major failure-prone systems. For starters there is the track-laying mechanism upon which the tank travels. Hit an obstacle at the wrong angle and the track falls off. This is a common problem of inexperienced drivers. It takes a few hours to get the track back on. Tracks also wear out. After anywhere from 1000 to 3000 km, it's replacement time. All those wheels and rollers associated with the tracks require lubrication and inspection for wear and tear. More so than with an automobile, the driving controls, transmission, brakes, and so on must be inspected frequently and maintained to avoid complete failure, usually when you can least afford it. The tank engine is also in a class by itself. Usually a diesel (although the U.S. M1 has a gas turbine) generating 500 to 1500 horsepower, it is often under stress.

Although there's plenty of work required just to keep moving, all is for nothing if the tank's weapons are not maintained. The turret mechanism weighs in at ten or more tons. It revolves on a set of bearings and is driven by electric motors. The tank's main gun and the machine guns must be periodically resighted, particularly after hard movement or firing; otherwise they will become progressively inaccurate. Finally, there is maintaining the electronics: the radio, the internal intercom, the electrical system in general and the fire-control system, which usually includes a computer. For all maintenance procedures, checklists and manuals are provided. Modern tanks are nearly as complicated and more difficult to keep functioning than most commercial aircraft. The more skillful and attentive the tank crew, the more likely the tank will run.

All these tank no-no's add up to tank losses, a lot of them. Combat losses can be very heavy. World War II and recent experience indicate that tank losses will be five to six times personnel losses. If a tank division takes 10 percent personnel losses in combat, it can expect to lose 55 percent of its tanks. Similar rates apply for other armored vehicles, particularly APC's. However, 60 percent of combat losses and 95 percent of noncombat losses can be repaired. Depending on the repair facilities available to a division, they should be recouped in five days, 20 percent of the recoverable losses returning each day.

If the surviving crewmen are well trained in vehicle maintenance and are mechanically inclined, they are a big help with the repair work. As this assumes that the wrecks have not been captured by the enemy, possession of an armored battlefield is critical.

Tank Units

Although tanks are usually combined with infantry units, they still exist administratively as separate battalions. Only the Russians and their imitators still employ large all-tank units. Almost all tank units in the world are now, or soon will be, organized in platoons of three or four tanks, companies of three platoons (plus one or two headquarters tanks) and battalions of three companies. A tank battalion thus contains 33 to 40 tanks. The U.S. Army still uses the 5-tank platoon, the 17-tank company and the 54-tank battalion.

Tactical tests by the U.S. Army have shown that the three-tank platoon is the most efficient. In this respect tanks are used like fighter aircraft. In the "loose deuce" formation, tanks operate in pairs, or in groups of three in those rare cases of full-strength platoons. One tank is the main fighting element while being covered by its partner(s).

In addition to the tanks themselves, some armies (usually non-Russian) attach ATGM, reconnaissance and mortar platoons to the tank battalion. Russian tank battalions are usually organized into regiments of three battalions plus the specialist units. All armies attach headquarters, ammunition resupply, refueling and mainte-nance units to the tank battalion. This adds 50 to 200 men above the actual tank crew strength.

Tank Tactics

The differences between infantry and tank tactics stem from the tank's larger size, greater firepower, better protection and lesser ability to detect what is going on around it.

A tank cannot sneak up on you. Even during an artillery barrage, all that noise should alert infantry to the possibility of tanks appearing out of the dust and smoke. Thus, a tank moving forward has announced it is coming and must move accordingly. Normally tanks move forward in battle up to 100 meters in front of the infantry. The APC's are 100 or more meters behind the dismounted infantry. The tanks assume either a column or a line formation, depending on the terrain. The tank platoon commander assigns sectors for each tank to observe. An infantry squad or platoon is assigned to each tank. The infantry communicates by hand signals, leather lungs, or, in Western armies, by a telephone on the rear of the tank. The tank commander should have at least his head sticking out of the turret hatch so that he can respond to whatever the infantry uncovers. Even if the unit is advancing with the infantry in their APC's, the tank commander should be outside the turret. This is rough on tank commanders, but not nearly as rough as getting hit by the antitank weapon they couldn't see from inside. The natural reluctance of tank commanders to expose themselves to enemy fire is one of the major reasons tanks are so vulnerable. Without the ability to detect the enemy, they are just large targets. In attacks by pure tank units, the speed of advance is faster (up to 20 km or more per hour), and preattack reconnaissance becomes even more critical, especially of minefields.

Without the ability to detect the enemy, the tank's enormous firepower is useless. A tank's protection will help somewhat, but the killing power of most modern antitank weapons renders it marginal.

The tank is an ideal defensive weapon, especially against other tanks. Although it is difficult to entrench, it can defilade to present a smaller target. The tank goes to the other side of a hill, ridge or rise, and the main gun is depressed as far as it will go. The enemy tank therefore sees little more than the main gun and part of the turret. In this respect Western tanks have a significant advantage (see the tank characteristics chart) because they can depress their guns by a few additional degrees. Russian tanks must expose nearly half of their frontal profile even in defilade.

Defilade is not the only way to defend a tank. If there is cover like trees or shrubs, a tank can be concealed so it can get the first few shots off before discovery. It can then move in reverse to a new firing position. In defense, the tank's mobility, firepower and protection are decisive advantages. Using prepared "fields of fire" (sighting the gun along likely enemy approaches) as well as the usual infantry defense techniques (mines, preplotted artillery barrages), the tank commander can stay inside the tank without missing anything. In defense, this is more important than in attacking, as there will be more artillery fire.

During defense, the infantry usually deploys in *front* of the tanks, primarily to prevent the enemy infantry from getting too close to them. But here the tank truly becomes an infantry support weapon. The tank's main gun and machine guns support the infantry while the infantry provides security and information for the tanks.

Normally most of the tank's 40 to 60 rounds of gun ammunition are antiarmor. Not more than a dozen high-explosive antipersonnel shells are carried. Two machine guns, one in the turret alongside the main gun and the other on top of the turret, are the primary antipersonnel weapons.

Some tanks are fitted with flamethrowers in place of the main gun (or simply use the gun barrel). Tanks also can generate their own smoke screens, either through smoke-shell dischargers on the turret or by a mechanism that squirts fuel oil over hot parts of the engine. In this case, the leading tanks are unprotected by smoke. But this gets us into some of the more interesting aspects of antitank warfare.

Antitank Tactics

Although tanks were initially invented as infantry support weapons, it was soon realized that they were also the best antitank weapons, largely because of their maneuverability and firepower. But they are expensive and difficult to maintain. Other antitank weapons are more abundant and almost as effective. Until ten years ago tank guns without the tank were towed and sometimes mounted on lightly armored vehicles. Towed, these cost one tenth as much as a tank; self-propelled cost one third.

Then the ATGM (antitank guided missile) changed the rules for antitank warfare. More numerous, more accurate, more powerful, longer ranged and much lighter than previous such weapons, the ATGM's have some weaknesses. They have a slower rate of fire (2 to 4 rounds per minute), an often prominent backblast that is visible from the target and a need for some operator guidance to the target (although more recent models are less dependent on this). Many ATGM's, particularly the early Russian ones, are inaccurate at ranges under 500 meters. These are all fatal defects, considering that in temperate climates like Europe's the average visibility ranges for seeing a tank are under 500 meters 40 percent of the time (500 to 1000 meters, 20 percent; 1000 to 2000 meters, 25 percent; 2000 or more meters, 15 percent), based on World War II experience and recent West German tests. Even in open terrain like Middle Eastern deserts, there are sufficient undulations to cover approaching tanks somewhat.

Consider the following situation: Armored vehicles are first seen advancing from 800 meters away at a rate of 12 km an hour (10 meters every 3 seconds). Your ATGM's can get off three or four shots each before the enemy is on top of them. Each time an ATGM is fired, the enemy vehicle crews are ready to pour machine-gun fire on the location of the missile launch, which is often where the missile operator is. Even if the operator is working from the new U.S. armored ATGM launchers (in defilade, with only the launch barrel and periscope showing), the exposed periscope may be hit, leaving the missile without guidance. Worse yet, the attacker may be able to lay down a smoke screen right on the defender's position. This leaves antitank defense up to the infantry rocket launchers, which cannot penetrate the frontal armor of most tanks. In a situation like this, the defending commander has to decide quickly whether or not to pull back to the next set of positions. This would leave the enemy advancing without benefit of smoke cover or knowledge of probable enemy positions. On the other hand, moving back to new positions while under attack is tricky. This is another situation that calls for troop and leadership quality.

If the advancing enemy is primarily tanks, this shoot-fall back-shoot technique can work well. The advancing enemy (for example, a Russian tank regiment with 100 tanks and only 10 APC's full of infantry) will have limited observation capability, particularly if artillery and infantry machine guns disable the enemy's exposed tank commanders or force them to stay inside the tanks. This works less well with a mixed infantry-tank force, as the infantry is either on foot or observing from its APC's.

The most feared antitank weapons are not ATGM's or other tanks but aircraft, self-guided missiles and mines. At least the tank can shoot back at ground systems. Mines can't be discovered until vehicles are stopped by them. Aircraft appear, do their damage and disappear too fast for the armored vehicles to shoot back. Self-guided missiles (artillery shells or "fire-and-forget" missiles) are the most difficult to defend against.

There *are* measures that can be taken against all these weapons. For mines there is careful reconnaissance. Alas, the most recent mine developments make even this difficult. Currently deployed are under-two-pound "track buster" mines delivered by helicopter, aircraft, infantry, rocket or artillery. A ton of these little nasties (about 1200 mines) can cover an area 1000 meters by 100 meters. Each tank would have about a 70 percent chance of hitting one and getting a track blown off. These mines are painted in camouflage colors and are difficult to detect from a vehicle. If the area is under enemy fire, the tank is not only stopped, but it very likely will be destroyed by some other antitank weapon. Crews require a lot of discipline to stay on board in such circumstances. Some armies (for example, North Vietnam's) even chain crewmen to their positions.

Aircraft come in two flavors: fixed wing and helicopters. The only effective fixed-wing aircraft have the 23mm to 30mm cannon to penetrate the thinner top armor of tanks. They travel at tree-top level at 100 or so meters per second. You might hear them coming 20 to 30 seconds before they arrive, but you will never be sure of the exact direction. If you are moving and generating your own noise, you won't have even that much warning. Your antiaircraft weapons will usually not be able to engage aircraft flying so low and fast. After the first pass, the best bet is to run for cover. If there isn't any, man the turret machine guns and try for a lucky shot before you get hit or the aircraft run out of ammunition.

Fixed-wing aircraft can also be deadly if they carry cluster bomb units (CBU's) filled with antitank munitions. See chart on air weapons.

Then there are helicopters armed with ATGM's. These are fast-moving gunships, lurking just below the horizon, popping up just long enough to get off a missile, guide it in and duck. Doing so at ranges up to 6000 meters, they are difficult to see and hit with antiaircraft weapons. Their biggest danger is inadvertently moving too close to an undetected enemy machine gun or helicopter. For this reason, helicopters using ATGM's prefer neat battlefields, firing from behind their own troops if possible. The latest ATGM's are self-guided ("fire and forget") or laser homing. Homing is not a panacea, as countermeasures can easily be developed. Also, missiles that home in on large hunks of metal can't discriminate between disabled and functional tanks. Microprocessors could make them more discriminating but more expensive. Laser designators can be seen with special viewers and then destroyed.

There are no perfect weapons, only more destructive ones. One trend is certain in antitank warfare: It is becoming increasingly difficult for armored vehicles to stay intact on the battlefield.

The Life of a Tanker

Historically, 25 percent of tanker casualties occurred when the tank crewman was outside his tank. Tankers spend over 90 percent of their time outside their tanks, APC's and other armored vehicles. Armored crewmen are in many respects servants of their vehicles. At least eight man-hours a day of maintenance must be given to these cranky mechanical monsters. If the climate or geography is bad for people, it's worse for tanks. And the vehicles must always be in peak combat condition. During combat the crew is enclosed in a very small space. Russian and French tanks are so small that the crews are selected first on the basis of height. The shortest 5 percent of the population is preferred. Anyone over 5 feet 6 inches is at a grave disadvantage in a Russian tank.

As the turret slews around, the main gun recoils and shells of 50 pounds or more are tossed about. Fractures, lacerations and amputations regularly occur among careless, fatigued or untrained crewmen. In the best of times the driver is squeezed into the front of the tank, working the direction and speed controls while straining to see through a few viewing slits. Even with his hatch open, only his head protrudes. No wonder tanks in combat seem to move blindly.

The other three crewmen are all in the turret. The gunner sits in a small seat next to the main gun with his face pressed against the viewing devices. These are fitted with a range finder and display information about the gun's bearing and the range of the viewed objects. The quality of fire-control systems varies greatly. The skill of the gunners varies also. Skillful gunners working quality equipment are magnificently effective. The other extreme, however, is a waste of ammunition. Assisting the gunner is the loader who, next to the tank commander, is most frequently injured as he moves about the cramped space to get the next shell into the gun. If the gun is not loaded, it does not shoot. The latest series of Russian tanks (T-64/72/80) have automatic loaders and three-man crews, but the mechanical loader doesn't always work properly. With only three men to maintain more machinery, you can imagine the effect on vehicle readiness. Western armies aren't the only ones in love with new gadgets.

Another critical crew skill is speedy restocking of ammunition and fuel. With some tanks this can take over an hour. Most tanks carry up to a ton of munitions and nearly as much fuel. Run low on these supplies during a battle, and the speed with which the crew is able to replenish can be critical. Israeli experience during the 1973 Arab-Israeli War led to the design of the Merkava tank. This vehicle has large doors in the rear, not for infantry, as was first thought, but for rapid loading of larger quantities of ammunition.

The tank commander can slip down into a seat in the turret, but he can't see enough from that position to command the tank properly. So an effective commander sticks his head and shoulders outside the turret until he is wounded. When that happens, everyone gets upset until the wounded man quiets down or the corpse is allowed to fall to the floor of the tank or is tossed overboard. At that point the gunner nominally takes over. The results can be imagined.

In the defense the tank can stay buttoned up with less loss of control. A good crew will have surveyed the surrounding terrain carefully and will be able to manage without the commander exposing himself to enemy artillery and small-arms fire. At this point the biggest danger often comes from fatigue and nausea caused by the usual engine gases that leak into the fighting compartment as well as by the propellant gases from each fired shell. Russian tanks are not as well ventilated as Western tanks, and the crews suffer accordingly. In hot climates the crews become, for all practical purposes, nonfunctional after less than an hour of combat. Fortunately, combat usually doesn't last that long. As soon as a tank finds itself in a position that is exposed to enemy fire, either it gets hit or it gets out of the battle. Tanks may spend a long time waiting for the other fellow to make a false move, but they get the brutal business of fighting over with quickly.

At night, tanks can still fight, using infrared searchlights or devices that electronically multiply available light. The dangers then become ambushes and unexpected hits. A major disadvantage of infrared gear is that the engine must be started to supply power. The noise then gives you away.

Infantrymen in APC's don't live much differently from the tank crews. The maintenance load is lighter because nine to twelve men are assigned to the mechanically simpler vehicles. Also when combat is most intense, the majority of the APC troops are outside. One final note on APC's: Because they are lighter and less stable than tanks, they cannot move as quickly across broken ground without severely injuring their passengers. The tank crews are securely strapped into their heavier and more stable vehicles. Depending on the quality of the suspension system, a modern tank can move 30 to 40 km per hour, but most APC's can move about half that speed safely. So you can see there's one problem after another.

Principal Main Battle Tanks

VEHICLE. This is the vehicle's official designation.

BUILT BY. This shows the nation that originally built the vehicle. In some cases the vehicle is also manufactured in other countries. Only six nations are *original* designers and producers of tanks: the United States, Russia, Britain, West Germany, France and Sweden. Fourteen other nations produce tanks based on other nations' designs: West Germany, United Kingdom, France, Sweden, Israel and Japan.

FIREPOWER. This is a numerical evaluation of the vehicle's firepower. It is arrived at by taking into account the following factors:
The "proving ground" performance of the vehicle's main gun and the various ammunition. New types of ammunition have increased the lethality of tank guns. A good example is the latest versions of the discarding-sabot ammunition for the 105mm gun. Particularly effective also is the HESH (high explosive squash head). This shell hits a tank, flattens out and explodes. This transmits a shock wave through the armor, which shakes things loose and causes pieces of armor and equipment inside the tank to fly about. The effect is usually fatal to the crew and disabling for the vehicle. The HESH is expensive, but does not lose velocity at long range or require a precise angle of hit like a HEAT (shaped-charge) shell.
The fire-control system. This includes the type of range finder (see below) as well as the computing system. The more recent electronic fire-control computers have proved to be much more effective than the older mechanical types. Unfortunately, the fancy gadgets are usually less reliable, although they are improving.

PRINCIPAL MAIN BATTLE TANKS

Russia has the most tanks, but also the oldest and not always the best.

Vehicle	Built by	Fire-power	Pro-tect	Range (km)	Grnd Pres	HP: Wt	Gun Dprs	Wght (tons)	Max Spd (km)	Range Finder Sys	Hght (m)	Main Gun	Rnds on Board	Rnds per Min	Max Range	MG 1	MG 2	In Use	Intro-duced
T-80	Russia	10	9	400	11	24	4	42	60	Laser	2.3	125	55	8	3000	12.7	7.62	3000	1981
T-72	Russia	9	8	500	12	25	4	40	60	Laser	2.5	125	50	8	2000	12.7	7.62	15000	1972
T-62	Russia	8	6	480	11	19	4	37	59	StadR	2.4	115	40	4	1500		7.62	45000	1962
T-55	Russia	6	4	300	12	16	4	36	50	StadG	2.4	100	43	3	1000	12.7	7.62	28000	1957
PT-76	Russia	3	2	260	7	17	4	14	44	StadG	2.2	76	40	4	1000	7.62	7.62	4000	1955
																	Total	95000	
M1	US	11	10	560	14	26	10	58	65	Laser	2.4	105	55	6	3000	12.7	7.62	1200	1981
M60A3	US	10	7	300	11	19	10	48	48	Laser	3.2	105	63	6	3000	12.7	7.62	1500	1977
M48A5	US	10	7	290	13	19	10	47	48	Coin	3.1	105	57	6	3000	12.7	7.62	1900	1976
M60A1	US	9	7	310	12	18	10	49	48	Coin	3.3	105	63	6	2500	12.7	7.62	6700	1962
M48A3	US	7	6	287	13	16	9	47	48	Coin	3.1	90	62	6	2000	12.7	7.62	3800	1952
																	Total	15100	
Leopard II	WGer	12	10	350	13	30	9	50	68	Laser	2.5	120	60	8	3500	7.62	7.62	1200	1978
Leopard I	WGer	9	7	375	13	23	9	40	65	StroC	2.6	105	60	6	2500	7.62	7.62	5800	1965
																	Total	7000	
Chieftian 5	UK	11	9	310	14	14	10	55	48	Laser	2.9	120	64	8	3500	12.7	7.62	1800	1966
Centurion 13	UK	9	7	250	14	14	10	52	35	RMG	3	105	64	6	2500	12.7	7.62 / 7.62	4000	1959
																	Total	5800	
AMX-30bis	Fr	9	6	400	12	19	8	36	65	OptC	2.8	105	50	8	2500	12.7	7.62	2000	1967
S-Tank	Swed	10	7	250	14	6	10	39	50	Laser	2.4	105	50	15	3000	7.62	7.62	300	1968
Merkava	Isr	10	8	500	15	16	10	58	45	Laser	2.7	105	50	6	3000	7.62	7.62	300	1978
Type 74	Jap	9	7		13	20	5	38	60	Laser	2.7		50	6	3000	12.7	7.62	280	1972
Type 61	Jap	7	6		14	17	5	35	45	Coin	3.2		50	6	2000	12.7	7.62	500	1962
																	Total	780	
										"West"							Total	31280	

The internal layout and organization of the tank. The cramped Russian tanks suffer in this respect. Crew performance is quickly degraded in them.

Gun stabilization and platform stability are the ability of the tank to provide the firing gun with sufficient stability on the move or immediately after a halt.

Ammunition carried. The more you have the more you can use to hit the other fellow.

Rate of fire. Especially the ability to get off the first shot and—in these days of better armor—the second shot. Experienced crews can fire faster than the number indicated for each tank. The chart merely indicates the tank's intrinsic rate of fire.

PROTECT is the numerical evaluation of the vehicle's ability to defend itself against enemy fire. This is a combination of the following factors:

Quantity and quality of armor. Armor thickness counts for less today than the composition of the armor. The recently introduced Chobham (composite) armor gives excellent if expensive protection against antitank weapons. This armor has layers of metals, ceramics and plastics which absorb and break up kinetic (high-velocity) and HEAT (shaped-charge) shells. Composite armor does not make the vehicle invulnerable, but it does allow it to survive otherwise fatal or disabling hits. The M1, Leopard II and Chieftain use composite armor. The T-80 apparently uses a Russian version.

Speed of the vehicle. This is a combination of actual top speed, vehicle power (see HP:WT below), ground pressure and quality of suspension system. Power and speed enable a vehicle to get out of the way quickly. Higher ground pressure makes it more likely that the vehicle will become stuck. The better suspension systems prevent the crew and the vehicle from being knocked about during high-speed cross-country movement.

Ability to lay smoke. Some vehicles have smoke-grenade dischargers. Others can form smoke by spraying fuel oil on hot parts of the engine. Some vehicles cannot produce any smoke to hide in.

Size. All armored vehicles are large. Height is the best indicator of a vehicle's ability to remain unseen.

Main gun depression. The greater this is, the less the tank is exposed when it goes into defilade behind a hill, with just its own gun and top of the turret visible to the enemy.

Viewing devices from inside the tank. Ideally the tank commander should have his head outside the tank. But this is not always possible. Various arrangements are made in tanks to provide viewing slits protected by bullet-proof glass. The quality of the gunner's sight is also considered.

RANGE is the unrefueled range of the vehicle in kilometers. Generally, in combat, 100 km of range equals 3 to 5 hours of running time (assuming 40 percent time off the road, 20 percent on the road and 40 percent not moving with the engine idling). This will vary with the seasons: more time in the summer with dry ground, less in the winter either during severe cold or mud. Cruising speed is usually 30 to 40 km per hour on the road.

GRND PRES is the ground pressure in pounds of pressure per square inch. The lower this is, the more easily the vehicle is able to cross soft ground like mud or sand. An infantryman's weight averages 2 to 10 pounds per square inch.

HP:WT is the horsepower-to-weight ratio (the horsepower of the engine divided by the vehicle weight). The higher this is, the more "lively" the vehicle will move. This is more important for acceleration than for pure speed.

GUN DPRS is the gun depression in degrees. The greater the depression, the better. A tank best defends from defilade. That is, it backs up behind a hill as far as it can go without being able to sight its gun over the hill. Depending on the slope of the hill, very little of the vehicle is visible to the enemy. At best, all the enemy sees are the gun and the top of the turret. On gentle slopes a small depression is adequate. Steeper slopes require more depression, unless you want to expose more of the tank.

WGHT is the full load weight of the vehicle in metric tons.

MAX SPD (in kilometers) is the maximum speed of the vehicle. Going cross-country, this is limited by the suspension and weight of the vehicle. Heavier vehicles have an easier time. For this reason APC's usually cannot keep up with tanks without permanently damaging their passengers.

RANGE FINDER SYS is the type of range-finder system used. Laser is the easiest and most accurate. *StadR* (stadia reticle) is quite primitive. *Coin* (coincidence), *StroC* (stereo coincidence) and *OptC* (optical coincidence) are based on more elaborate optics and get much better as the gunner's experience increases. RMG (ranging machine gun) fires 12.7mm machine-gun rounds (with tracer) until the target is hit. Then the main gun, mounted along the same axis as the machine gun, is fired.

HGHT is the height of the vehicle in meters (1 meter = 3.3 feet). Measured to the top of the turret.

MAIN GUN is the caliber of the main gun in millimeters.

RNDS ON BOARD is the number of main gun rounds of ammunition carried. The more the better. Normally a mix of armor-piercing and antipersonnel rounds are carried. Usually over 75 percent of the rounds are armor piercing.

RNDS PER MIN is the nominal number of rounds per minute the main gun can fire. The higher the better.

MAX RANGE is the maximum effective range of the main gun in meters. The farther the better.

MG 1, MG 2 are the machine guns carried in addition to the main gun. The caliber of the machine guns is given in millimeters. One machine gun is usually mounted next to the main gun and can be fired in its place. The second machine gun is usually on the top of the turret for defense against aircraft or nonarmored ground targets.

IN USE is the number of this type in use as of 1982. The BMP, BMD, XM2, XM3, T-80, T-72, M1 and Leopard II are still in production. The annual production and number on order for each of these vehicles is: BMP—2000 per year (12,000); BMD—200 to 300 per year (2000); XM2/XM3—1400 per year (6900); T-80—1200 per year (25,000); T-72—1400 per year (10,000); M1—800 per year (9000); Leopard II—400 per year (2300). Because Western nations do not produce vehicles at full production rates as the Russians do, they could therefore quickly achieve annual rates two to three times those given.

INTRODUCED is the year the vehicle was first put into use. It gives you an indication of how old some vehicles are. Any vehicle over ten years old is living on borrowed time.

Principal Armored Personnel Carriers

APC's (armored personnel carriers) are also referred to as IFV's (infantry fighting vehicles). They are also used widely as reconnaissance vehicles, often with more weapons mounted and fewer men carried.

Many of the terms are the same as those of the tank charts (see above).

COMFORT is the relative "livability" of the vehicle for passengers. The higher the number, the more livable it is. This is an important consideration because low livability tires the passengers and renders them less effective.

PASSENGERS is the number of passengers carried comfortably in an APC. You can carry 30 percent to 40 percent more with a lot of crowding.

FLOAT indicates whether or not the vehicle can float. The number indicates the speed (in kilometers per hour) in the water.

GUN PORTS is the number of gun ports in an APC that the passengers can use to engage targets with their rifles.

WEAP 1, 2, 3 represents the weapons mounted on an APC. Given in caliber (millimeters). All are machine guns except the 73mm guns on the BMP and BMD. Recent versions of the BMP have an automatic (23mm) cannon mounted in place of the 73mm gun. ATGM is an antitank guided missile that can be fired from the vehicle.

Crew size for tanks is four, except for the S-Tank, T-72 and T-80. These vehicles have automatic loaders and thus have only a vehicle commander, gunner and driver. APC's usually have a crew of only two (commander-gunner and driver). Some have a third man assigned as a gunner if there are sufficient onboard weapons.

PRINCIPAL ARMORED PERSONNEL CARRIERS

Russia has the edge in quantity and quality of APC's.

Vehicle	Built by	Fire-power	Pro-tect	Com-fort	Grnd Pres	HP: Wt	Pass-engers	Wght (tons)	Max Spd (km)	Hght (m)	Float	# Gun Ports	Max Range (km)	Weap 1	Weap 2	Weap 3	In Use	First Used
BMP	Russia	6	3	3	9	23	11	12	55	2	8	9	300	73	7.62	ATGM	25000	1967
BMD	Russia	6	3	4	9	28	9	10	55	1.9	6	0	300	73	7.62	ATGM	3000	1969
BRDM	Russia	2	2	5	Wheel	20	3	7	100	2.3	10	0	750	14.5	7.62		15000	1966
BTR-60	Russia	2	2	5	Wheel	18	16	10	80	2.3	10	6	500	14.5	7.62		15000	1961
BTR-50	Russia	2	2	5	7	17	16	15	44	2	10	0	260		7.62		20000	1957
																Total	78000	
XM-2	US	8	5	6	7	23	9	21.4	68	2.6	7	6	480	25	7.62	ATGM	3500	1981
XM-3	US	8	5	6	7	23	5	21.4	68	2.6	7	0	480	25	7.62	ATGM	3400	1981
LVTP-7	US	3	3	7	9	17	28	24	60	3.3	13.5	0	480	12.7	7.62		940	1972
M-113	US	2	2	6	8	20	13	11	65	2.5	5.8	0	480	12.7	7.62	ATGM	45000	1960
																Total	52840	
AMX-10P	Fr	6	4	6	8	20	11	14	65	2.5	7.9	2	600	20	7.62		2000	1973
Marder	WGer	7	5	6	12	21	10	28	75	2.9	No	2	520	20	7.62	7.62	3000	1971

VEHICLE NOTES

Each nation tends to have a "philosophy" on armored warfare which it applies consistently to its vehicle designs.

Russia has gone for massive numbers of effective yet expendable tanks. The tanks have effective guns but light armor. They make up for this lack of armor by being low and wide with a well-sloped shape to increase the effectiveness of armor. The tanks are fast. The Russians were the first to accept the fact that a tank could run about the battlefield like a sports car. Western drivers were amazed when they first tested them. They are cramped, uncomfortable, difficult to maintain (especially with all the new gadgets they keep putting in them) and numerous. The Russians have built about 50,000 tanks in the past sixteen years. The oldest one still in service, the T-55, is a direct descendant of the famous T-34 of World War II. First produced in the early 1950's (as the T-54), the T-55 is still produced in non-Russian countries. China, in particular, makes a T-55 variant called the T-59. The T-55 caught fire too easily (fuel tanks in the front of the vehicle did not help) and had inferior fire control. Still, it was an improvement over the T-34.

In the early 1960's the T-62 became the main tank of the Russians. This had a more powerful gun than the T-55 and was the only successful improvement. Beset by numerous mechanical problems, the T-62 appears to have been replaced sooner than planned.

In the late 1960's the T-72 came along, first as a model called the T-64. A radical departure, it contained a laser range finder and a more complex fire-control system (with more maintenance problems). In addition, an automatic loader was installed, reducing the crew size to three. This meant that there was 25 percent less manpower to perform 25 percent more maintenance. The gun is a slightly improved version of the one found in the T-62. Not entirely satisfied with the T-72, the Russians are starting just now mass production of the T-80, which apparently tries to digest properly the technological advances of the T-72. In addition, it features composite armor.

The PT-76 persists in service because it is basically a good design for reconnaissance.

Right after World War II, the Russians quickly went from wheeled APC's to fully tracked APC's to provide infantry transport that could keep up with their tanks. They were not entirely successful. The BMP is so cramped and uncomfortable that the infantry is in poor shape to do much fighting after a high-speed romp with the tanks. The vehicle looks impressive as hell, though. The earlier APC's are still used. In addition, divisions in the Arctic use an unarmored tracked vehicle specially designed to keep going in the snow and the cold.

The *United States* never developed a reputation for outstanding armored-vehicle design. After World War II, the United States built tanks that tried to have the best of everything: thick armor, heavy firepower, high crew comfort, advanced fire control. In most particulars, these objectives were achieved. The biggest problem was maintenance. Although much comment is made about the larger size of U.S. tanks, this does not appear to have adversely affected their combat performance. United States tanks were built to last, particularly through extensive peacetime use. The M48 was a development of the heavy tanks developed toward the end of World War II. The M60 was basically an upgrade of the M48 design. A recent upgrade of the M48 (M48A5) is considered the equal of the earliest M60. The new M1 tank is, for all practical purposes, a development of the M60. The only radical addition is composite armor.

The U.S. APC's were built just for transport. The primary stimulus for building a U.S. IFV was the Russian BMP. The IFV concept (fighting from the vehicle, in particular) has yet to be proved in combat. Evidence to date is not encouraging. The U.S. XM-2 is similar to the BMP with the 23mm cannon; the XM-3 is the reconnaissance version. The M-113 continues to be widely used, as are a few thousand World War II era half-track APC's. The LVTP-7 is the U.S. Marine Corps' amphibious APC.

Other nations adhered to slightly different tank-design formulas. *Germany,* with loads of World War II experience, came out with a series of tanks somewhere between those of the United States and Russia in philosophy. The Germans stressed quality, high firepower and speed, and accepted lower weight and protection. The *French* went for an even lighter tank, while the *British* opted for higher firepower and protection at the expense of speed. The *Swedes* developed a turret-less tank with an automatic loader. This was more of a defensive weapon, well suited to their war policy.

Japan simply adopted what it needed from Western designs in order to produce a good defensive tank. *Israel,* after years of using other people's tanks, designed its own, the Merkava, for defensive warfare. It is heavily protected, and its larger-than-usual storage compartment is easy to resupply through large doors.

Non-U.S.–Russian APC's usually follow the Russian model. The AMX-10P and the Marder are both similar to the BMP.

Switzerland, Austria, Argentina and India all produce tanks with designs similar to those shown in this chart.

With a few exceptions, nations that manufacture tanks do not use those of another nation. Germany still has thousands of U.S. tanks. It is only in the last twenty years that the German tank-building capability was revived after it was stopped at the end of World War II.

Nonproducing nations tend to have more than one type of tank. Purchasing armored vehicles appears to have more to do with political arrangements than with technical merit.

Portable Antitank Weapons: Missiles and Rockets

There are three general types of antitank weapons today: missiles and rockets, guns and mines. All can use shaped-charge warheads (see below for explanation). Guns are usually found on tanks or other vehicles. Some antitank guns, particularly in the Russian Army, are still towed. Most guns rely primarily on *kinetic* (high-velocity) shells, which allow for quick shooting. These shells travel at over 1000 meters a second.

Mines, one of the more effective weapons, are discussed elsewhere. Mines are passive; you must place them in the way of the enemy. For most troops not in tanks, their main hope against mechanized combat troops lies in the weapons shown in this chart, which are also the most numerous antitank weapons. While

ANTITANK WEAPONS

More than at any previous time, antitank weapons are numerous and effective enough to stop tanks.

Maker	Name	Accuracy % Probability of hit					Armor Pen (mm)	Effective Range (meters)		Speed (mps)	Back-blast	Missile Weight (lbs)	Launch System Weight (lbs)
		>300m	>500m	>1000m	>1500	>max		Min	Max				
US	TOW	80	90	90	90	90	750	65	3000	200	Yes	40	184
US	TOW2	80	90	90	90	90	750	65	3750	200	Yes	40	
US	TOW3	80	90	90	90	90	1500	65	3750	200	Yes	40	
US	DRAGON	0	90	90	0	0	500	300	1000	100	No	27	32
US	Hellfire	0	80	90	90	90	900	500	6000	300	No	95	
US	LAW	50	0	0	0	0	200	5	75	100	No	5.5	
Russia	AT-3 Sagger	0	50	70	90	90	430	300	2000	150	No	25	57
Russia	AT-4 Spigot	50	90	90	90	90	550	50	2000	200	No	30	
Russia	AT-6	0	0	80	80	80	750	1000	4000	300	No	60	
Russia	RPG-7V	80	30	0	0	0	320	5	500	200	No	5	15.4
Russia	RPG-LAW	60	0	0	0	0	250	5	200	100	No	6	12
Russia	RPG-16	80	70	0	0	0	500	5	500	200	No	16	
France	Milan	70	90	90	90	90	500	25	2000	180	No	26	
France	HOT	70	80	90	90	90	750	75	4000	200	Yes	46	
Britain	Swingfire	50	70	90	90	90	500	150	4000	150	No	75	
US	106mm RR	90	80	70	0	0	500	10	1100	300	Yes	37	460
Sweden	Carl Gustav	80	40	0	0	0	400	5	450	200	No	5.7	34
W Ger.	Armbrust	80	0	0	0	0	300	5	300	200	No	2.4	13.9
	Averages	56	58	57	49	49	575	146	2146	193		32	101

there are only some 100,000 tanks in use, there are over 100,000 guided-missile systems and millions of antitank rocket launchers. It is debatable which system will ultimately more effectively destroy tanks and other armored vehicles. Each tank carries at least 30 antitank shells, with 4 or 5 reloads. Each missile launcher has only about 10 missiles. Tanks are expected to last longer than missile launchers. Keep in mind that the weapons described below have not been fully tested in combat. Since they were last used extensively in 1973, there have been considerable technical improvements.

MAKER is the country of manufacture.

NAME is the official designation of the weapon. The chart does not contain all the weapon types currently in use. It does contain the vast majority. For example, for the TOW system about 300,000 missiles have been produced, with production continuing. The TOW2 has just entered production in 1981, and the TOW3 will start being made in 1983. The DRAGON is a smaller and less effective missile, and therefore probably fewer than 100,000 missiles will be produced. Hellfire may begin production in 1982 and may eventually be made in large quantities (over 100,000 missiles). The LAW is a one-shot rocket launcher that has been manufactured in the hundreds of thousands. The Sagger has probably been produced in quantities as large as 250,000 but is no longer being made. It is being gradually replaced by the Spigot. This points out a peculiar problem with missiles: they are expensive, as much as $10,000 each. The Russians, in particular, do not readily discard obsolete weapons. Although nearly 15,000 TOW missiles have been fired, the majority in training, far fewer Saggers have been expended. Already many of these missiles' propellant, explosive and guidance components have deteriorated to the point of uselessness. In sum, many Russian and not a few Western missiles will not perform in combat because of old age.

The RPG-LAW is a one-shot rocket launcher. The RPG series are all rocket launchers. With the exception of the one-shot RPG-LAW, a recent development, they can be reloaded. The Milan, HOT and Swingfire are all European missiles similar to the US TOW (HOT, Swingfire) and DRAGON (Milan). Only the Milan has been produced in large quantities (over 130,000 missiles). The 106mm RR (recoil-less rifle) has been included because it is still used by many nations. It is usually mounted on a light, unarmored vehicle, or on no vehicle at all, and has a prominent backblast that exposes its position. It's better than nothing.

The Carl Gustav and Armbrust are particularly effective rocket launchers that are used by a number of countries.

ACCURACY is the percentage probability of hitting a target at various ranges. Destroying the target depends on armor penetration. The percentage probability of obtaining a hit is given for each distance shown at ideal conditions. Many older missiles require a few seconds after launch for coordinating the guidance system and the recently launched missiles. This accounts for the sometimes lengthy minimum range. The maximum range is often a function of the reach of the guidance system. Many systems have a thin wire trailing from the missiles to the guidance system. When you're out of wire, you're out of control. Missiles also run out of momentum. The missile propellant is usually burned up within seconds of launch. This, plus visual limitations, prevents hits at farther distances.

Other factors degrade accuracy. To obtain the listed hit probabilities, you need a stationary or slowly moving target in plain sight, in clear weather and not shooting back. A rapidly maneuvering target (30 to 40 km per hour; that is, 8 to 11 meters a second), heading for cover will be more difficult to hit, particularly at longer ranges where the missiles will take up to 20 seconds or more to get there. Foggy weather, artificial smoke or the usual battle dust make seeing and hitting the target more difficult. Finally, the enemy shoots back if given a chance. Awareness of missiles causes an experienced enemy to maintain a missile watch. Even a missile with little backblast generates some dust under dry conditions. The missiles themselves can be seen. Once the experienced enemy detects incoming missiles, he will head for cover while shooting back at the source of the launch. This has a tendency to disturb the aim of the soldier guiding the missile.

There are four ways to get the missile on the target. The most primitive is used by the rocket launchers (LAW, RPG, Armbrust, Carl Gustav). Since the projectiles are unguided, you simply *aim the weapon,* pull the trigger and hope for the best. The earliest versions of the ATGM (antitank guided missile) allowed the operator to *maneuver the missile to the target* with a joystick. Speed could not be controlled, only altitude and direction. If the operator got nervous or lacked skill, accuracy suffered. Only the Sagger and Swingfire still use this system. The next generation (TOW, DRAGON, Milan, HOT, AT-4) use a system whereby the operator need only *keep the target in his sights and the missile will home in on it.* Enemy fire can still make the operator wince and, as the sight shifts, so does the aiming point of the missile. The Hellfire and AT-6 use a terminal homing system. The missiles are launched in the general direction of the targets. The *missile has a seeker in it that homes in on target* once the missile is 1000 to 2000 meters from the target.

The targets are identified several ways. Most dependable is a ground or air-based laser designator, which illuminates the target with laser light invisible to the naked eye. Other seekers can home in on heat sources (infrared), images (any object that fits the profile of the desired target), large masses of metal, or large moving objects.

Each method has shortcomings. Lasers cannot easily penetrate smoke or fog. The seekers are also blinded by smoke and fog, but this plagues any guidance system. Special sighting devices also make the laser light visible, thus allowing the defender to bring the designator under fire. If no ground designators are available, the attacker can use a designator at the point of launch. This is not much better than the third system explained above. Infrared seekers can be distracted by false targets like burning vehicles or even enemy flares. Microprocessors will eventually provide a degree of discrimination, but they are not available now. Image recognition and metal detection are also a few years away.

The biggest advantage of the seeker approach is that the airborne or land launch platforms don't need to hang around to guide the missile in. They survive much longer.

Some missiles (Hellfire and AT-6) home in on targets without guidance, but most require human assistance. Shot and shell falling close by has been shown to degrade human guidance. Combat experience against antitank missiles has established a set of defensive tactics—rapid movement for cover and firing on the missile's operator—which, in many cases, has reduced missile effectiveness by ten or more times. An instinct for self-preservation and a little experience usually degrades the most lethal weapons.

ARMOR PEN. is the number of millimeters of armor the warhead will penetrate if the armor is hit *directly* at a *right angle*. Shaped-charge warheads operate by "focused explosion." The front half is a hollow shell, the rear half explosive with a cone-shaped depression. At the tip of the warhead is the detonator. When the warhead strikes the target, this detonator ignites the rear of the explosive. The cone-shaped depression causes the explosive to form a metal-penetrating stream of superhot gas. This plasma jet burns a small hole in the armor and once inside the tank, will usually ignite something else like the ammunition, fuel or crew. It is not always fatal, as the plasma jet is only 10 percent to 20 percent the width of the warhead and dissipates quickly. The rule of thumb is that a shaped charge penetrates five times its warhead diameter (a 100mm-wide warhead penetrates 500mm of armor).

There are several ways to defeat shaped charges:

Standoff distance. Once the plasma jet is formed, a fraction of a second after impact, it burns through whatever is in front of it and is gone. If sheets of thin metal are rigged about 300mm from the tank's armor, the warhead explodes and the plasma jet burns harmlessly through the air space. Because of the need for standoff distance, shaped-charge missiles must move slowly. If they are too fast, they will not allow a proper time for the formation of the plasma-jet explosion. Thus, it is not possible to use a higher-velocity missile that will crash through such shields.

Sloped armor. This is also used to ricochet kinetic shells off the tank. Shaped-charge rounds will do the same. Even if they ignite at a bad angle, the warhead may be jarred and a less effective plasma jet will be formed. Most modern tanks have relatively flat surfaces only on the turret sides. The turret is only a small portion of the tank's target area, so the average slope degrades the average shaped-charge hit from 25 percent to 50 percent. A normal penetration of 500mm becomes 375mm or 250mm.

Composite armor. Instead of just 50mm to 200mm of armor, an equal or greater thickness of lighter, layered materials is used. This combination of metal, plastic and ceramic layers absorbs much of the plasma jet's heat without melting. Good composite (or Chobham) armor degrades a shaped charge's penetration by a factor of three or more.

The above defenses, especially the sloped armor present on almost all tanks, combine to make shaped charges much less effective when they hit. Tanks like the U.S. M1 (with some 200mm of sloped, composite armor plus some thin metal skirts) can withstand most hits by warheads capable of penetrating over 1000mm of normal armor. This does not produce immunity. Even nonpenetrating hits can damage components of the tank, like the running gear, engine, weapons. One or more hits may make the tank ineffective without destroying it. By comparison, the maximum armor thickness of other modern tanks is: T-55/T-62—170mm; M-48/M-60—120mm; Leopard I—70mm; AMX-30—50mm; Centurion—152mm. Western tanks have thicker armor on the sides and rear. Russian tanks are thinly armored in these areas and depend on not letting the enemy shoot at anything but their fronts. The above thicknesses should be multiplied by 1.3 to 1.5, depending on how good the armor slope is. Using composite armor adds 60 percent more thickness without additional weight but with additional bulk.

Meanwhile, the technology of shaped-charge warheads goes forward. One improvement is combining two shaped charges in the same warhead plus a metal "penetrator." These two charges go off one behind the other, thus increasing the penetration to as much as ten times the warhead diameter.

EFFECTIVE RANGE is expressed in minimum and maximum. The minimum is necessary to arm the warhead and get the missile under operator control after launch. This control time varies with the sophistication of the missile system. Maximum range is also a matter of control. A shaped charge is just as effective at any range. The critical factor is being able to hit something. Unguided rockets become much less accurate as they run out of speed. A guided missile is controllable only for so long. The wire-guided missiles are limited by the length of their wire. Newer types do not use wire, but these are limited by either the ability of the missiles to stay airborne or by the fire-control equipment's ability to keep the missile under control.

SPEED (meters per second). This is limited primarily by the missile's need to hit the target at relatively low speed to allow the plasma jet to form. Some missiles, because of technological superiority, can hit their targets at a higher speed. Ideally these missiles should travel at the highest speed possible. This reduces the time in flight and makes spotting the missiles more difficult. At long ranges (over 3000 meters) a 200-meter-per-second missile takes 15 seconds to reach the target. That's a long time—enough time for the enemy to react.

BACKBLAST. All missiles have backblast. Some have a very prominent backblast and are indicated by a *Yes* label. Such missiles will be spotted more easily by their intended victims. Even the Sagger, with a very small backblast, was detected easily by Israeli troops once they learned vigilance and effective evasive tactics.

MISSILE WEIGHT. This is the missile, rocket or projectile weight itself and shows the relative portability of the system.

LAUNCH SYSTEM WEIGHT. Many systems have a reusable launcher containing the guidance system. This usually includes a power supply. The LAW-type weapons are self-contained. The launcher is thrown away once the rocket is launched. The heavier systems are used only in vehicles. Hellfire and the AT-6 are meant primarily for use in aircraft.

4

ARTILLERY: THE KILLER

ARTILLERY IS large-caliber guns firing explosive shells. From the user's point of view, artillery is a humane weapon. It does enormous destruction to the enemy and poses little risk to friendly troops. Even better, the users are usually out of sight of their victims. Artillery is a rich man's weapon, a less wealthy army can be just as destructive without artillery. All it has to do is trade lives, rather than guns and shells, for lives.

When asked which weapon they fear most, soldiers put artillery at the top of the list. The reason isn't just that artillery causes the most casualties—artillery is unpredictable. You can't fight it. Even tanks can be fought, but artillery is out of sight and always there with death and mutilation.

During World War II, artillery accounted for 58 percent of all casualties. In open plains and deserts, 75 percent of the casualties were artillery caused; in mixed terrain, 63 percent; and in forests and towns, 50 percent. Today most combat troops have armored transport, but artillery guns, munitions and their methods of employment have improved considerably. In addition, armies are more motorized and more dependent on supplies of fuel, ammunition and spares, all carried by unarmored vehicles. Even heavily armored Russian divisions consist of 68 percent unarmored vehicles, perfect targets for the new, improved artillery. The infantry, and everyone else, has more to fear.

Artillery Fire and Missions

There are two primary types of artillery fire: *barrage* and *concentration.*

A barrage is literally a wall of fire—shells exploding in a line—that is usually employed to screen troops from enemy observation or to prevent enemy movement. A rolling barrage moves forward at preplanned speed in front of an advance. If this is done properly, the advancing troops will appear out of the shellfire in front of the defending positions.

A concentration is high-density fire for the purpose of destroying a specific target. Barrages and concentrations are fired at three levels of intensity: harassment (up to 10 percent destruction, enough to keep the target troops' heads down), neutralization (about 30 percent destruction, causing a temporary inability of the target to perform), destruction (50 percent to 60 percent destruction, resulting in the permanent inability of the unit to perform). "Temporary" means about a day. "Permanent" means until the unit is pulled out of combat, rebuilt and rested.

There are a number of basic artillery missions. Each is either *preplanned* (guns are assigned to fire a specific number of shells at a specific target according to a schedule), or *on call* (similar to a preplanned mission except fired only when called for), or *target of opportunity* (an observer registers the fire on the spot).

Offensive and defensive barrages are preplanned fires to assist attacking or defending troops by providing a wall of fire. Usually of neutralization or destruction intensity, they may be either stationary or rolling (moving every few minutes). They often use smoke and high-explosive shells together.

A *standing barrage* is a screen to prevent enemy movement or observation. Done at harassment intensity, it often includes smoke shells and poison gas. It guards a flank of an advance or cuts off the enemy's retreat or route for reinforcements. It is almost always preplanned or on call.

A *fire assault* concentrates against specific targets for the purpose of destruction. It is usually preplanned, but is also available to observers during attacks. This is the heaviest intensity of fire.

Harassment is random fire on enemy positions to keep the

enemy from functioning normally or sleeping at night. It is usually very light fire, causing only a few percent destruction.

Interdiction is similar to, but heavier than, harassment fire. Usually employed on roads or routes behind enemy lines for the purpose of impeding movement, it has varying degrees of intensity depending on the ammunition supply or how badly you want to interdict.

Counterbattery is fire at enemy artillery to suppress or destroy.

Techniques of Artillery Use

Artillery is warfare by the numbers. Even aircraft, for all their technical sophistication, are guided to the target by the skill and talent of a human pilot. Artillery is almost wholly mathematics and formulas. References on a map guide the artillery crews as they service their guns. Human, or more frequently, electronic, surveyors plot the position of the guns. Previously surveyed maps provide the location of the targets. Artillery observers at the front radio back the approximate location of targets that don't show on a map. Even this can now be done with a laser device that either computes the distance and bearing of the target from the guns or paints the target with its rays. The artillery shells then home in on the rays reflected from the target. The electronic battlefield is closer than you think, and, as usual, the artillery leads the way.

Beginning with World War I, artillery fire became almost primarily indirect; that is, the gunners could not see their targets. Instead, trigonometry, ballistics, map surveys, electronic communications, observers and registration by fire directed the shells from behind friendly lines. Out of the way of enemy observation and fire, the guns were less likely to be hit. Thus, the guns survived longer in combat while their targets survived less.

The techniques of modern artillery are quite simple. First, the ballistic flight of the artillery shells is fairly predictable. Adjustments are made for unpredictable elements, such as the minute differences in propellant from one batch of shells to another and certain aspects of the weather like wind and humidity. Second, map making has reached the point where by using traditional survey techniques, the position of the guns can be calculated accurately. To locate the target, triangulation is used. The position of the guns is surveyed accurately, and the observer estimates the distance and bearing of the target in relation to a fixed point (his own location or a prominent geographical reference). Then trigonometry is used to compute the bearing and barrel elevation (for range) of the guns. Shells are fired, one at a time, and corrections are made until the shell is on target. Then all the other guns let loose on the correct bearings. This registration process often takes place before a battle. After that, the observer or commander simply calls down fire on a preregistered point. This fire is called a "concentration" because it concentrates the fire of a certain number of guns on a certain area for a certain amount of time.

During World War II, communications techniques first allowed each observer to control hundreds of guns, "every gun within range," as the saying went. Before that time, each observer talked to one unit of guns. Much more complex communications and plotting systems were required to tie in a large number of artillery units and observers. The U.S. Army first developed this system in the late 1930's, and continues to lead its development.

Because the artillery units themselves will be spread all over the place, determining the bearing and elevation is not only a very complex problem but one that must be solved quickly before the target moves. Today Western armies can have the shells on the way in less than a minute (sometimes in less than 15 seconds). The major flaw in all this is the increasing use of electronic jamming. Without dependable communications, the guns might as well go back to direct fire, although good training and imagination can overcome the damage caused by electronic warfare.

The Russian-trained armies employ about half their artillery for direct fire. This solves the electronic-warfare problem. It also allows the enemy to shoot back more easily at the artillery. Although the artillery in direct fire is two or three times as effective as in indirect fire, its losses increase ten times or more. This is not surprising considering these direct-fire guns must be used 1500 meters or less from their targets. Any farther away, the gunners can't see the target, because of terrain, climate or the smoke of battle. Also, even though an increasing proportion of Russian artillery is self-propelled and lightly armored, the majority is still towed. This makes the guns and crews much more vulnerable to enemy fire. Although the Russian artillery is becoming more sophisticated, it still does not operate as spontaneously as the Western armies'. The Russians still place greater faith in mass than in gadgets and fancy footwork.

In the recent past, counterbattery—artillery firing at artillery—was a tricky and expensive business. Enemy artillery could be only approximately located through reconnaissance or sound and flash. The latter literally used the roar and/or flash of the firing guns to calculate their location and range. As this technique wasn't accurate, large quantities of shell had to be fired in order to destroy or uproot the defending artillery.

Western armies now use computer-assisted radars to locate enemy guns in seconds and with great precision. Russian techniques, although older and cruder, still make it difficult for a gun to stay in one place for hours of firing. The age of shoot and scoot has arrived, but only if the troops can use and maintain the gadgets competently.

Electronic warfare has complicated this picture. If the guns are running around shooting and scooting, they are very dependent on radio communication with their observers. If the enemy is blasting away with its jammers, the guns have unreliable communications. At best, this slows down the process of requesting fire and getting it. This can be circumvented a number of ways. The unit can prepare the firing positions ahead of time and be prepared to have the infantry signal with flares for unscheduled fires. Or telephone wire can be run to the various positions, only a few of which are occupied by guns at any one time. Otherwise fire can be delivered on schedule to previously selected targets. You can also use antiradiation missiles or other ECM against the artillery-tracking radars. Things never become impossible, just more difficult.

And accidents happen. The infantry takes a dim view of getting hit by friendly fire. This is accepted more philosophically in the Russian Army, whose artillery fire tends to be rigid, preplanned and massive, and very closely coordinated with the infantry and tanks. Losses from friendly fire are considered preferable to leaving the infantry unprotected too long by a barrage.

LIFE ON THE GUN CREW

Like tank crews, gunners do a lot of maintenance, especially if their guns are self-propelled. The self-propelled guns are larger than tanks and have armor more like that of APC's. They usually mount a larger gun (152mm or 155mm being the most common caliber) and have a larger fighting compartment to allow for the greater amount of activity associated with firing a larger number of shells.

Field artillery guns are, if anything, easier to maintain than tank guns, mainly because of their lower shell velocity. Guns, as opposed to howitzers, have longer barrels, which, because of the higher velocity of their shells, undergo wear and tear.

Gunners have a lot more precombat work than tankers. To avoid counterbattery fire, or simply to support a complex fire plan for a major defensive or offensive operation, they usually have to prepare a number of alternate firing sites. Teams of gunners and surveyors are sent to spots chosen from the map to mark the locations for the guns. Arrangements may also be made to store ammunition at the spot. When the location is used, the guns are rolled in and lined up at the preselected location, their gunsights are aligned on the survey stakes, and the guns are trained on the given bearings. Then it's load, fire, load, fire until the required number of shells have been sent on their way. Most guns can fire off six or more shells a minute for a few minutes. After that they have to slow down to two or three a minute to avoid overheating the barrel.

Firing the gun is a well-trained drill. The gunner keeps the gun lined up properly, shifting the alignment according to the fire plan so that the proper number of shells fall on the right targets in the right sequence. The loader gets the shells into the gun. The ammo crew, the "gun bunnies," keeps the supply of shells (weighing 90 pounds for 155mm shells, plus the lighter propellant charge that is loaded separately) moving. The gun chief keeps checking that the right type of shell is being loaded, that the right type of fuse and/or fuse setting are being used, and that things are functioning in general. He also keeps the gun log, which is important for maintenance and for adjusting gun alignment for barrel wear.

During a major operation, as many as 500 shells per gun per day may be fired. That could be over four hours of steady firing, but it's never all at once. Usually there are bursts of a few shells, or a few dozen, interrupted by displacement to new positions, maintenance and, if the front is close or the enemy has broken through, defense against ground attack. There is always the danger of air attack and the dreaded counterbattery fire.

Particularly in a defensive situation, the guns may be on call at all hours. You just wait, day and night, for the call "fire mission." Then you scramble through the drill as quickly as possible. An infantryman's life depends on the gunner's speed and accuracy.

ORGANIZATION OF ARTILLERY

Almost all artillery is organized into battalions—usually of 18 guns, but sometimes less—containing guns of the same type and caliber. The typical battalion has 3 *firing batteries,* each with one third of the battalion's guns. In addition, there is always a headquarters battery containing the communications and fire-control specialists and their equipment. Some Western armies have a fifth *detail battery* that takes care of ammunition resupply, maintenance and other details. The actual gun crews of an 18-gun battalion number under 200 men. Ammunition supply troops number another 100, while fire-control and support personnel are another 100 to 200 men. The Western armies average some 500 men per battalion, the Russians about 300, although they place the equivalent of another 100 men per battalion under the control of divisional artillery headquarters.

Beyond the battalion, artillery is divided into two broad types: divisional and nondivisional. Divisional artillery uses lighter guns. The nondivisional artillery is under the control of higher headquarters and is assigned to divisions according to need. Divisional artillery consists of all the guns assigned to a division. In the U.S. Army this usually means 3 battalions of 155mm guns, 1 battalion of 203mm guns, and a battalion of longer-range rockets (either free flight or internally guided). In the Russian Army there are 2 battalions of 122mm guns, 1 of 152mm, 1 of 18 122mm, 40-tube rocket launchers, and 1 of large rockets. All other armies have various combinations of the above; for example, the total artillery of a U.S. division is 72 pieces; of a Russian division, 76 pieces; of a West German division, 92 pieces. But this is only the divisional artillery that forms the artillery under direct divisional control. It gets up to 80 percent by weight of the ammunition allocated to the division. But a division has far more artillery. Everything from an 81mm mortar to a 125mm tank gun is artillery. If you count all this other "artillery," there are 400 to 500 pieces for each U.S., Russian or West German division. Only 100 years ago, few armies had as many as 6 guns per 1000 men, while today the average is about 30 per 1000. The standard of dying appears to have gone up faster than the standard of living.

Nondivisional artillery is generally of the same caliber as divisional artillery, with a larger proportion of the heavier guns. Few armies use anything larger than 203mm anymore. Nondivisional artillery does contain the long-range (up to 800 km) missiles. These are usually armed with nuclear or chemical warheads. In wartime, armies strive to have as many guns assigned to nondivisional artillery as to all the division's battalions. For example, a corps with 3 divisions and 12 battalions of divisional artillery would have an additional 12 battalions of divisional artillery. This ratio is rarely achieved, particularly in peacetime.

Artillery assigned to smaller units in a division has much less available ammunition. It is not practical to deliver masses of ammunition to front-line units. Every shot must count. Direct support artillery is much more liable to be destroyed than divisional artillery, even though much of it uses the same indirect-fire techniques. This is the price paid for responsiveness to the requirements of supported units.

SHELLS SMART AND DUMB

There are many kinds of artillery shells:

High explosive (HE). Still the standard artillery shell; basically a shell container with an explosive charge of 5 to 20 or more pounds, depending on the shell's caliber.

Smoke. Creates a smoke cloud lasting from 10 to 20 minutes or more.

Star shell. A big illuminating flare with a parachute to delay its fall so that the light will last for 5 to 10 minutes or more.

Chemical. Loaded with various poison gases.

Nuclear. 152mm or larger, with a yield of up to 5 kilotons.

HEAT. High-Explosive Antitank shell.

Beehive. A shotgun type with thousands of darts.

There are also a number of fuse types. These devices are usually in the front of the shell and ignite the explosive.

Contact. The simplest; it ignites the shell when it hits anything.

Delayed action. Delays ignition for up to a few moments after contact so that the shell may penetrate; used for destroying fortifications, creating craters, etc.

Proximity. Has a radar range finder that ignites the shell a preset distance from a solid object. Good for getting an airburst. This is necessary for improved conventional munitions (ICM) with submunitions and increases the effectiveness of most other shells, as it prevents much of the effect of just hitting the ground. Originally designed for use against aircraft, and still used for that purpose.

VT (variable time). Poor man's proximity fuse. Gunners can preset the shell to ignite a certain number of seconds after leaving the gun. If the calculations are correct, this has the same effect as a proximity fuse, unless you are shooting at moving targets, especially aircraft.

There is also a trend toward ICM's, which reduce the number of guns while increasing their effectiveness. These shells are much more expensive than conventional shells, up to $30,000 versus under $500 for the most expensive conventional shell. Their features include:

Rocket boosters. Up to 50 percent more range, although some accuracy and payload are lost.

Guidance systems. Homing in on point targets, like tanks and hard enemy positions.

Submunitions. Many smaller shells that spread the damage over a wider area or many small mines. May also be equipped with delayed-action fuses.

Otherwise improved payloads. Incendiary, poison gas, smoke, illumination devices, etc.

The number of guns in the divisional artillery is less important than the supply and quality of ammunition and fire control. Divisional artillery is basically a delivery service. It delivers ammunition in large quantities to targets acquired and designated by the division's combat units.

SELF-PROPELLED VS. TOWED

For most of artillery's history, there has been a "horse artillery"—everyone rode, and the entire unit was organized and equipped for speed of movement and action. Modern self-propelled artillery follows that tradition, with a few twists.

Originally tanks were intended to be little more than self-propelled artillery. The armor was added to give essential protection against infantry weapons and shell fragments. The track-laying movement system was selected over the more efficient wheels because a tank's initial function was to move over the shot-up World War I battlefields in support of the stalemated infantry.

Soon the tanks became tanks as we now know them. Fighting each other pushed their original artillery function into the background. The armor became heavier as tank armament became more powerful. During World War II, self-propelled artillery was rediscovered by mounting conventional guns on tank chassis.

Today self-propelled artillery is conventional artillery mounted on a lightly armored chassis. Its advantages over towed artillery are better cross-country mobility, faster emplacement and displacement, better protection against enemy fire, an ability to keep up with armored units (especially when under fire), and a superior ability to employ direct fire and survive.

Towed artillery takes about 30 minutes to get ready to fire and 15 minutes to get on the road again. Self-propelled artillery takes about half that time to get into fire order (fire control is not yet fully automated, especially in Russian-type armies) and a few minutes to get on the road again.

There are drawbacks. Being tracked vehicles, they are more prone to breakdowns. Their direct-fire mode exposes them to more enemy fire and, armor protection notwithstanding, heavier losses. Finally, although the guns are self-propelled, not all the support vehicles are. Ammunition resupply is still dependent on wheeled vehicles. Overall, however, self-propelled artillery is superior to towed artillery, if you can afford the expense. Western armies generally can. The Russians are introducing self-propelled guns, particularly for regimental artillery.

ROCKETS AND MISSILES

Modern field artillery rockets were a Russian innovation that most European armies have adopted. Originally a replacement for Russian-style mass artillery fire when enough conventional artillery pieces weren't available, rockets became the right weapon at the right time. With the constant threat of deadly accurate counterbattery fire (especially against the Russians), rockets can provide massed fire quickly. In addition, the chemical warfare, first accepted by the Russians, found in rockets an ideal means of delivery. Chemicals are an area weapon; they needn't be delivered with great accuracy. Chemicals should be delivered in massive doses over the shortest possible period of time. With either chemical or high-explosive rounds, rockets are ideal for hitting enemy troops assembling for an attack. Surprise is important, as these troops will be unprepared and out of their APC's or positions. Finally, given the more crude artillery location techniques of the Russians, rockets provide the ideal counterbattery weapon.

A battalion of 18 BM-21 rocket launchers can fire 720 rounds as far as 20 km in 30 seconds, and be on their way in a few minutes, since the launchers are mounted on trucks. In their new position the launchers can be reloaded with 35 tons of rockets in 10 minutes.

The 720-round volley will devastate an area as large as 2000 by 500 meters in 30 seconds. It would take an equal number of conventional artillery 6 minutes at rapid fire rate to do the same damage. In addition, rockets, being lower-velocity projectiles, do not require heavy shell casings, thus allowing a higher proportion of high explosive. Their effect on impact may not produce as many lethal fragments as a conventional shell, but the shattering of troop morale is unmatched.

Field artillery missiles are a post-World War II development. They were developed for one primary purpose: the delivery of nuclear warheads. Most now use inertial guidance systems and have ranges approaching 1000 km. This puts them in the class of strategic weapons. The shorter-range missiles (100 to 200 km) will probably be used in nonnuclear warfare to deliver chemical warheads at rear-area targets. Western missiles will also be used to deliver certain ICM, like mines. These can be particularly devastating against truck and armored convoys moving up through the "safe" rear area.

HELICOPTERS

A unique weapon, gunship helicopters are used as artillery and should be considered as such. Perhaps the ultimate self-propelled artillery, helicopters carry a wide range of weapons: primarily rockets, high-speed (3000 rounds per minute) machine guns, automatic grenade (40mm) launchers, and, most important, ATGM's. Unarmored transport helicopters are also equipped to drop mines. United States, British and some other divisions have gunships. Most armies use gunships as nondivisional artillery. They are used like the horse artillery of old, the rapid reserve—to be sent wherever the danger is greatest.

Principal Artillery in Use

The artillery weapons shown represent over 90 percent currently in use worldwide. The United States and Russia export nearly all of it. Some nations, such as Britain and France, manufacture all or most of their own artillery, but their holdings are minor. China manufactures copies of Russian equipment and exports some.

CALIBER is the diameter of the shell in millimeters. NAME is the designation of the artillery piece. The 105mm guns were standard during World War II, but now

PRINCIPAL ARTILLERY IN USE

The only differences in artillery are created by the demands for longer range, lighter weight or mobility.

Caliber (mm)	Name	Origin	Range (km)	ROF per min	Radius (m)	Shell (kg)	Protec- tion	AT Cap mm pen	Mobility	Weight (tons)
105	M102	US	11.5	3	175	15	0	102	Towed	1.15
105	M101A1	US	11	3	175	15	0	102	Towed	2.26
122	M55/D74	Russia	24	6	210	21.8	0	460	Towed	5.5
122	M63/D30	Russia	15.3	8	240	25.5	0	230	Towed	3.2
122	M1974	Russia	15.3	8	240	25.5	5	460	SP	16
122	BM-21	Russia	20.5	4	2000	45.9	0	0	SP	11.5
130	M46	Russia	27	6	280	33.4	0	230	Towed	7.7
140	RPU-14	Russia	9.8	4	750	39.6	0	0	Towed	1.2
152	M55/D20	Russia	24	1	350	43.6	0	800	Towed	5.7
152	M1973	Russia	24	2	350	43.6	3	800	SP	28
155	XM198	US	22	2	360	43.5	0	800	Towed	7.8
155	FH70	NATO	30	2	360	43.5	0	800	Towed	9.3
155	M114A1	US	14.6	2	360	43.5	0	800	Towed	5.8
155	M109A1	US	18	2	360	43.5	7	800	SP	23.8
175	M107	US	32.7	.5	520	66.6	0	0	SP	28.2
203	M110A2	US	29	.5	470	90.6	0	0	SP	28.2
180	S23	Russia	30	1	700	88	0	0	Towed	21.4
240	BM-24	Russia	10.2	3	750	112.5	0	0	SP	15.2

are used only by Western armies in airborne units. The 122mm weapons are all of Russian origin. The D74 is a long-range gun, while the D30 is their standard towed howitzer. The M1974 is the towed version of the D30. The BM-21 is the standard, although not the only, rocket launcher. The 130mm M46 is the most widely used Russian long-range gun. The RPU-14 is the standard rocket launcher in airborne units. The Russian D20 and M1973 are the standard 152mm artillery and are identical except that the M1973 is self-propelled. The M114A1 is out of production but still widely used and will be replaced by the (155mm) M-198 and the FH70. The M110A2 (203mm) is the standard U.S. heavy artillery, with the M107 (175mm) being phased out. The S23 is the most widely used Russian heavy gun. In place of heavy artillery the Russians use missiles and rockets like the BM-24.

ORIGIN is the manufacturing country. NATO indicates a group effort by Britain, Germany and Italy.

RANGE is the extreme range of the gun in kilometers. Generally, best accuracy is achieved at about two thirds of the maximum range. Although the chart does not state it, you can see which artillery is howitzers, and which is guns. Howitzers have relatively short barrels and can fire accurately at high angles in order to hit behind obstacles like hills. Guns have longer barrels; thus, a higher shell velocity, a longer range and a flatter trajectory make them less able to hit behind hills. Guns also require a new barrel after as few as 400 rounds, whereas howitzer barrels are good for ten to twenty times that number. For example, the U.S. 175mm gun barrel has to be replaced every 400 rounds. Not only that, as replacement time approaches, barrel wear affects accuracy and must be calculated along with other factors like range, wind and humidity. Range can be extended 40 percent to 50 percent with RAP (rocket-assisted propellant) shells. Accuracy suffers, but this is not important if many shells are fired at a target, as even at extreme range, shells will fall within a 150-meter circle.

ROF PER MIN is the sustained rate of fire per minute. Most guns can fire double to triple that rate for a minute or so. They are restricted from firing so fast over a longer time by barrel overheating or, in the case of the larger weapons, by the ability of the crew to get the heavier shells into the gun. For rocket launchers the net number of rockets per reload cycle is given. The rocket launchers fire all their shells in less than a minute. The BM-21 fires 40 rockets, the RPU-14 fires 16 and the BM-24 fires 12.

RADIUS (in meters) is the area covered by a battery volley (one shell each from a battery of six guns) of HE (high explosive). In this area there is a 50 percent chance of being hit. ICM (improved conventional munitions) used by Western armies—and soon by Russian armies—increases this area by two or three times while increasing the probability of getting hit to over 90 percent.

SHELL is the weight of the standard HE shell in kilograms. A complete round, which includes propellant and packing material, weighs 30 percent to 50 percent more. An increasing number of shell types are becoming available. Their weight varies by only 20 percent either way.

PROTECTION is the level of protection the gun has when mounted on an armored chassis.

AT CAP MM PEN (antitank capability millimeters penetration) is the armor-piercing capability of that gun when using a HEAT-shaped-charge shell.

MOBILITY indicates whether the weapon is towed or self-propelled.

WEIGHT is given in metric tons. For self-propelled guns this includes the vehicle. For towed pieces only the gun weight is given.

DESTRUCTION TABLE FOR "AVERAGE" ARTILLERY UNIT

Enormous quantities of munitions are needed to destroy combat units, particularly those with armored vehicles.

Activity of Defending Unit	Area Covered (sq km)	% Casualties per 100 tons of Ammo		Tons of Ammunition Expended to: Neutralize		Destroy	
		Armor Unit	Soft Unit	Armor Unit	Soft Unit	Armor Unit	Soft Unit
Hasty Attack	1	31	109	96	28	160	46
Prepared Attack	1	21	75	143	40	239	67
Assembly	1.7	30	49	99	61	165	101
Hasty Defense	3.6	9	18	345	169	576	282
Prepared Defense	3.6	3	6	1043	517	1739	862
Dispersed Defense	7	1	3	2029	1005	3381	1676

Artillery Destruction Table

This chart shows how much artillery ammunition must be used to inflict various levels of damage on armored or unarmored units. Keep in mind that ICM (improved conventional munitions) change these calculations considerably, but not without cost, literally. A conventional 155mm HE shell costs about $500. Some ICM cost twenty times as much. But the savings (in theory, as many of these ICM have not been used in combat) compare favorably. For example, the M712 Copperhead round costs $30,000. It contains a guidance system that homes in on reflected laser rays painted on the target by a forward observer. One hit can be expected for every five shells. That's $150,000, which isn't a bad exchange rate, since a Russian tank costs over $1,000,000. You can buy 300 unguided $500 HE shells for $150,000. According to Russian practice, it takes over 500 HE rounds to knock out a tank. And there are additional savings. One is time. It takes an artillery battalion nearly half an hour to fire those 500 shells. The tactical battle will be over before half that time has elapsed. Therefore, the only way artillery can be an antitank weapon is with ICM. Other types of ICM can deliver antitank or antipersonnel mines or "bomblets." Both cover a much wider area with one half to one tenth as much ammunition.

ACTIVITY OF THE DEFENDING UNIT indicates men and equipment deployment. The units involved are generally reinforced battalions. HASTY ATTACK is a quick deployment from the road as the unit lines up to attack. PREPARED ATTACK is more deliberate, the troops having taken more precautions against artillery effects by staying inside vehicles and closer to protecting terrain. ASSEMBLY is troops assembled in an area for redeployment, rest, etc. HASTY DEFENSE is when the unit quickly seeks protective terrain; not much time for digging or other preparations. PREPARED DEFENSE is doing all that can be done to mitigate the effects of artillery fire. DISPERSED DEFENSE is similar to prepared defense except the troops are spread over a wider area. This is done to minimize the effects of artillery, particularly nuclear weapons. It also helps defend more ground.

AREA COVERED is the area occupied by the unit under attack by the artillery. It is the area into which all the artillery fire falls. It is roughly square-shaped and is measured in square kilometers.

% CASUALTIES PER 100 TONS OF AMMO is the percentage of the unit's troops and/or vehicles that will be killed, destroyed or disabled. The two classes of targets are: ARMOR, a unit consisting primarily of tanks, APC's and self-propelled artillery, and SOFT, a unit consisting of only troops and/or nonarmored vehicles.

Hitting the target is difficult enough. You must also hit it with sufficient ammunition to do significant damage. Supply becomes a critical problem. It takes a lot of ammunition to hurt a unit. A good rule of thumb is 1 105mm to 155mm shell per 100 square meters to neutralize armored vehicles or dug in infantry. One shell per 1000 square meters will suffice for exposed troops and unarmored weapons. You can only carry so much ammunition with you. A Western division can have 3000 tons (about 75,000 shells) available, a Russian-type division at most about two thirds that amount.

Then there is the problem of getting the ammunition on the target quickly enough. It takes one artillery battalion over sixteen hours to neutralize a mechanized infantry battalion in hasty defense. By that time the defender has had ample opportunity to bring up another battalion to prevent a breakthrough. This was an essential problem in World War I: To ensure destruction, the artillery barrages were so long that they eliminated any possibility of surprise.

TONS OF AMMUNITION EXPENDED TO NEUTRALIZE OR DESTROY indicates how much ammunition will be needed to neutralize (destroy 30 percent) or destroy (take out 50 percent) of the unit.

Time is critical when using artillery. Often the target is moving. More often there is counterbattery fire. Anytime you inflict an extensive shelling, the enemy suspects that an advance by your ground forces will follow.

For major operations it is not unusual for each tank or infantry battalion to be supported by three to five or more artillery battalions.

On the average, each 18-gun battalion can fire 100 tons of shell in 69 minutes. This is far too long; usually more guns are used to deliver high tonnages of ammo. A hundred guns could fire 100 tons of shell in 12.5 minutes. This need for speed, in order to avoid counterbattery fire, is another reason for the more effective, and expensive, ICM.

5

COMBAT SUPPORT

THE INFANTRY, tank crews and artillerymen are not up at the front all by themselves. They have plenty of company in the form of combat support troops. And much support is needed.

THE MULTIPLIER EFFECT OF COMBAT SUPPORT

Combat support makes the combat troops' efforts much more effective. Reduce combat support and consider the consequences: Without *engineer support,* enemy and natural obstacles become insurmountable. Rivers and minefields cannot economically be crossed. Fortifications become much less effective. With no *signal support,* communications become spotty at best and will most likely break down entirely under the stress of combat. With no *transportation support,* units run out of ammunition, fuel, food and everything else within days. Without *military police,* the traffic problems soon become intolerable, prohibiting any movement. Combat troops must be detailed to take care of prisoners of war and perform other security duties. Without *maintenance support,* the equipment will probably be in no condition even to enter combat, much less continue for any length of time. Without *chemical troops,* enemy use of chemical or nuclear weapons will stop troops cold. Without *electronic-warfare troops,* the enemy will be able to eavesdrop on communications and disrupt them at will, without retaliation. Without *headquarters,* you will not be able to respond to enemy activity or initiate any effective action yourself.

NATIONAL DIFFERENCES

There is a fundamental difference in the use of combat support units between Western and Russian-style armies. The Russians recognize that support units are expensive and available only to armies created by technologically powerful and well-organized societies. Less well-endowed armies can compensate by devoting more of their resources to pure combat power and by planning to fight a short war, a very short war. Without proper combat support, a modern army can lose over half its strength as a result of equipment wearing out and/or breaking down in the course of a month's operations. Combat only increases this loss rate. Western armies attempt to keep more equipment operational longer. Russian armies attempt to win the war in a month, before their equipment ceases to function.

Each division has separate units of support troops. In addition, each army has additional nondivisional support units, often in proportion to the number of divisions in the army. Western armies usually have six to twelve divisions, Russian armies usually half that number. Below is a description of each combat support unit's function. The size of the units shows the differences between various nations. The larger units are generally for Western armies.

ENGINEERS

Every division has at least a battalion (400 to 1100 men) of engineers. Often regiments or brigades have their own engineer company (100 to 300 men). Nondivisional units usually exist on a ratio of two or more engineer battalions per division in an army.

Whenever something has to be built or torn down, the engineers are called in. Part of the engineer company is actually a combat branch, doing demolition or construction work under enemy fire.

Bridges. Most engineers are builders. The combat-engineer battalions in divisions devote over 30 percent of their personnel to river-crossing operations. They use various self-propelled and truck-carried bridging equipment. The former are 15- to 20-meter bridges mounted on an unarmored tank chassis or a truck. They can support 50 tons (a tank). Longer bridge sections, which can also double as ferries, are carried on trucks.

Digging and minelaying. Engineers also supervise, or in Russian-type armies, control all construction equipment. The excavation equipment enables a Russian division to entrench one of their regiments. Each regiment has enough equipment to entrench one battalion a day. The minelayers, limited by the number of mines available, can lay up to 10 km of mines to defend a regiment, at a rate of 800 mines weighing up to 8 tons per kilometer. This performance is generally unlikely except after long preparations. In a major offensive operation, a division is not likely to carry more than 100 tons of mines. Western armies depend more on "track busters" which are often delivered by air. Stored at air bases, they weigh under 3 pounds, one tenth as much as conventional mines.

Defensive positions. Many kinds of engineer resources are devoted to obstructing enemy movement and preparing defensive positions. This is the most important construction work engineers do.

Mine clearing. Most infantrymen are trained to clear mines when engineers are not available, but the engineers are faster with their specialized equipment. Capabilities vary considerably. The Western armies have the edge in difficult-to-clear mines and sophisticated mine-clearing devices. The Russians have much less sophisticated matériel.

Demolition. What goes up, the engineers bring down. Handling tons of explosive to bring down large structures requires expertise. Bridges, in particular, are difficult to destroy. Airfields, roads, railroads, structures of all kinds may have to be destroyed by engineer troops. Each job normally takes a few hours if the proper equipment is at hand. For example, clearing a 75-meter abatis (a large earthen obstacle laden with wire, logs and booby traps) takes 4 hours, using a squad of engineers with explosives and a combat-engineer vehicle (CEV), a tank-bulldozer. With just chain-saws and hand tools, it takes the same crew 16 hours.

Construction and repair. Aside from field fortifications dug out of the ground, engineers also can put up one-story, prefabricated buildings and large tents, and build roads, runways and railroads. Special construction battalions build them, and regular engineers do the repairs. Repairing a large road crater (30 feet by 20 feet by 10 feet) takes an hour with a combat-engineer vehicle. Filling in runway craters takes more time because they are usually smaller but more numerous.

Maps. Normally the engineers are responsible for preparing, reproducing and distributing maps. Nobody can fight a war without good maps.

Utilities. Engineers are responsible for generating power in the field. Any large plant—a field bakery, decontamination equipment, field baths—is maintained, if not operated, by engineers. It simplifies looking for help when something breaks down.

SIGNAL

Every division has a signal battalion (400 to 1000 men). Often every unit down to battalion size has its own signal unit, representing about 5 percent of its strength. Nondivisional units exist on a ratio of one signal battalion per division in an army.

Communications start with a radio in practically every vehicle and the ever-present field telephones. The radios have to be maintained and the wire strung from one field phone to another.

Nondivisional signal battalions have several responsibilities. They maintain long-range communications systems, especially satellite links. They set up dozens of radio and telephone "nets" (party lines). They ensure maximum security for communications. Finally, they give technical assistance to electronic-warfare and intelligence troops.

The dozens of radio and field phone nets in a division can easily fall into chaos without the efficient efforts of the signal troops. A net (either phone or radio) can include dozens of parties. There are nets for tactical control of units, for staffs, for supply units, for the air force, for artillery observers. At headquarters the nets are linked with one another through multiple radios or phone switchboards. Sometimes radios can switch nets by changing frequencies, but often the range of frequencies available is too small. "Net discipline" as well as "signal discipline" in general must be strictly enforced: no useless chatter, unauthorized breaking in on another net or transmitting in other than the rigid, authorized format.

CHEMICAL

The trend is for divisions to have a chemical battalion and for much smaller detachments to have lesser units (a chemical platoon for a regiment or brigade).

These troops are a recent development. Their primary responsibility is the detection of chemical, nuclear and biological weapons and the decontamination of their effects. More on their functions is given in Chapter 21.

TRANSPORT

Every division has a transport battalion, combat battalions usually have platoons and nondivisional units number two or three transport battalions per division in an army.

Transport units move supplies. They consist primarily of trucks and trailers. A typical medium truck battalion has about 200 5-ton trucks, each capable of pulling a 10-ton trailer. This gives the unit a maximum practical carrying capacity of 2000 tons, allowing for out-of-service vehicles and variable load size. A transport company has a 500-ton capacity, a platoon 150 tons. Most units are general purpose. A few specialize. For example, tank transporter units consist of 30 to 60 heavy tractor trailers, each with a 50- to 60-ton capacity. Other heavy units may only be able to carry fuel.

Also included are air-transport units, controlled directly by the army, which contain helicopters and light fixed-wing aircraft. The trend is toward one aviation battalion per division, although this unit commands and controls more than it transports. Additional battalions are controlled by army or other higher headquarters.

For more details see the chapters on logistics, air transport and naval transport.

MILITARY POLICE

Most divisions have at least a company (100 to 200 men) of MP's, military police. At army level there is often a battalion or more. The main function of the MP's in wartime is traffic control. In rear areas, particularly behind those occupied by fighting troops, they are also used to maintain order, including guarding prisoners.

MEDICAL

Each division has a medical battalion (300 to 900 men). Smaller combat units, down to the battalion level, have smaller detachments, 2 percent to 5 percent of their strength. Nondivisional units usually number one battalion per division in an army.

At and below the division level the medical troops give first aid and evacuate wounded troops as quickly as possible. Because about two thirds of combat losses can be returned to duty eventually, it is essential to prevent wounds from worsening and evacuate the wounded to rear areas for recuperation as quickly as possible.

Medical units are also responsible for supervising preventive medicine. In cold climates this means treating and monitoring exposure casualties. The chief medical officer alerts the unit commanders when these losses start to get out of hand. In any area with a lot of civilians, venereal disease will be a major problem, often the number one cause of days lost to noncombat causes. In disease-prone locales, like the tropics, the medical troops distribute medicines and eradicate pests. At all times the medical troops monitor the purity of food and water and living conditions in general.

MAINTENANCE

Each division has a maintenance battalion (400 to 1000 men) and smaller units devoting up to 10 percent of their strength to maintenance detachments. Nondivisional units exist on the same ratio, up to 10 percent of nondivisional unit strength.

Because of the high breakdown rate of equipment under field conditions, the maintenance units wage, at best, a holding action. (See the chapter on attrition.) Western armies make a greater attempt to keep everything moving. In Russian armies most broken-down equipment is abandoned so that the follow-on maintenance units can recover and rebuild them.

Like medical units, maintenance units devote a lot of their efforts to performing and supervising regular maintenance of equipment. This function, in the long run, has more impact than the ability to perform many repairs quickly during combat.

HEADQUARTERS

Each division has, in effect, a headquarters battalion, plus one or more headquarters-support units, like finance and intelligence. Smaller units devote about 5 percent of their manpower to headquarter functions. Nondivisional units follow the same ratio.

Headquarters administer, plan and coordinate. They include intelligence units that often send detachments to subordinate units to screen prisoners, examine captured matériel or just look at things first-hand.

Headquarters' primary job is collecting information from all subordinate units, analyzing it and issuing appropriate orders. Without a functioning headquarters, subordinate units become independent of each other. And much less effective.

ELECTRONICS

The trend is toward one electronic-warfare battalion (400 to 700 men) per division. Each army may also have an additional battalion or two.

See the chapter on the electronic battlefield for more detail. The electronic-warfare battalion monitors enemy signal traffic, and, when appropriate, jams enemy communications.

6

PARAMILITARY FORCES AND RESERVES

NOT ALL SOLDIERS serve in the armed forces. In many nations paramilitary forces, police organized into combat units, account for substantial military power. In addition, reserves of former and part-time soldiers are an essential supplement to active forces.

Police Armies

Americans do not normally equate police units with military power, yet in wartime the demands on police forces increase dramatically. This is especially true if a nation is invaded or possesses a number of disloyal occupants. Even if those two conditions are absent, the police are called upon to help mobilize manpower, move extraordinary quantities of military personnel and equipment and supervise shifting civilian populations. European nations, long faced with these problems, historically have looked upon peacetime police forces as paramilitary forces. Many European nations have organized and equipped their police accordingly.

Russia, an extreme example of police strength and dependence, possesses 200,000 KGB border police. This "army" is equipped with armored units, armed patrol boats and patrol aircraft. In reality, the KGB troops serve the same function as the U.S. Coast Guard and Border Patrol, but there are fewer than 50,000 people in the latter two. Russia also has 260,000 MVD internal security troops organized into combat units. There is no comparable force in Western nations, aside from some relatively minuscule riot-police troops. Many countries maintain national police forces, and in the United States most states also have state police forces. But none compare with the size and mandate of the MVD. With one MVD trooper per 1000 population, Russia has a reserve to back up the local police should anyone get out of hand in an emergency, or at any other time.

These paramilitary forces are important not just because they are a standing force but also because their personnel are carefully chosen, especially for political reliability. In authoritarian nations this is a critical element in keeping the countries functioning under stress. In wartime it is not unusual for paramilitary forces to be sent into combat as light infantry. They are usually rapidly destroyed because they lack training and experience in heavy combat. Paramilitary forces are more accustomed to fighting those who cannot offer an effective resistance.

Military Reserves

Reserves are civilians who have recently served in the active armed forces and/or civilians who spend a few weeks a year in training to maintain military skills. These reserves are used in four ways:

Fillers to bring peacetime combat units up to full strength. This is an essential element of the Russian reserve system. Of the 173 divisions in the Russian Army, only 54 are at full strength in the same sense as the United States' 20 divisions (17 army, 3 marine). About 500,000 reserves are required to bring the other divisions up to war strength, plus another 500,000 to fill out nondivisional units (artillery, transport, engineer and support). Some training of reserves takes place, but the primary objective is equipment maintenance and the division's ability to absorb up to 10,000 or more reservists on short notice.

Maintenance of complete units. The U.S. system maintains many units in the reserves with full equipment and *active* reserve personnel. This is significantly different from the Russian system. The competence of the U.S. reserve units is not on the same level as that of the active units. However, the U.S. reserves, for all their faults, are probably superior to the Russian reserves. One reason is that a Russian reservist is assigned to a position in a unit he may never see. Instead of regular training, as in the U.S. system, he is obligated to serve no more than three months a year. Most of the forces that originally went into Afghanistan were composed of reservists. They were quickly withdrawn and replaced by active-duty troops.

Replacements for combat losses. These are mostly infantrymen, artillery specialists and tank crews. In heavy combat, total losses less wounded returned to service may amount to 3 percent per division per day. With 100 divisions in combat, that amounts to 10,000 men per day, or 300,000 per month if you want a large number.

Formation of new units. To form a new division requires not just the personnel strength of the division but also a certain number of nondivisional troops, at least 20,000 per new division. Manpower will not be the restraining factor; equipment and specialists will be. Both military and civilian skills are needed. Civilians can be conscripted directly into the military. Military specialists (especially the infantry, tank and artillery crews, etc., specialties for which there is no civilian equivalent) can only be obtained from those with prior military service.

THE RUSSIAN SYSTEM

Russia keeps nearly every veteran in the reserve until age fifty. Most other nations use the same general concept. As these men receive virtually no refresher training, only those who have left active service in the last few years retain a usable amount of experience. This reduces Russia's useful reserve to a million men times the number of years you want to go back—say two to five million men.

The Russian system, originally developed in nineteenth-century Germany, is suitable for a nation lacking great wealth. In Russia a reservist may not be called up for more than ninety days a year unless a national emergency is declared. This is done not out of regard for the reservist, but in recognition of the labor shortage situation. Extended call-ups would severely disrupt the economy. Additionally, a shortage of managerial manpower prevents the establishment of extensive, fully manned reserve units.

An example of the problems of this system was in the Russian mobilization against Poland in 1980. In areas adjacent to Poland,

Russia had 57 divisions. At least 40 would be needed to guarantee the quick conquest of Poland. Of the 57 divisions, only 28 were fully active, 19 in East Germany, five in Czechoslovakia, two in Poland and two in Russia. For political reasons Russia would probably decline to commit all its divisions in East Germany. This would leave no more than 20 available active divisions. The remaining 20 divisions would have to come from the reserves. Calling up the reserves to fill out those 20 divisions would affect over 200,000 men. This would have a noticeable effect on the local economy, especially if maintained for the full 90 days. Twenty million man-days would be lost. The cost would be not just lost manpower, but also the expense of maintaining the troops and civilian transport. The strain on the local economy was one of the critical, but not mentioned, factors causing Russia to demobilize.

Israel provides another example. Mobilization calls up over 15 percent of the Jewish population and severely disrupts the economy.

More critical than disruption is the lack of quality. Most Russian-style armed forces rely more on conscripts than do Western armies. The Russian Army is 77 percent conscript. Most of the noncommissioned officers are appointed conscripts. Russian officers, all volunteer and professional, perform much of the supervision done by NCO's in Western armies. Supervision, management and leadership, already spread thin in the regular Russian Army, is spread thinner still when the reserves are called up. When fully mobilized the Russian Army is 85 percent conscript, 95 percent in about a third of the divisions. If history is any guide, this third of the Russian Army will be less than half as effective as the first third.

There is only one solution for Russia's quality problem: training. Most Western armies train their reserves, or attempt to. A trained combat soldier is very much a technician. The care and use of weapons are complex skills only maintained through constant practice. Even full-time troops develop their combat skills only so far, because only combat itself trains a fully effective combat soldier. Reserves that do not regularly train are therefore little more than prescreened raw material. They are important, but the skills they retain consist largely of a knowledge of military routine, which is not critical and can be quickly learned on the job by civilians. A larger training problem is getting everyone conditioned physically for the intense demands of wartime operations.

The Russian reserve system provides high numbers but lower effectiveness. The Russians are not unaware of this. They are diligent students of past experience. Their solution is to prepare for a short war, short enough so that their deficiencies will not catch up with them. Thus, their reserves are more aptly designated veterans. The reserves tabulated in Chapter 29 reflect this use of reserves.

Navies and air forces also use reserves, but not as extensively as the armies. They need skilled personnel to man their complex equipment. Reserves are used primarily as laborers and other unskilled support workers.

THE U.S. SYSTEM

The U.S. National Guard and reserve systems are usually explained as incorrectly in the press as is the Russian system. The Russian system does not, as the press and the Pentagon would have you believe, produce an army of 173 combat-ready divisions. At the same time, the United States does not maintain only 20 combat-ready divisions. The U.S. total is actually 30. The additional units are composed of reservists who train monthly and during large-scale summer exercises. These units are generally full strength and perform comparably with active units. This may not sound impressive until you stop to think of the shape in which nations using the Russian system must be.

Almost all the additional U.S. combat units, 40 percent of the infantry and the armor battalions, come from the National Guard, which is what other countries call the militia. After many years of the National Guard operating more as a social club than as a combat force, the two World Wars of this century turned it into a very professional force.

The U.S. reserves are primarily support units, the specialist units that are so difficult to create quickly in wartime. The National Guard and the reserves account for over 50 percent of the infantry and armor battalions, 58 percent of the artillery and over 60 percent of the combat support units.

Most nations have small air and naval reserves. Again, the United States is unique in having hundreds of combat aircraft manned by reservists. Many navies have older combat and support ships in reserve. Only the United States carries out such an ambitious training program for its naval reservists.

The U.S. reserve system is a recent development. Only a wealthy economy can provide enough skilled people with enough leisure time to become part-time soldiers. Western nations tend to develop reserve systems like that of the United States.

The chief advantages of the U.S. system are the existence of fully formed units and the quality of these units. The quality derives primarily from continued training with the same personnel. In this way, the reserves often achieve a degree of efficiency even active units cannot match. Also, under the Russian system many troops are thrown together ineffectively just before going into action.

The Uncounted Reserves

When wars break out, a lot of civilians find themselves in uniform doing much the same work they did in peacetime. As warfare becomes more technological, the skills of the support soldier become relatively more important. Skills, particularly complex ones, only survive through practice. A reservist who learned a technical skill in the service and went on to another career as a civilian is no longer useful for his military specialty, but is quite valuable in his regularly practiced civilian skill. A radio repairman, for example, could be put to work in uniform immediately.

The ability of a nation to do this depends on the quantity and quality of its skilled civilians. The Western nations have a distinct advantage. In addition, the countries with high living standards can draw on a larger surplus of skills. Poor nations live closer to the threshold of existence. Current examples are African nations which lost their thin reserve of technicians in the post-colonial era. Starvation, economic collapse and a general inability to make any massive efforts (like a major war) resulted. Russia suffered similar deprivation during World War II. Much of the aid given Russia during the war was not war matériel but industrial goods, raw materials and other supplies lacking because too many Russian technicians were at the front getting killed. Russians remember this trauma better than most Western nations. It inspires more profoundly antiwar feelings than most Westerners realize exist.

Part 2
AIR
OPERATIONS

7

THE AIR FORCE: FIGHTERS, BOMBERS AND SNOOPERS

AIR OPERATIONS revolve around the gathering of information. They always have; they still do. Air forces cannot occupy ground. They cannot seize or even defend cities. Their accuracy is questionable. Their effective nonnuclear firepower is less than that of ground forces. Air forces are best at intelligence gathering. Air forces are also the best way to prevent the enemy from snooping around your own backyard. There are also the more visible aspects of air operations: air superiority, bombing and air transport. Air forces are very flexible and can be concentrated quickly. Yet the majority of air resources are devoted to gathering information or preventing the enemy from doing the same.

The techniques of gathering information from the air have changed little in the past seventy years. Reconnaissance aircraft, usually combat planes stripped of most armament and equipped with cameras and sensors, speed over enemy territory. They are protected by other planes from enemy aircraft and ground-based antiaircraft defenses. All this protection has the fringe benefit of allowing friendly bombers in to do their own work. The bombers also operate as "recon" craft, for they must be able to spot their targets precisely in order to shoot or bomb them.

There are three main air force missions: reconnaissance, intercept, and strike.

Ground Control and Support

All air forces support their aircraft with ground-based radars and control. The Russian-trained air forces attempt to trade off pilot skill and aircraft quality with more aircraft and more complete "positive ground control." Western air forces put more radar-detection equipment aboard their planes and train their pilots to a higher level of skill so that their aircraft become self-contained weapons systems.

From the very beginning of air warfare, air or ground observers spotted enemy air activity. All this intelligence would be sent to a central command center, which would send friendly planes to the area where they could do the most damage. Pilots would have little more to do than take off, follow instructions from the ground controllers vectoring them to the enemy aircraft, fire their weapons and return to base. The only extraordinary maneuvers needed would be quick getaways in case this neat little weapon system ran into trouble.

Unfortunately for the "positive ground-control" school of air operations, it has a number of critical weaknesses. Everything depends on the continued functioning of the ground-based detection and control centers. In a future war Western and Russian air forces have every intention of going after enemy ground control early and often. Ground control depends on the continued functioning of radars, which are vulnerable to destruction by antiradiation missiles (ARM) or interference from jammers.

The Russians took to the positive ground-control system as a solution to their other problems, which are:

Technical inferiority. The Russian aircraft industry has not been able to produce machines that are as efficient as Western ones. They try to compensate by building more aircraft, but this leads to additional problems.

Personnel inferiority. Given a smaller base of technically competent personnel, the Russians are unable to man their aircraft or staff their maintenance forces to the same level of performance as the West's. This problem is compounded by the larger number of aircraft, and their crude design and materials make them more difficult to maintain. The maintenance load is kept down by flying the aircraft fewer hours, less than half as many as are flown by Western aircraft. This leads however to less experienced pilots.

Doctrine. Russian doctrine, developed with an eye toward minimizing shortcomings, concentrates on the massive attack. This means putting more aircraft against the enemy early in the war. No matter if the maintenance resources are quickly overwhelmed. Hit the enemy hard enough and early enough, and nothing else will matter. This the Russians learned from their own experience during World War II. It was also an element of the German *blitzkrieg,* or lightning war, concept. Hit the enemy hard enough in a short space of time, and you paralyze him. This is a gamble and not necessarily a foolish one. It all comes down to whether or not the first blow is a telling one. If not, the war is lost. The Russians learned this lesson by having it done to them many times. They have yet to pull it off themselves.

Reconnaissance Missions

Reconnaissance is the primary mission of air power. Looking at reconnaissance from the viewpoint of the ground-forces commander, its primary user, it appears as follows:

Tactical reconnaissance. Much is done by low-performance, nonjet aircraft, including helicopters. It involves constant monitoring of an enemy in contact with friendly forces. It consists primarily of photography. Infrared photography can detect camouflaged positions through the unnatural pattern of heat from dead camouflage material and hidden vehicles. Visual observation is still used at this level, but less so because of the intensity of ground fire. Unpiloted aircraft (remotely piloted vehicles) are replacing human observers to some extent. In place of film cameras there are sometimes TV cameras or sensors to detect heat, electronic signals, movement, large metal objects. Increasingly aircraft drop passive sensors that detect sound, heat, large metal objects, and broadcast the data for evaluation or warning. Finally, long-range reconnaissance patrols or espionage agents may be dropped behind enemy lines.

Reconnaissance missions are usually flown very close to the ground to increase the quality of information. Sometimes this is done purposely to draw enemy fire and reveal its position.

The information obtained includes the location, strength, identification and activities of enemy units. It is particularly important to evaluate the results of an air strike, a ground attack, or an artillery barrage. All the photographs and tapes are passed back to intelligence units for sorting out and analysis. Intelligence then goes over the results with the combat units and, in turn, requests further information.

Operational reconnaissance. This is longer range and less urgent than tactical reconnaissance. The latter usually restricts itself to the area from the point of contact to perhaps 20 km into the enemy's rear position, as well as your own rear area if the situation is fluid. Operational reconnaissance may extend hundreds of kilometers into the enemy's rear area. The information required is the same as for tactical reconnaissance. Because the enemy ground forces are not yet in contact, there is less immediacy, at least for the friendly ground forces. Such is not the case for friendly air forces, or for more senior ground commanders. Enemy air forces are scouted out during operational reconnaissance. Also, enemy missile forces must be watched, especially for telltale signs of imminent use.

Specially equipped fighter and strike aircraft perform these missions deep into enemy territory. The opposition is more intense, if only because more enemy-defended territory is covered. The aircraft usually carry no weapons and little additional weight beyond their cameras, sensors and perhaps an ECM (electronic countermeasures) pod. A reconnaissance aircraft depends primarily on speed and ECM to avoid detection and damage from enemy ground or air forces. The reconnaissance aircraft usually flies low and fast to avoid enemy radar detection. Radio and radar will usually not be used, as they give the enemy an electronic signature to home in on. Using onboard computers to monitor passive, or listening, sensors, the reconnaissance pilot roars low and fast into enemy terrain. The computer automatically chooses the safest path through the thicket of enemy radars. Random, or just lucky, ground fire often causes the most damage. Overall, however, the reconnaissance pilot should not encounter too much danger. He must possess all the flying skill of a good combat pilot but little of the killer instinct. The only thing a reconnaissance pilot fights is detection. He runs toward his target, and away from danger. But it requires no little courage to go in alone and overcome numerous obstacles in order to take pictures. And without this information the armed forces are blind.

Strategic reconnaissance. This form covers information on a nation's total means of waging war: its armed forces, economic strength and resources, etc. In addition, strategic reconnaissance gathers the same types of information as tactical and operational reconnaissance. While strategic reconnaissance is increasingly taken care of by satellites (see Chapter 20), either long-range bombers are used or specially designed planes like the U.S.'s famous U-2 or the SR-71 (Blackbird), which can overfly just about any area on earth without harm. They can be detected though, so there are political restrictions on them in peacetime. In wartime strategic reconnaissance becomes a combat mission and is often carried out by specialized aircraft (U-2, SR-71) or, in the case of long-range missions, by regular reconnaissance aircraft. The slower bombers may perform these types of missions only if their considerable ECM can defend them. The bombers often lack the speed and low-flying ability of smaller aircraft. Each mission must be evaluated separately.

Interception Missions

Fighters. Air-to-air combat. Soon after reconnaissance was discovered, air-to-air combat followed. If reconnaissance was so useful, it followed that denying it to the enemy was almost as important. If that were achieved, you could wage war with the twin advantages of knowledge of the enemy's activities and secrecy for your own.

With air-to-air combat, some things never change. Seventy years ago, during World War I, pilots discovered that the easiest way to destroy an enemy aircraft was to ambush it from behind. Dogfighting—high maneuverability, tighter turning and greater speed—was a poor second choice and usually ended up in a stalemate. Today the same basic tactics apply, with a number of important additions and modifications.

Modern air combat is more a matter of teamwork and technology. Longer-range weapons and better communications allow pilots to detect and attack the enemy at longer distances. A 30mm automatic cannon allows 800-meter shots compared with 100 meters with 7.62mm machine guns 68 years ago. Missiles allow kills at ranges of over 200 km.

But then, as now, spotting the other fellow first is the key to success. *Some 80 percent of air kills are the result of the attacker surprising the defender. The victim never even sees his attacker.* The cardinal rule of air-to-air combat is: Hit the other fellow while he isn't looking. Obtain the favorable position (usually high and behind the enemy), and get in the first shot.

When fighter goes up against fighter, the orderly, planned routine of other air operations goes out the window. To intercept the other side's missions and to defend against interception, you must depend on technology and skill. Technology manifests itself in the form of technically superior aircraft. Skill is what you want your fighter pilots to have.

All pilots require a wide range of flying skills. Some are obvious, like knowing how to efficiently take an aircraft through a wide range of maneuvers, from tricky landings and takeoffs under bad weather conditions to reacting to unexpected changes in flying conditions, like air pockets and component failures. Some skills are seemingly mundane, like how to do a thorough preflight check on your aircraft. A loose component or suspicious instrument can lead to flying problems. Flying problems during combat can be fatal.

Another potentially fatal flying problem is poor fuel management. As a rule of thumb, a fighter can take its total flying range and divide it into thirds: one third for going out ("operating radius"), one third for coming back and one third for combat. A typical modern fighter can cruise at 900 km per hour. It might have an extreme range of 2700 km. That gives it a theoretical flying time of three hours. However, a high-performance fighter obtains its speed by having an engine that can increase its fuel consumption enormously for short periods. For example, at cruise speed the fighter burns about .56 percent of its fuel per minute. By kicking in the afterburner, cruise speed can be more than tripled, and fuel consumption increased more than twenty times. At full war power an F-15 can burn up its combat reserve in less than three minutes. It can also escape from most unfavorable situations because of its sudden speed. A less experienced pilot could abuse the high performance of his aircraft to get him out of an emergency. Once a fighter reaches BINGO fuel, only enough remaining to get home, combat must cease. Otherwise the aircraft will crash and be just as useless as if lost to enemy action.

There are several decisive factors in interception missions. First, the side with the superior detection devices and ECM often gets the superior position. Second, everything being equal, the side with superior tactics gains the advantage or mitigates the other side's electronic superiority. Tactics include the ability to obtain strategic and tactical surprise. Pearl Harbor and the other opening battles of the 1939–1945 war are examples of surprise. It worked, didn't it? Third, everything being equal, the side with the more skilled and resourceful pilots gains the superior position.

A fourth factor, everything again being equal, is sheer numbers.

If one side has three or four of these elements in its favor, it can win a virtually bloodless victory. A recent example was the 1967 Arab-Israeli War. The Israelis had everything going for them except numbers. They utterly destroyed the numerically superior Arab air forces with minimal loss. When things are more equal, various levels of attrition set in. It then becomes a matter of who still has something flying when the dust settles.

Electronic warfare is becoming an increasingly critical element. The "low and fast" flying techniques of strike and recon aircraft protect not only against ground-to-air weapons but also against interceptors. The current way to detect low-flying aircraft is by airborne "look-down" radars either in the interceptors or in larger air warning and control aircraft (AWAC) circling hundreds of kilometers inside friendly territory. The look-down radar employs computers and generally reliable electronic components to separate the low-flying aircraft from all the ground clutter. The Western air forces have the edge in this area. The Russian Air Force attempts to counter with its own less capable AWACS and with less efficient but still potentially effective patrols of interceptors ready to go to the area to which a low-flying enemy aircraft was reported headed and attempt to pick up the target visually. This puts the Russian interceptor at the mercy of higher-flying Western interceptors.

This maneuver can be countered in turn by the long-standing Russian tactic of operating patrols on two levels: high, to jump enemy interceptors and low, to jump the low-flying intruders. Given a large enough superiority of numbers, a willingness to take higher losses, and good weather for visual contact, the superior numbers approach will work. The interceptors will take heavy losses as they fight superior Western aircraft from disadvantageous positions. This is the most that a technically inferior air force can hope.

A numerically inferior air force normally aborts missions when faced with the prospect of taking losses, unless the mission is so important, which is often the case, that its successful completion is worth the losses.

As will be explained below, the attacker can counter this Russian technique by sending in large groups of aircraft composed of many different types, each with a specific offensive or defensive function. It can all get very complicated. Yet air combat still comes down to pilot skill, courage and quick reflexes, just as it did sixty-eight years ago.

Strike Missions

Air-to-ground combat can be offensive or defensive, but because of the limited endurance of aircraft, it is usually offensive. Intelligence and planning staffs select the most cost-effective targets. The favorite targets for strike aircraft are airfields, enemy aircraft on the ground (where they can't strike back), radar, and ground-control stations, ammo dumps and transportation lines.

Flying into enemy airspace is a risky business. High- and low-altitude antiaircraft defenses are numerous. Radars are all over the place. The best approach is to sneak in one plane at a time, a few hundred meters up. More experienced or audacious pilots fly low enough to singe the treetops. Coming in that low, at 500 to 600 km per hour, gives the enemy little time to react before you have gone past him. Using radar mapping and other electronic navigation devices, the target is found, the ammunition released and an equally rapid exit made.

That's one way of doing it. One aircraft at a time. But if the weather is bad, which it is most of the time in Central Europe, and not all your aircraft have sufficiently sophisticated electronics to get you in and out without good visibility, then you go in as a group with specialist aircraft. You also do this if the enemy has functioning AWACS.

The group consists of electronic-warfare aircraft, or Wild Weasels, and fighters. If the target is heavily defended by anti-aircraft guns and missiles, the group includes flak (antiaircraft) suppression aircraft. The Wild Weasels will confuse the enemy radars. If you can't evade all the enemy aircraft, your own fighters would defend you. Everything depends on the Weasels, as you can't take enough fighters in to fend off everything the enemy can throw at you. The Weasels also befuddle the enemy flak radars. If some guns are pre-positioned to fire visually at the approaching aircraft, it's flak suppression time. If all goes according to plan, the Weasels guide and protect the electronically less sophisticated strike aircraft to the target, where they release their loads. Everyone then fights their way home.

Although they are striving mightily to catch up, the Russians still cannot match Western aircraft in overall sophistication. They do not yet have functioning AWACS aircraft. Their solitary strike aircraft are untried, and their Wild Weasels are somewhat tame compared with Western models. Russian pilots, by and large, are not as expert, experienced or audacious as their Western counterparts.

This results in Russian strike missions being more massive and willing to succeed by attrition. Masses of fighters swarm ahead to secure air superiority. Further masses of strike aircraft go after the radars. And then come other strike aircraft, which take out a large range of targets while they can.

It's no wonder the Russians have so many guided missiles in Eastern Europe. These missiles are not reliable either, but the ones that do get moving can't be stopped. So there is great concern that the Russians won't take a very sporting attitude toward their air force's inability to hit anything. They may just go nuclear and launch the missiles. There is a lesser, but not insignificant, chance that they will use only chemical warheads on the missiles. But this is a very expensive way to temporarily knock out enemy rear-area installations, while nuclear weapons will do it permanently.

Drones

Drones and remotely piloted vehicles (RPV's) are pilotless aircraft. A drone flies a preprogrammed course, sometimes with onboard navigation equipment to correct any flight deviations. An RPV is controlled from the ground. With electronic warfare becoming ever more intense, the advantages of the drones over RPV's have increased. An RPV's link with its ground controller can be jammed. A drone is impervious to such jamming.

The rationale for such aircraft is simple; you don't lose a pilot if a drone is shot down. Further, because they don't carry a pilot much weight is saved. Finally, if they are used primarily for noncombat missions, they can be lighter and cheaper still.

Originally drones were used largely as *targets* to perfect aircraft and antiaircraft weapons. Under ground control, RPV's could match many of the maneuvers of piloted aircraft.

Reconnaissance, including electronic warfare, is fast becoming *the* primary mission of drones. Sensors are much lighter than normal bombloads, and are reusable. Flying low and slow, and of small size, drones are difficult to detect.

Target acquisition, while a form of reconnaissance, is becoming a distinct mission. Advances in electronics have made it possible for the artillery simultaneously to see what the drone sees. High flight endurance of some drones allows them to loiter over the battlefield long enough to give the artillery an ongoing view of targets. In addition, the artillery receives immediate information on the effectiveness of its fire.

Drones and RPV's also have an important role in *air combat,* even without being armed. Drones can carry electronic gear that mimics a larger aircraft or even groups of aircraft. In addition to acting as decoys, they can spearhead strike missions. Carrying electronic-warfare gear, they can detect enemy radars. Piloted aircraft can then fire antiradiation missiles, or other munitions, to destroy the enemy defenses.

In the past most of the development work was done by the United States. Other nations are now developing their own, because rapid advances in electronic technology have made it possible to make very capable drones at relatively low cost. In this respect the West has a considerable advantage.

However, there is a major problem. One man's technological breakthrough is another man's threat. Drones threaten to take away pilot jobs. Few people in the air forces will come right out and say this. But halfhearted enthusiasm for drones can be traced back to pilots' unease over their becoming *too* effective. This is ironic, as the air forces themselves had to fight similar prejudice in their early years.

Fighters, Bombers and Recon Aircraft

COMMON NAME is what people usually call the aircraft. Often, but not always, this name is officially recognized. Examples of unofficial names are Warthog, Aardvark and BUFF (Big Ugly Fat Fellow). The Russian names are those assigned by NATO.

DESIGNATION is an American and Russian convention. The European aircraft usually just use their names. The Tornado is a British-German-Italian joint effort. The Jaguar is a French-British cooperative venture. The Alpha is French-British. The F-6 is a Chinese copy of the MiG-19. The American designations indicate the primary function of the aircraft (F = Fighter, A = Attack). Russian designations refer to the design bureau responsible for developing the aircraft. The Russians

FIGHTERS, BOMBERS AND RECON AIRCRAFT

Western aircraft are generally more capable, and just as abundant, as Russian models.

Common Name	Desig-nation	Capability Ratings				Normal Combat Radius (km)	Year Intro-duced	No. in Use 1983	Main-tain-ability	Max Weight	Wing Load-ing	Thrust:Weight		Max Load	Max Speed (Mach)	Number Built or Planned
		Inter-cept	Strike	Recon	All Weather							Loaded	Clean			
West																
Phantom	F-4E	7	16	7	6	1100	1963	3200	15	28	0.57	1279	1721	7.2	2	5177
F-6/A-5	China	5	1	4	2	680	1962	2400	20	9	0.35	1667	1768	.5	1.2	2600
Tiger	F-5E	6	8	6	3	1000	1972	2200	8	11	0.61	901	1250	3.1	1.6	2500
Mirage III	France	6	1	6	3	1300	1963	1300	15	14	0.39	993	1115	1.5	2.2	1500
Starfighter	F-104	6	5	5	5	1200	1958	1000	25	14	0.77	1279	1689	3.4	2.2	2541
Mirage F1	France	6	5	6	3	1000	1973	700	16	15	0.61	1039	1411	4	2.2	700
Corsair	A-7E	5	7	2	5	550	1966	700	12	19	0.55	789	1230	6.8	.9	1534
Eagle	F-15A	12	10	9	6	990	1977	660	11	25	0.44	2000	2809	7.2	2.5	1130
Skyhawk	A-4	4	8	2	6	1500	1960	600	10	20	0.80	465	571	3.7	.92	2960
Jaguar	NATO	5	12	5	4	1300	1972	550	10	18	0.75	811	1081	4.5	1.5	550
Alpha	NATO	1	5	3	1	520	1979	450	8	8	0.43	795	1125	2.2	.85	600
Falcon	F-16A	10	6	6	4	900	1980	410	10	16	0.57	1563	2747	6.9	2	2900
Warthog	A-10	5	18	2	3	500	1977	380	7	23	0.49	783	1139	7.2	.56	735
BUFF	B-52	1	30	7	9	16000	1955	340	65	225	0.60	604	688	27.2	.9	465
Aardvark	F-111F	6	20	8	9	2000	1967	300	30	45	0.65	871	1120	10	2.2	501
Harrier	AV-8A	4	6	5	2	400	1969	300	15	11	0.59	1818	2703	3.6	.9	330
Tomcat	F-14A	8	10	6	8	1000	1970	300	25	34	0.64	1228	1522	6.5	2.4	597
Intruder	A-6E	4	10	4	8	750	1963	190	20	27	0.55	689	984	8.1	1.05	600
Kfir C2	Israel	7	6	6	2	780	1974	150	15	15	0.42	1218	1721	4.3	2.3	240
Tornado	NATO	9	14	8	9	1300	1980	100	24	24	0.67	1333	1905	7.2	2	660
Hornet	F-18A	9	12	6	7	1000	1982	50	12	22	0.59	1455	2238	7.7	1.8	1500
Mirage 2000	France	7	6	4	5	1600	1983	6	15	16.5	0.40	1333	2268	6.8	2.3	570
	B1B	3	45	8	9	5800	1984	2	30	217	1.17	553	723	51	2.2	100
Total								18271								30990

Russia															
Fishbed	MiG-21J 5	2	2	2	1100	1956	4000	15	9	0.41	1548	1842	1.5	2.1	10000
Flogger-B	MiG-23 8	1	5	4	960	1971	1900	25	17	0.46	1474	1565	1	2.3	2000
Flogger-D	MiG-27 7	5	6	6	700	1973	1400	30	18	0.49	1408	1690	3	1.7	1400
Fitter	SU-7 3	1	1	2	480	1960	1100	12	15	0.47	1467	1571	1	1.2	4400
Fitter-C	SU-17 4	4	4	4	600	1972	1000	20	18	0.55	1384	1929	5	2.1	1100
Flagon E/F	SC-15 5	1	3	5	725	1973	850	40	21	0.57	1415	1487	1	2.3	1100
Fencer	SU-24 5	9	6	7	1600	1974	600	35	40	0.82	1268	1536	7	2.1	550
Fishpot	SU-11 4	1	3	4	400	1967	400	15	18	0.69	1225	1297	1	1.8	650
Backfire	TU-22M 4	10	6	5	2500	1974	400	50	110	0.67	800	871	9	2.01	420
Badger	TU-16 2	6	4	3	1500	1955	325	40	75	0.45	560	636	9	.77	2000
Foxbat	MiG-25 5	1	7	4	500	1970	300	40	38	0.67	1293	1406	3	2.8	500
Blinder	TU-22 2	5	5	5	1500	1962	200	50	83	0.60	647	680	4	1.4	250
Firebar	Yak-28 5	1	6	6	928	1961	150	35	16	0.44	1646	1757	1	1.1	400
Forger	Yak-36 3	3	4	1	300	1975	100	25	10	0.63	1750	2011	1.3	.9	100
Foxhound	MiG-25M 7	1	5	6	1800	1982	70	50	33	0.60	1557	2131	9	2.3	200
Fiddler	TU-28P 4	1	4	6	1600	1962	70	40	45	0.80	1200	1227	1	1.7	130
Frogfoot	SU-25 0	11	1	1	500	1983	70	20	17	0.40	882	1364	6	.6	400
Fulcrum	MiG-29 5	7	7	5	900	1984	10	30	17	0.38	1412	2182	6	2.3	5000
Ram K	SU-27 6	5	5	4	800	1983	10	40	27	1.04	1444	1696	4	2.3	4000
Blackjack	Ram-P 2	32	6	8	5200	1985	2	50	250	0.71	560	636	30	2.3	100
Total							12957								34700
Average	West 6	11	5	5	1877	1971	761	18	37	0.59	1107	1545	8	1.68	1347
Average	Russia 4	4	4	4	1230	1971	648	33	44	0.59	1247	1476	5	1.80	1735
Average	All 5	8	5	5	1576	1971	710	25	40	0.59	1172	1513	7	1.74	1528

also use odd numbers to indicate fighters and even numbers for bombers. The Russian names are those assigned by NATO. The SU-17 is exported in less capable versions designated SU-20 and SU-22.

CAPABILITY RATINGS are numerical evaluations of the aircraft's effectiveness in performing interception, strike, and reconnaissance missions. Each calls for a different combination of equipment capabilities.

INTERCEPT is the ability to detect and destroy enemy aircraft. This rating is determined by evaluating the following aircraft capabilities:

Electronics. Onboard radar is important, especially look-down radar that can detect targets flying close to the ground. Other electronics are necessary to control long-range missile weapons. Even the traditional dogfight is enhanced with electronic equipment. The HUD (head up display) is a see-through computer display in front of the pilot. On this device the computer displays information on aircraft speed, altitude and bearing as well as fire-control information, cross hairs for targets, terrain horizon outlines for navigation, weapons status, tracking data from the radar, ECM data, etc. Using an HUD the pilot can keep his head up and be in touch with the situation while still taking advantage of numerous electronic aids.

Speed refers primarily to the ability of the aircraft's power plant to achieve top speed as quickly as possible. This is because top speed wastes so much fuel—10 percent to 15 percent of internal fuel per minute—that it can only be used briefly for maneuvering or climbing. Cruise speed allows the pilot to intercept the target and still have fuel left for high-speed maneuvering into a firing position. Typically, a cruise speed of 900 km per hour consumes about .5 percent of internal fuel per minute, or 250 meters per second. At a typical top speed of Mach 2, the craft moves 600 meters per second. This speed advantage over a cruising target can only be used if you have the fuel.

Maneuverability. In order to bring weapons to bear on the enemy aircraft, you must be able to get close enough to use them. You must also be able to maneuver into a position where the enemy will not detect you. If the enemy does detect you, you will need maneuverability to avoid becoming the victim. This maneuverability consists of engine power (thrust-to-weight ratio), wing loading and general mechanical superiority.

Weapons. Superior missiles allow destroying the detected intercept fast enough, and far enough away, to avoid any counterattack.

STRIKE is the ability to deliver bombs to ground targets. This is evaluated by the following aircraft capabilities:

Maximum load and munitions quality. This is the amount of weapons the aircraft can carry. The quality of these weapons acts as a multiplier. Guided or self-guided bombs and missiles are much more effective than unguided iron bombs, which just drop and explode. Western air forces tend to have more sophisticated munitions.

Electronics. Ground-attack radars and computers take much of the inaccuracy out of bombing. These devices find the target and then assist the pilot in releasing his bombs at the right point. Bombing can be done through pilot skill alone, but such skill takes time to develop and even more so when flying modern high-speed jet bombers. This is why electronic navigation and bombing aids have become so essential.

Maneuverability. A different type of maneuverability is required for bombing than for interception. Coming in low (down to less than 100 meters), loaded (over 5 tons of bombs) and fast (600 to 700 km per hour), the aircraft's ability to maneuver accurately is essential. Moving at 170 meters per second at an altitude of 100 meters is tricky. The target may be an area less (often much less) than 100 meters square. Depending on the size and prominence of the target, you often can't make your final approach until you are a mile or two away. This gives you somewhat less than 15 seconds to get the bombs on the target before the enemy gets you.

Aircraft durability. Bomber aircraft take more sustained punishment than higher-flying interceptors and recon aircraft. Ground weapons may not be as deadly, on a one-for-one basis, as air-to-air weapons, but a lot more people will be firing—from 200 to 1000 armed soldiers per square kilometer in the area passed over. For every few hundred troops armed with machine guns, there is a heavier antiaircraft gun or a few portable surface-to-air missiles. None of these weapons is normally capable of knocking you down by itself. It is the cumulative effect that makes life interesting. All that minor damage adds up. Worse yet, damage that would be a minor inconvenience at 2000 meters up looms more ominously when the altitude is 100 meters, the speed 200 meters per second, and the consequences of a slight loss of aircraft control or altitude critical. Effective ground-attack aircraft cannot be built like tanks because of armor weight. Instead, the aircraft are built to absorb much damage and still be capable of returning by providing armor for the pilot and some other vital components. Other components are provided in duplicate or triplicate so that you can afford to lose things. On the U.S. A-10 bomber, the engines are mounted over the rear fuselage to make them more difficult to hit and to reduce the infrared signature homed in on by many missiles. Thus, a combination of some lightweight armor, overbuilding and clever design gives a ground-attack bomber the durability to survive more than one pass.

RECON is the ability to take pictures over enemy territory without getting shot down. A single interceptor-type aircraft operating clean, without weapons load, will often be fast and agile enough to avoid enemy defenses. Recon ability was derived by evaluating the following aircraft capabilities:

Speed and thrust-to-weight ratio allow an aircraft to fly high and fast, or low and fast. Either approach is acceptable. The high approach is good for photographing large areas. If there are not a lot of high-altitude SAM systems, you can get away with flying fast and high. Generally enemy aircraft won't be able to get at you. American specialized reconnaissance aircraft like the TR-1 (U-2) and SR-71 can avoid even the missiles (these are covered under naval patrol aircraft). But other aircraft must go in low if too many SAM's are present. Here, thrust is important, as the denser atmosphere at low altitude requires more engine power to attain the same high speeds. Also, at lower altitudes the drag from extra fuel tanks is much higher than at higher altitudes. The result is that you can go about twice as far at high altitude than you can on the deck. And you can't go anywhere unless you go fast.

ECM (electronic countermeasures) are needed not only to penetrate enemy detection systems, but also to assist the collection of information. In these days of ever-increasing electronic warfare, recon aircraft are often sent in to take pictures of the enemy electronic defenses. The ECM equipment consists of sensors that detect various transmissions and transmitters which broadcast signals that obliterate or confuse enemy reception. Its computers study the enemy transmissions and warn the pilot of a potentially dangerous one. Some ECM computers also suggest the least dangerous path through upcoming transmitters. For reconnaissance you need only add a recording device to bring back a picture. Not all ECM loads contain all the above equipment. The minimum load is sensors and transmitters.

Navigation equipment. You can't get from here to there if you don't know the way. Often, the recon mission involves finding a very small target. If you are coming in very low and/or the target is heavily defended, there may be one safer approach that requires precise flying. Precise navigation is required not only to complete the mission but to survive it. On a recon mission, radio silence is usually maintained to give the enemy one less electronic transmission to home in on. So you can't call for directions either. You are on your own.

ALL WEATHER is the ability to navigate an aircraft in all-weather conditions. This ability is determined by evaluating the capabilities of the aircraft navigation equipment. A rating of O indicates an inability to fly safely in anything but clear-weather conditions.

NORMAL COMBAT RADIUS represents the normal maximum distance from the aircraft base to the area where it performs its mission. The usual rule of thumb is that an aircraft's radius will be one third the maximum distance it can fly with available fuel. This formula must take into account the need to fly out and back as well as the considerable fuel consumed while completing the mission.

For interceptors this means allowing for air combat at high speed. Fuel consumption of .5 percent per minute during cruise rises to 10 percent or more per minute during combat. Thus, 2 to 3 hours are spent going to and from a 10- to 15-minute battle.

Bombers have the same problem. They must also crank up the engine power to do their job. The fuel needs are not as high as for interceptors. Bombers do not always use an afterburner. Thus, going to maximum military power burns 2 percent to 3 percent fuel per minute. This power is applied only as the bomber makes its pass on the target. Depending on weapon load, a bomber makes from one to three or more passes. More fuel is required for loitering. Bombers, more than other aircraft, cannot force a battle. They must wait upon the ground operations by flying in circles.

Recon aircraft also need that one third fuel for the heart of the mission, which often involves using it for high-speed evasion of enemy aircraft or missiles. In this case, sufficient fuel is an excellent defensive measure. If the reconnaissance aircraft have more fuel than the interceptors, they can generally outrun them. Recon aircraft often carry drop tanks, which are let go as they make their final approach to the target. Flying clean increases speed and the efficiency of fuel use. The final approach to the target will usually be taken at the highest possible speed to avoid detection.

Bombers and recon aircraft can extend their range if they are certain that they will not have to burn additional fuel for loitering or for high-speed operations.

YEAR INTRODUCED is the year the aircraft type was first delivered to a combat unit. The longer an aircraft type is in service, the more it will be improved. Later versions have substantially improved performance. Aircraft must be overhauled frequently, and have a flight life of only 500 to 3000 or more hours.

NO. IN USE 1982 is the number of the type of aircraft in use during 1982. As this figure had to be projected in late 1981, there may be some small errors due to increased accident/war losses or production changes. Some air forces may also have decided to retire some older aircraft early.

MAINTAINABILITY is the relative ease of maintenance for the aircraft. The number is the man-hours of maintenance required for each hour of flight. High-performance aircraft wear out quickly, as the pilots push operating limits of the equipment. It will not fail every time, parts must be replaced very frequently.

Modern aircraft require from 7 to 50 man-hours of maintenance per flight hour. A decrease in the maintenance per man-hour makes a big difference in the number of technicians needed. For example, if a peacetime squadron of 24 aircraft flew each 40 hours per month, this would equal 960 flight hours. At 20 man-hours per flight hour, 19,200 man-hours per month would be needed for maintenance. With a peacetime availability of 118 man-hours per man per month, 163 maintenance troops would be needed to keep 24 aircraft flying. In wartime the available number of man-hours can be tripled by cutting out leave and other duties while following a 12-hour-day, 7-day-week schedule. But aircraft use will increase by far more, something on the order of 2 to 3 sorties per day, or 7 hours a day. That's 3360 maintenance hours per day, nearly twice the daily man-hours. What happens? Maintenance can be deferred, which will produce more frequent failure during use, resulting either in lost aircraft or a return to base without completion of the mission. Lost aircraft reduce the maintenance load, but repair of damaged aircraft takes up some of this slack. Much battle damage can be repaired during routine maintenance.

Maintenance demands increase with the complexity of the aircraft. Most Western aircraft devote over half their maintenance to electronic components, which continue to proliferate. Even so, aircraft components are constantly made simpler or more easily maintained. For example, the J-79 engine in the F-4 has 22,000 parts and requires 1.7 man-hours per flight hour. The F-100 engine in the F-16 has 31,100 parts and requires 4.2 hours. The F-101X will replace the F-100 and have only 20,000 parts. Its maintenance man-hours requirement will be under 2 hours per flight hour.

Russian equipment generally requires more maintenance. Yet the manpower pool for technicians is smaller and less well trained. Russian aircraft have only one third as much peacetime airtime as Western aircraft. Still, they have a higher peacetime accident rate, partially as a result of less qualified pilots, partly because of poor maintenance. In wartime the Russian equipment's maintenance system will be strained even more than those in Western air forces.

MAX WEIGHT is the maximum takeoff weight in metric tons, usually with a full load of weapons. Aircraft often operate with less than this weight, particularly on intercept missions. Bombers and recon aircraft more likely carry a full load.

WING LOADING is the ratio of full load weight to square meter of wing surface. This is one of the chief determinants of an aircraft's maneuverability. An aircraft with high wing loading will be stiffer and less agile in flight. However, wing loading is not the only determinant of maneuverability. Equally important are engine power and the general design of the aircraft, particularly the mechanical arrangement of the wing flaps, slats, fences, spoilers and sundry other mechanical devices. Look carefully at the wing of a commercial aircraft during landing or takeoff. Combat aircraft often have even busier wings. Some aircraft also have swing wings, which can be swept back for faster speed or extended for greater maneuverability. The swing-wing aircraft are Tomcat, Eagle, Tornado, Flogger, Backfire and Fitter-C. There are more components to fail but also higher performance.

THRUST:WEIGHT ratio is the pounds of engine thrust per ton of aircraft weight at full (afterburner) power in two configurations: LOADED is at full load (MAX WEIGHT); CLEAN, at full load minus the maximum load, in its most combat-worthy condition. The aircraft would still have some internal fuel and weapons.

This ratio is the best indicator of an aircraft's ability to move. A low ratio indicates a relatively sluggish aircraft. A high ratio, particularly one over 2200, indicates a highly maneuverable and lively aircraft. Vertical takeoff aircraft (Harrier and Forger) have a misleadingly high ratio because much of their thrust is dedicated to getting them off the ground straight up like a helicopter.

MAX LOAD is the maximum weapon load (in tons) hung outside the aircraft from the wings and fuselage.

This load is usually bombs and missiles, but it can also be fuel tanks or pods (aerodynamically streamlined containers) carrying cannon, rockets, ECM gear, recon equipment. Anything carried this way can be dropped in flight. These loads do impair speed and maneuverability.

Some aircraft also have internal cannon and a few hundred rounds of ammunition. This is reflected by their capability ratings.

MAX SPEED is the maximum speed in Mach numbers. Mach 1 equals the speed of sound at the most efficient height, which is about 10,000 meters. Supersonic (Mach 1+) speed is attained only by full engine power with afterburner. Normal cruise speed is about 900 km per hour at 10,000 meters. Flying at lower altitudes uses more fuel and strains the pilot because the aircraft is usually more difficult to control in the thicker lower atmosphere. Flying on the deck, 100 meters altitude, causes even more strain because of the constant need to avoid obstacles.

NUMBER BUILT OR PLANNED is the figure as of 1982. For Russian aircraft the figure is the total number built. For non-Russian aircraft it is the number on order.

Air Weapons

AIR-TO-AIR MISSILES are of two basic types: *infrared homing* (IR) and *radar homing* (RH). By far the most widely used are the Sidewinder and the Sparrow. Over 100,000 Sidewinders have been built so far. In addition, most other IR-type missiles are copies of the Sidewinder. The Sparrow has also become standard. The Sky Flash was designed as an improved Sparrow. Russian missiles are, for the most part, derivatives of the Sidewinder and Sparrow. The Phoenix is ahead of its time in having its own radar, which, when 16 km from its target, seeks out the enemy on its own. The next generation of RH missiles will employ this technique, but the next generation of RH won't arrive until the middle of the decade, because a Phoenix currently costs $2.7 million, compared with $85,000 for a Sidewinder and $170,000 for a Sparrow. Even an air-to-surface Harpoon missile costs over $1 million, which is why it is designed to be used against ships only. The Maverick, originally to be used against tanks, now costs $440,000 per missile. Tanks only cost two to three times that amount. Obtaining advanced technology ahead of its time is expensive.

AIR-TO-SURFACE MISSILES come in four types:

Homing missiles are sent by the launching aircraft in the general direction of the target. Thereafter, the missiles' own sensors take over. They are most often used against ships or other large targets, which are easier to hit than other aircraft. For this reason, most air-to-surface homing missiles are somewhat cheaper than the Phoenix. Examples are: Harpoon, Exocet, Kormoran, Maverick, ALCM, SRAM, AS-2, AS-3, AS-4, AS-5, AS-6 and AS-10. The Russian missiles have crude but effective homing devices when not faced with any countermeasures. The AS-10, like one version of the Maverick, is a tactical missile that homes in on reflected laser energy produced by an air- or ground-based target designator.

Antiradiation missiles (ARM) are primarily used to hit or shut down enemy radars by homing in on the enemy transmitter. The ARM is an electronic-warfare weapon. It has an onboard computer and considerable intelligence capability. In order to keep one step ahead of enemy efforts to preserve its radars, the newer models are smarter and faster. They can be used against most transmissions and will make the use of RH missiles dangerous. Examples are: HARM, Shrike, AS-9.

Guided missiles are controlled by an operator in the aircraft via a TV camera in the missile. The operator either literally flies the missile to the target or simply gets the target onto the TV screen. At that point the guidance system homes in on the TV image. Examples are versions of the Maverick.

Guided bombs are unpowered bombs using the same techniques as guided missiles. There is a wide variety, and more are being developed by the United States. These "smart bombs" allow aircraft to stay away from defending antiaircraft fire while achieving accurate hits on difficult targets. A guidance system and control surfaces, wings and fins, are attached to a bomb. An example is the Walleye.

AIR WEAPONS

The simpler weapons, like cannon, survive because of their greater re-liability.

Weapon	Made by	Target Detection	Aspect	Range (km)	Weight (lbs)	Speed (mps)	Guid-ance	Rank	Used On
Air-to-Air Missiles									
Phoenix	US	Active	All	200	975	1600	9	1	F-14
Sky Flash	Britain	Active	All	45	425	1000	7	2	F-4, Tornado
Sparrow AIM-7F	US	Active	All	40	500	1200	6	3	F-4, F-14, F-15, F-18
Sidewinder AIM-9L	US	Passive	Chase	14	190	820	5	4	Most
Shafrir	Israel	Passive	Rear	5	205	700	5	5	Most
Magic R550	France	Passive	Rear	10	200	1000	5	6	Most
Sidewinder AIM-9J	US	Passive	Rear	10	185	820	4	7	Most
AA-7RH Apex	Russia	Active	All	30	700	1100	4	8	MiG-23
AA-8RH Aphid	Russia	Active	All	15	350	800	4	9	MiG-23
AA-6RH Acrid	Russia	Active	All	40	1700	1400	3	10	MiG-25
AA-8IR Aphid	Russia	Passive	Rear	7	250	800	3	11	MiG-23, most
AA-7IR Apex	Russia	Passive	Rear	16	600	1100	3	12	MiG-23
Sidewinder AIM-9B	US	Passive	Rear	4	159	650	2	13	Most
AA-6IR Acrid	Russia	Passive	Rear	20	1300	1400	2	14	MiG-25
AA-3 Anab	Russia	Passive	Rear	10	500	800	2	15	SU-9,SU-11,SU-15
AA-2 Atoll	Russia	Passive	Rear	7	155	700	1	16	Most
Air-to-Surface Missiles									
Harpoon	US	Active	All	110	1200	280	7	1	S-3, P-3, A-6,A-7
Exocet	France	Active	All	60	1442	300	7	2	F-4
Kormoran	W Ger	Both	All	37	1320	300	7	3	F-104, Tornado
HARM AGM-88A	US	Passive	All	20	798	1200	7	4	F-4, A-7, Wild Weasels
Shrike AGM-45A	US	Passive	All	16	400	650	6	5	F-4, A-4, A-6,A-7, Wild Weasels
ALCM AGM-86	US	Active	All	2500	2800	240	5	6	B-52
AS-6 Kingfish	Russia	Active	All	250	10600	800	5	7	Tu-22M, Tu-16
AS-4 Replacement	Russia	Active	All	800	12100	1150	5	8	Tu-22M, Tu-22
AS-5 Kelt	Russia	Active	All	180	7700	300	4	9	Tu-16

AS-9 ARM	Russia	Passive	All	80	2200	260	4	10	Su-24
Maverick AGM-65	US	Passive	All	20	462	600	4	11	Most
AS-4 Kitchen	Russia	Passive	All	300	13200	400	3	12	Tu-22M, TU-22, Tu-95
Bullpup AGM-12	US	Passive	All	17	1786	650	3	13	Most
SRAM AGM-69A	US	Active	All	>20	2240	1000	3	14	B-52
Walleye AGM-62A	US	Passive	All	4	1100	200	2	15	Most
AS-10	Russia	Passive	All	10	2600	200	2	16	Su-24
AS-7 Kerry	Russia	Passive	All	10	2600	300	1	17	Su-24, Su-17, MiG-27
AS-3 Kangaroo	Russia	Passive	All	650	24200	550	1	18	Tu-95
AS-2 Kipper	Russia	Passive	All	210	9300	400	1	19	Tu-16

Cannon									
GAU-8 30mm	US	Both	Chase	1.2	57	1020	3	1	A-10
M-61A1 20mm	US	Both	Chase	.8	28	1036	3	2	Most
ADEN 30mm	Britain	Both	Chase	.8	10	790	2	3	Most
Gsh-23 23mm	Russia	Both	Chase	.8	27	950	1	4	Most
NR-30 30mm	Russia	Both	Chase	.8	14	780	1	5	SU-17, SU-7

Drones and RPV's									
AQM-34 series	US	Both	All	3000	6200	200	4	1	Does Not Apply
MQM-74 Chukar	US	Both	All	454	492	265	3	2	Does Not Apply
Scout	Israel	Both	All	450	250	28	2	3	Does Not Apply
La-17	Russia	Both	All	280	4000	280	1	4	Does Not Apply

CANNON shown are representative of the more common types. The GAU-8 is the only cannon designed solely for destroying tanks. It could be devastating against aircraft except for the fact that it weighs over 3800 pounds, ten times as much as any other fighter cannon. Next to the name of each cannon is its caliber.

DRONES AND RPV'S. The former are self-guided aircraft; the latter are controlled from the ground.

MADE BY is the nation that designed the weapon and, in most cases, is the sole source of that weapon.

TARGET DETECTION indicates whether the using aircraft must emit an electronic signal in order to guide the weapon. *Active* means that a signal is sent and can be detected or defeated by countermeasures. *Passive* means that no signals are transmitted, making the weapon much more difficult to defeat with countermeasures. *Both* indicates that both passive and active means are used.

ASPECT shows the direction from which the aircraft may make an attack with that weapon. *All* means that an enemy aircraft, or ground target, may be attacked from any direction. *Rear* means that enemy aircraft may only be attacked from the rear. *Chase* indicates that although the weapon can be used to attack from all directions, it is far more effective when used from the rear during a chase of the target aircraft. This is particularly the case with cannon.

RANGE (in kilometers) is the maximum effective range. This will be more—or less—for exceptionally large or small targets. Longer range is largely a function of the size and efficiency of the propulsion system and of the range of the radar. Cannon range is longer when these same cannon are used in ground-based vehicles because of the great loss of accuracy from a rapidly moving platform.

WEIGHT (in pounds) of the missile. There are four components in a missile: airframe (shell), propulsion system, guidance system and warhead. The warhead usually comprises 15 percent of missile weight. Propulsion-system weight is a function of range, while airframe weight is a function of missile size. The largest variable is the guidance system, including the fins and other control surfaces as well as the flight computer, radio gear and sensors. Western missiles' higher level of technology allows lighter and more capable guidance systems. The weight given for cannon is the weight of shell the cannon fires per second.

SPEED (in meters per second). Higher speed is always desirable. It can be obtained only when the guidance system can handle it. Most air-to-air missiles have solid-fuel motors that burn out quickly (in 2 to 10 seconds), leaving only momentum to carry them to the end of their mission. This high initial speed thus accounts for the minimum range of missiles, as the higher speeds are more difficult to control. It also limits the range of highly maneuverable missiles intended for use at close range. Too high a speed and the minimum range will be too long. Ideally it should be no more than a few hundred meters. Using cannon, aircraft have been brought down at under a hundred meters. This is dangerous to do with missiles, as the destruction is generally more catastrophic and the debris has been known to take the attacker down with it. The speed given is the highest speed attained at motor burnout.

GUIDANCE is an evaluation—from 1 (poor) to 9 (excellent)—of the missile's guidance-system quality. The evaluation includes the accuracy and dependability of the system as well as its resistance to countermeasures. Even passive systems, especially infrared ones, have a number of weaknesses. Because they home in on heat, they can be confused by such natural phenomena as the sun or the hot surface of a desert. Also, clouds can mask the heat source. Radar-guided systems are very sensitive to jamming. This is usually circumvented by having the missiles home on the jamming source.

RANK indicates comparative effectiveness within the classes, taking all factors discussed into account.

AIRCRAFT USED ON is the aircraft that uses each of these weapons. *Most* indicates a simple missile that can easily be mounted on most aircraft. In the West the majority of the aircraft can easily be equipped to use passive-guidance missiles. The pilot need only turn on the missile, listen for the audio signal that the missile is "locked on" to something and then launch it. For missiles dependent on aircraft radar, a more elaborate setup is required. In this case a few hundred pounds of equipment must be installed and integrated with the missile's electronics.

Russian aircraft are unique in that until recently each new Russian aircraft had a special missile. Another Russian habit was equipping these missiles with either a passive (IR = Infrared) or an active (RH = Radar Homing) system. The RH version is heavier and has a longer range. This solves some inventory problems, but is wasteful because the IR version uses a larger motor than necessary.

Russian aircraft usually carry two of each version. Russian aircraft missiles have been noticeably unsuccessful in combat. An example of their inefficiency was the incident several years ago when a South Korean airliner strayed over northern Russian airspace. A Russian interceptor eventually found the airliner and attempted to shoot it down with several missiles. The airliner, a Boeing 707, was able to crash land on a frozen lake. The Russian missiles were unable to destroy a subsonic civilian airliner flying a straight course. The Russians still equip most of their aircraft with cannon, an intrinsically more reliable weapon.

NOTES ON IRON BOMBS AND UNGUIDED MISSILES

Air-to-ground operations have traditionally used a wide range of gravity bombs and unguided missiles. These have not been eclipsed by guided air-to-ground munitions. As the list below demonstrates, iron bombs have become a lot smarter and deadlier.

Iron bombs are the traditional high-explosive-filled containers dropped by aircraft. Largely unchanged since World War II, they come in numerous sizes, weighing from a few pounds to over a ton.

CBU (cluster bomb units) saw widespread and effective use during the Vietnam War. They are containers of smaller bombs. When the container is dropped, it breaks open and distributes the smaller bombs over a wide area. A typical 600-pound CBU contains 150 smaller (3-pound) bombs which would fall over an area 50 meters wide and 200 meters long. Often the pilot can select a smaller, or a larger, pattern. The 50-by-200 meter pattern might result in a better-than-50-percent chance of any unprotected personnel being injured. The CBU's carry a variety of loads: antitank, incendiary and chemical. Their weight varies from a few ounces to 20 or more pounds. In addition, the bombs can be equipped with timers or sensors that turn them into land mines.

FAE (fuel air explosive) is a variation on the napalm bomb (thickened gasoline). An FAE hits the ground, breaks open and creates a mist of flammable liquid. A small delayed-action explosive then goes off, causing the cloud to ignite. The pressure of the blast is sufficient to wreck vehicles, ships and equipment as well as being fatal to any personnel. The only other devices to produce similar blast effects are nuclear weapons. Pound for pound, FAE.bombs are three to five times as destructive as high explosive. For example, an 1100-pound FAE would destroy most equipment and injure all personnel within 250 meters of the impact point. These devices are also used in CBU's. Because of their area effect, FAE bombs have been successful in clearing mines. In Vietnam large FAE's were used to clear helicopter landing sites in the jungle.

Incendiaries are the familiar napalm plus an assortment of other flammable items. These are being supplanted by FAE.

Special-purpose bombs fulfill a variety of specialized needs. Concrete-piercing bombs are used to crater airfields and destroy heavy structures. Also in this category are chemical and nuclear weapons.

Unguided missiles, or rockets, are still used. They have a variety of loads: fragmentation, illumination, smoke, armor piercing. Their range is several thousand meters. Usually carried in pods containing 7 to 32 rockets each.

8

AIR DEFENSE

ON THE FACE of it, aircraft appear to be sitting ducks for ground fire. At the beginning of World War II, the Germans thought that one aircraft would be shot down for every fifty 88mm shells fired. But Allied aircraft were found to be made of sterner stuff, requiring 12,000 shells for each kill. The advent of missiles didn't much improve the ratio. Over Vietnam and in the Middle East it was found that about fifty surface-to-air missiles were needed to bring down one aircraft.

Also, the aircraft developed an unpleasant habit of attacking the air-defense units. During World War II and ever since, air defense has proved to be an impediment, at times even a deterrent, to aircraft attacks, but air defense cannot stop aircraft in the long run. The primary cause for aircraft having failed to destroy targets has not been air defense, but shortcomings in aircraft weapons.

Air-defense forces strive for attrition and deterrence. They attempt to force aircraft either to abort their mission or take losses by making high-altitude flight unsafe but not impossible. They force aircraft to fly low enough to be engaged by antiaircraft guns; at least the gunners can see what they are shooting at. Indeed, this is where most aircraft losses to air-defense forces take place. In North Vietnam over 80 percent of losses were to antiaircraft guns, while in the 1973 Arab-Israeli War, antiaircraft guns inflicted 30 percent to 50 percent or more of the losses (the Israelis claim the higher number). On the other hand, over 10,000 cannon shells were required for each of these aircraft losses. Everything depends on the quantity and quality of the air-defense forces, and on the quality of the aircraft they oppose.

Detect, Acquire, Track, Destroy

That's the four-step process by which air-defense weapons attempt to destroy aircraft. Each step is fraught with insoluble problems. Pilots have a keen sense of self-preservation and diligently respond to whatever advantage the air-defense people may gain.

Detection. Aircraft, no matter how large, are small objects in the vastness of the airspace in which they operate. Early air-defense forces learned that to detect aircraft effectively they must prepare ambushes by establishing fields of fire in preselected areas. This approach can use visual or radar sighting, sometimes both, particularly for tactical air defense. Pilots, not unaware of this situation, take pains to make detection as difficult as possible through electronic warfare. If the electronic warfare is successful, as it was over North Vietnam, the surface-to-air missiles are fired much like unguided rockets, if at all. The attacking aircraft know when they are being painted by the detection radar and, if they have the electronic-warfare equipment, proceed to "play Space Invaders on the enemy radar screens," as one U.S. pilot put it. Lacking electronic-warfare resources, the aircraft can go on the deck. At an altitude of 100 meters most detection radars cannot spot it. For this reason, airborne detection systems are becoming more popular: the U.S. AWACS (airborne warning and control system) and the Russian MOSS aircraft. The U.S. system works, the Russian MOSS appears less successful. At least the Russians are not building many of them, a sure sign of trouble. Other reports indicate that the MOSS only works some of the time under optimal conditions.

Acquire. Once the individual aircraft has been successfully detected, it must be "acquired," that is, confirmed to be an enemy aircraft and to be flying a bearing that will bring it within range of your weapons. Detection radar always has a longer range than missiles or antiaircraft guns to allow you to destroy the aircraft as far away as possible. This is important for two reasons. First, the air-defense system deploys behind the fighting front. The aircraft target may be many kilometers from the air-defense radar. Also, the aircraft may have its own standoff missiles, which will be released anywhere from a few to a few hundred kilometers from the target. Every kilometer counts. Second, the aircraft will most likely be traveling at high speed, from 200 to 700 meters per second. For a tactical air-defense system with a maximum range of under 20 km, every second counts.

Incorrect identification causes serious problems. During the 1973 war the Arabs fired 2100 missiles and destroyed 85 aircraft; 45 were Arab. All aircraft carry an IFF (identification, friend or foe) device that gives a coded electronic response when interrogated by radar. But codes can be broken, responses can be ignored or misinterpreted. The Arab air-defense people and at least 45 Arab pilots know all about this problem. It plagues all air forces.

Tracking. All of the above happens in less than a minute. Perhaps in seconds. If these first two steps are successfully completed, you must now track the target with fire-control radar and fire a missile or guns. By now the aircraft is probably aware that it is a target. Evasive measures can be taken even under fire. The most successful ones "break the track," usually by violent maneuvers. It's still not too late for electronic warfare either, as radar controls the firing of most missiles and antiaircraft. Other measures are ejecting flares (which provide a more tempting target for heat-seeking missiles) or chaff (strips of aluminum that create a "cloud" which confuses the tracking radar).

Destroy. Even if there is a hit or hits, there is still no assurance that the aircraft will be destroyed or even substantially damaged. Modern aircraft are built to last. They are overbuilt, with many duplicate or triplicate systems. It is somewhat presumptuous for the 15-pound warhead of the SA-9 missile to destroy a 20+-ton aircraft, especially when the warhead often does not achieve a direct hit but only explodes nearby, say, 10 meters away.

Air Defense Weapons

Small warheads. It is possible to destroy an aircraft with a small warhead. An aircraft moving along 100 meters from the ground at 600 km per hour is very susceptible to the slightest damage. More massive missile damage occurs at the higher altitudes, which gives the aircraft some distance in which to sort things out.

Smaller missiles, particularly the low-altitude portable ones, have only 5-pound warheads. Half the Israeli A-4 aircraft that were hit with them in the 1973 war returned to base and landed. The success of the smaller missiles is limited by both their small warheads and the lack of a proximity fuse, a radar device that ignites the warhead when a target is 1 to 20 meters away. Achieving a direct hit with these small missiles is mandatory and very difficult. The missiles can only be fired at the rear of the aircraft, where the heat source is sufficient to attract the heat seeker. As these aircraft come in low, fast and unexpectedly, you will have ten seconds at most to get off a shot. A sharp pilot immediately looks for a hill to slip behind to break the track of any pursuing missile. He can then come around for another pass and force you to burn up more missiles.

This is where guns come into their own. Their 20mm to 57mm shells either hit or they don't. When they do hit they inflict damage, not just through their high-explosive charges but also through their high-velocity impact. The shell weights vary considerably in this range. A 20mm shell weighs only 3.5 ounces. The 57mm weighs in at 100 ounces, over 6 pounds. The most commonly used Russian shell is the 23mm (7 ounces). The United States currently uses a 20mm, but plans on upgrading to 40mm (30 ounces). The basic tradeoff here is between the high rate of fire and the destructive capacity of the shell. The Swiss-developed Gepard (two 35mm cannon), used primarily by West Germany, can deliver eighteen 20-ounce shells a second. The Russian ZSU-23 (four 23mm cannon) can deliver sixty 7-ounce shells a second. On the basis of weight of shell fired, the Russians appear to have the edge, but Russian guns would be firing at heavier and better-protected Western aircraft, while the Germans would be firing at lighter and more vulnerable Russian aircraft. The sides should switch weapons.

The Russians did have helicopters in mind when they fielded the ZSU-23, and the lighter 23mm shell is excellent against older helicopters. More recent U.S. models have been designed specifically to withstand hits by 23mm shells. In the final analysis, the purpose of the antiaircraft guns is to damage the aircraft enough to send them home. The real destruction must be left to larger weapons.

Larger warheads. Larger missiles have warheads, which can be very elaborate, and weigh hundreds of pounds. The primary goal is to expand the range of destructiveness of these warheads as much as possible. Using shaped charges to direct the flight of high-velocity fragments, the warheads can be fatal to most aircraft a hundred or more meters away. Proximity fuses make near misses as effective as direct hits. Western missiles have more elaborate warheads than the Russian ones, but it's only a matter of time before all warheads are equally lethal.

Large caliber (75mm and up) antiaircraft guns also use proximity fuses and specially designed fragmentation explosives. They are employed less frequently for several reasons. A large number of shells are required to obtain a hit. Large surface-to-air missiles can cover higher and lower altitudes than large-caliber antiaircraft guns. Finally, large surface-to-air missiles are more capable of improvement than large-caliber antiaircraft guns.

Large-caliber antiaircraft guns are used only by Russia and her client states. They are utilized much as during World War II and North Vietnam (1966–1975). The guns basically barrage an area that radar has indicated the intruding aircraft will enter. In clear weather, with radars rendered inoperable by electronic warfare, antiaircraft guns can still serve quite well.

One area in which heavy-caliber antiaircraft guns are still effective is at sea. Most large-caliber naval guns are deployed against both aircraft and cruise missiles. Special smaller-caliber antiaircraft guns have also been deployed at sea. Once turned on, they automatically engage anything moving fast enough to be a cruise missile. As these missiles come in at up to 1000 meters per second, they don't leave much time for human intervention.

Very small-caliber antiaircraft guns. Although not decisive weapons, machine guns and rifles can have an effect. During World War II, the Russians developed the tactic of attacked ground troops firing blindly into the air. This not only damaged aircraft, but it also maintained the morale of the troops under air attack. Doing something, anything, to fight back is better than simply looking for a place to hide. Most tanks, and many other vehicles, are equipped with a heavy-caliber machine gun (12.7mm to 14.5mm) that is effective to an altitude of 1000 meters. Enough of this small-caliber fire in the air will not discourage aircraft—they usually won't even see it—but damage will be done. In very rare instances, aircraft have even been brought down. Usually, however, just more damage is created, and the damage adds up.

Tactical Deployment

Air-defense units are deployed according to their range and mobility. The shortest-range and most mobile equipment travels with combat units. The longer-range and less mobile equipment is set up as far back as 100 km behind the fighting front to protect rear installations and to give additional high-altitude protection to front-line units.

Ideally, air-defense units should be stationed on high ground to allow the greatest coverage by the radar-controlled or visually controlled missiles. This is often not practical, particularly in a mobile battle, where everyone will be on the road when the aircraft attack. In a defensive situation, or when attacking a rear area, the aircraft are more cautious and come in with the intention of destroying the air-defense units. As the latter are usually unarmored and full of explosive and flammable materials, they are very vulnerable targets. The new ICM (improved conventional munitions) include bombs that spread dozens of smaller bombs (88 from one 155mm shell or up to 200 incendiaries in an aircraft bomb) over a few hundred thousand square meters. Air-defense systems are fragile things.

The key to a successful system is layers of defense at multiple depth and altitude. The Russians, for example, put ZSU-23 (2-km range) cannon and SA-9 (8-km range) missiles right up with the leading combat troops. In addition, the troops have shoulder-fired surface-to-air missiles (the Russian SA-7, 3-km range). A few kilometers behind are self-propelled systems: the SA-8 (12-km range), SA-11 (25-km range) and SA-6 (30-km range). The SA-4 (75-km range) and SA-10 (50-km range) systems provide defense farther back, with only some of them extending their coverage beyond the leading units. Western nations use the same general principles. In addition, friendly aircraft are essential for successful air defense. For most Western nations, air superiority is seen as the senior air-defense technique. The Russians put more of their air-defense strength on the ground.

The Ultimate Air Defense: Air Superiority

The successful destruction of enemy air forces in World War II has made Western armies slow to realize that this may not always happen. The West is now making a belated attempt to develop air-defense strength on a par with the Russian system. It is still quite possible that Western forces will attain air superiority in a future war. But if they don't, or until they do, the burden of defense will be on the ground air-defense units.

Even if general air superiority is achieved, the enemy will be able to attain temporary air superiority. When this occurs, aircraft-delivered weapons will cause great destruction. As Israel has shown in its wars with the Arabs, it is possible to smother the other side's air force. But unless this can be assured, there will be a need for ground air defense. One sector may be stripped of air cover in order to achieve air superiority somewhere else. This leaves only ground-based air defense. For all these reasons, a doctrine of mutual support of ground air defense and air superiority operations has grown.

The major problem with combining ground air defense and air superiority operations is the difficulty of sorting out friendly and enemy aircraft. Operational experience has shown this to be a virtually insurmountable problem. Losses to friendly fire will probably encourage even more use of the exclusion rule whereby air defense fires on either anyone or on no one. The acceptability of this method depends on how many losses to friendly fire each side is willing to take.

Air-Defense Weapons

The following chart gives the characteristics of most of the world's air-defense weapons. Shown are the most widely used weapons, which are representative of the ones not indicated. As it is a relatively easy matter to manufacture air-defense weapons, a few nations do not have a monopoly on their manufacture, as is the case with aircraft, tanks and similar weapons.

NATION indicates the country that originally designed the weapon.

TYPE gives the designation of the weapon. For Russian weapons the NATO designation is given. See below for notes on the systems.

EFFECTIVENESS is a general evaluation of the relative capabilities of the weapon on a 1 to 100 scale. These are estimates, as many of these weapons have not been used in combat. Those that have been used against real targets have since undergone modification. All air-defense weapons constantly undergo upgrading.

AIR-DEFENSE WEAPONS

Air-defense weapons tend to be either low altitude/short range or high altitude/long range. All missiles are very sensitive to electronic counter-measures.

Type	Nation	Effec- tiveness	Effective Altitude (meters) Minimum	Maximum	Range (km)	Rate of Fire (rps)	All Weather Cap.	Caliber	Missiles Barrels	Naval Version	Mobil- ity
Patriot	US	100	100	24000	60	Missile	7	410	4	No	SP
Standard ER RIM67A	US	70	50	25000	110	Missile	6	343	2	Same	Ship
Improved Hawk	US	70	30	18000	40	Missile	6	370	3	No	Mobile
SA-12	Russia	67	100	30000	100	Missile	7	400	2	No	Mobile
SA-10	Russia	60	300	4500	50	Missile	6	450	4	SA-N-6	Mobile
Standard MR RIM66A	US	60	50	25000	32	Missile	6	305	2	Same	Ship
Hawk	US	60	100	11000	30	Missile	6	350	3	No	Mobile
Nike-Hercules	US	55	1000	50000	150	Missile	6	800	1	No	Portable
SA-N-3 Improved	Russia	51	150	25000	55	Missile	6	600	2	Same	Ship
Tartar RIM24B	US	50	50	20000	20	Missile	6	300	2	Same	Ship
SA-11	Russia	50	30	14000	28	Missile	6	335	4	SA-N-7	SP
SA-3	Russia	48	300	15000	35	Missile	4	450	2	SA-N-1	SP
SA-6	Russia	45	200	13000	35	Missile	6	350	3	No	SP
SA-N-3	Russia	45	150	25000	30	Missile	6	305	2	Same	Ship
Roland	W Ger.	42	10	3000	6.3	Missile	6	163	4	No	SP
SA-4	Russia	40	1000	25000	70	Missile	5	900	2	No	Mobile
Sea Sparrow RIM7H	US	38	15	5000	5	Missile	6	200	8	Same	Ship
Rapier	Britain	36	10	3000	6.5	Missile	4	133	4	No	SP
Crotale	France	33	50	3600	8.5	Missile	5	156	4	No	SP
SA-13	Russia	31	10	10000	7	Missile	0	110	4	No	SP
SA-2	Russia	30	1500	24000	44	Missile	4	500	1	SA-N-2	Portable

Phalanx	US	25	0	2000	2	50	7	20	6	Same	Ship
SA-8	Russia	25	50	10000	12	Missile	4	210	4	SA-N-4	SP
Twin 30	Russia	15	0	2000	3	8	5	30	2	Same	Ship
Chapparral	US	15	100	1000	5	Missile	0	127	4	Yes	SP
DIVAD	US	12	0	2000	4	10	6	40	2	Yes	SP
Gepard	Switzer.	12	0	2000	4	18	6	35	2	No	SP
SA-9	Russia	10	20	5000	7	Missile	0	110	4	No	SP
AMX-30SA	France	10	0	2000	3.5	21	4	30	2	No	SP
ZSU-23	Russia	8	0	2000	2.5	65	3	23	4	Yes	SP
Vulcan	US	5	0	2000	1.5	50	0	20	6	Yes	SP
ZSU-57	Russia	5	0	4000	6	4	3	57	2	Yes	SP
M-42	US	4	0	1500	3	4	0	40	2	No	SP
Stinger	US	4	0	4800	5	Missile	0	70	1	No	SP
SA-7	Russia	2	25	4200	3.5	Missile	0	70	1	SA-N-5	SP
ZPU-4	Russia	2	0	1400	1.4	40	0	14.5	4	Yes	SP
.50 Cal	Many	1	0	1000	1	10	0	12.7	1,2,3,4	Yes	SP

In addition, the countermeasures, primarily electronic, their potential targets possess have a considerable influence on their effectiveness. The ratings take into account such factors as system reliability, quality of target acquisition system and lethality of warhead.

EFFECTIVE ALTITUDE MAXIMUM is the maximum altitude (in meters) at which the weapon can reasonably be expected to hit a target. This limit is imposed primarily by the weapon's ability to reach that high. Missiles operate best at higher altitudes. Guns tend to do better closer to the ground.

EFFECTIVE ALTITUDE MINIMUM indicates the minimum altitude (in meters) at which the weapon can reasonably be expected to hit a target. For missiles this minimum represents the limits of its guidance-control systems. Because of this problem, the minimum is optimistic and depends on how ideal the situation is for the missile system. Missiles don't like surprises, such as aircraft doing violent maneuvers close to the ground.

RANGE is the maximum horizontal range of the weapon in kilometers. This limit is dictated by the range of the system's target acquisition ability.

RATE OF FIRE (RPS) indicates the number of projectiles the system can fire per second (rounds per second). Missile systems are indicated. Such systems can usually fire all the missiles on their launchers within a few seconds. It is common to fire two or more at each target in order better to assure a hit. Gun systems similarly rely on a high rate of fire in order to increase the chances of a hit.

ALL-WEATHER CAPABILITY is a numerical rating of the system's ability to operate at night and in bad weather. Even the most sophisticated systems often have manual backups. Missile and radar-guided gun systems are quite complex. Parts of them can fail without preventing the weapon from being effectively fired. But this requires clear weather so that the target can be visually acquired and tracked. The higher ratings indicate a better all-weather capability.

CALIBER is the diameter of the shell or missile in millimeters.

MISSILE BARRELS shows the number of barrels a gun system has or the number of missiles per launcher a missile system has. The more the better.

NAVAL VERSION indicates if the system is also used aboard ship. If so, this is indicated by "navy" or its navy name. The naval versions generally have the same operating characteristics as the land-based versions.

MOBILITY SP means self-propelled; the entire system moves on one or more vehicles and can be used while equipment is still on the vehicles.

Mobile means the entire system can be moved on vehicles and put into action after a minimum of unloading and preparation. *Portable* means that the system can be moved but requires extensive setup. *Fixed* means that the system usually operates from permanent sites. *Ship* means that the system is normally mounted on a ship. All naval versions operate from ships.

THE MISSILE SYSTEMS

Patriot is a recently introduced system that underwent twenty years of development. It will replace the *Hawk* and *Nike-Hercules.*

Standard ER (extended range) is the standard ship missile system for the U.S. Navy.

Improved Hawk is an extensively upgraded version of the original *Hawk.* This is the primary air-defense missile system for U.S. and many Western ground forces.

SA-12 is the newest Russian long-range system. It uses much improved radar and guidance systems that give it good capability against small, low-flying targets like cruise missiles.

SA-10 is the latest missile system for the defense of Russia itself. Its principal features are quick reaction time and high speed (2000 meters per second).

Standard MR (medium range) is a shorter-range version of the *Standard ER* (see above).

Hawk is the original U.S. missile system for defense against low-flying aircraft.

Nike-Hercules is the standard long-range air-defense missile system used by the U.S. and many other Western nations.

SA-N-3 Improved is the latest version of the standard heavy naval air-defense missile system, and is the Russian equivalent of the U.S. *Standard.*

Tartar is an older U.S. Navy missile system. Being replaced by the *Standard* systems.

SA-11 is a battlefield missile system intended initially to complement, and perhaps replace, the SA-6. Probably has increased reliability and accuracy against low-flying aircraft.

SA-3 is an older system, still used because the Russians never throw anything away. Easily defeated but still a potentially lethal nuisance.

SA-6 is the principal Russian battlefield air-defense missile system for which there is no Western equivalent.

SA-N-3 is an older version of the improved model (see above).

Roland is a battlefield missile system. Used by many Western nations.

SA-4 is the Russian heavy battlefield missile system. Primarily used for long-range, high-flying aircraft. Russian equivalent of the *Nike-Hercules.*

Sea Sparrow is an air-to-air missile used in an air-defense system adapted for shipboard use.

Rapier is a battlefield missile system. Used extensively for airfield defense against low-flying aircraft.

Crotale is similar to the *Rapier* and *Roland* systems.

SA-13 is yet another tactical system (in addition to SA-6, S-7, S-8, S-9). A clear-weather system with excellent ability against low-flying aircraft.

SA-2 is an obsolete missile system still used in the defense of Russian airspace.

Phalanx is a "last-chance" automatic defense system against surface-to-surface missiles and low-flying aircraft. Used on ships only.

SA-8 is the Russian equivalent of the *Roland.*

Twin 30 is the Russian equivalent of the *Phalanx.* Its capability is closer to older light flak, small-caliber air-defense cannon. Many navies still use such weapons, although it is unlikely that an aircraft will come that close. Unless automated, like the Phalanx, it will be difficult for such weapons successfully to engage missiles coming in at 500 or more meters a second.

Chaparral is an air-to-air missile (the Sidewinder) used as a battlefield missile system.

DIVAD (DIVisional Air Defense) is a recently introduced battlefield air-defense cannon.

Gepard is a battlefield air-defense cannon. Although originally developed in Switzerland, this system is primarily used by West Germany.

SA-9 is similar in concept to *Chaparral,* although it uses a missile more similar to the SA-7 than to an air-to-air type.

AMX 30SA is a battlefield air-defense cannon.

ZSU-23 is a battlefield air-defense cannon. This is a widely used Russian system and is reportedly to be updated shortly. The new version will be of larger-caliber (30mm?), with improved electronics and ammunition.

Vulcan is a battlefield air-defense cannon. It is being replaced by DIVAD.

ZSU-57 is an older battlefield air-defense cannon that has been largely replaced by the ZSU-23.

M-42 is a World War II era battlefield air-defense cannon that is still found in many armies.

Stinger is an air-defense missile system carried and fired by one man. This replaces the older Redeye missile.

SA-7 is a Russian version of the Stinger.

ZPU-4 is typical of the multiple machine guns used for air defense. Generally only used by less well-equipped armies or by reserve formations.

.50 Cal (12.7mm) is a machine gun commonly found mounted on armored vehicles and trucks for air defense.

Part 3
NAVAL OPERATIONS

9

THE NAVY: ON THE SURFACE

IF A FUTURE naval war occurs, the two divergent success formulas of the two major powers will be battling one another.

Lessons from the Past

There is nothing more instructive than defeat. The Russian Navy has taken lessons not only from its own defeats but from those of its enemies. Adopting the submarine doctrine of Germany, which failed in World Wars I and II, and the *kamikaze* doctrine of the Japanese, which failed in 1945, the Russians have developed a style of warfare widely regarded as potentially successful. This style depends on the stealth of submarines, sheer numbers of cruise missiles and vast minefields. This last element, mines, victimized the Russian Navy three times in this century (1905, 1914–1917 and 1941–1945).

Surprise is essential. Get in the first shot and make it count. The bulk of the Russian Navy is built and trained for this type of operation, and it is not organized for a long war. Anything beyond a few months will be beyond its planned capabilities.

Construct ships for maximum "one-shot" capability. The targets in order of decreasing importance are enemy missile-carrying submarines, aircraft carriers, nuclear attack submarines and enemy shipping.

Learn from the Japanese kamikaze *experience.* A multitude of aircraft that crash themselves into ships will overcome the most massive defensive systems. Thus, the large number of Russian cruise missiles.

Learn from the German experience with submarines. Send enough submarines against the West before it can mobilize its offense, and you will deny it use of the seas. The Germans almost succeeded. They began World War II with only 57 submarines to counter as many as 16,000 ships. If the one third of the ships controlled by Britain had been quickly decimated, victory would have been within sight. If the Russians start a war, they will go in with 200 submarines attacking over 20,000 ships.

The United States has also learned important lessons from World War II.

No matter what you thought before the war, new superior weapon systems will soon assert themselves. Before World War II it was still assumed that the battleship was the decisive naval weapon. The aircraft carrier was just another support system, and an untried one at that. Today the carriers are the battleships, and the nuclear attack submarines are the untried systems. The United States covers its bets by attempting to maintain strength in both areas. The submarines, however, are being given proportionately greater resources.

Oceanic powers that depend on merchant shipping can win only if they maintain control of the seas. The Western allies did this and were victorious. The Japanese were not able to withstand the onslaught of U.S. submarines and ended the war in economic collapse. Antisubmarine-warfare techniques from World War II have given way to more modern concepts.

Surprise will not guarantee Russia a victory, particularly if potential victims are wary. Pearl Harbor, and the numerous other attacks the Japanese launched simultaneously in 1941, destroyed numerous ships and aircraft, but America recovered. There is even less apprehension about surprise because everyone remembers Pearl Harbor. The Russians have had more Pearl Harbors than anyone and would like, for a change, to be the instigators, not the victims.

Superior information-gathering ability need not allow you to call the shots. Cracking the enemy's codes and keeping him under observation at all times will not allow you to get inside his head.

Too much information, incorrectly interpreted, can lead to fatally wrong conclusions.

Techniques of Modern Surface Warfare

DEPLOYMENT

About 85 percent of the Russian Navy and 60 percent of Western navies are in port during peacetime. Ships at sea are on exercises. An aircraft carrier task force consists of an aircraft carrier plus six to ten cruiser and frigate escorts, one or more submarines and a few replenishment ships. A noncarrier task force contains the same mix of ships, without the carrier. Here the key is complementary ships specialized for antisubmarine warfare, air defense or surface attack.

DETECTION

If you want to destroy the other fellow, first you must find him. The U.S. and Russian navies both use satellite reconnaissance and a worldwide network of intelligence-collection ships, ground stations and aircraft. In peacetime the Russians shadow all major Western task forces with combat or surveillance ships. If war comes suddenly, these shadows may be quickly blown away. If war comes gradually, the shadow ships will be joined by additional ships to assist in launching a surprise attack. Otherwise, the Western navies, with their extensive land- and carrier-based aircraft fleets, will have the edge in discovering the enemy first.

ATTACK

The biggest difference since World War II is *standoff* weapons, guided missiles released at distances up to 800 km. Submarines also use either missiles (up to 100 km) or long-range torpedoes (40-km range). Another effect of missiles is the reduction in explosive and kinetic energy versus that of World War II shells and bombs. The space taken up in the missile for propulsion and guidance is at the expense of explosive. For these reasons a World War II battleship is highly resistant to the much less destructive modern missile hits.

DEFENSE

Basic techniques vary, depending on the task force. An aircraft carrier task force uses a much more elaborate defense than a noncarrier task force. This defense system is described in the chapter on naval air operations. Noncarrier task forces use ship- or land-based aircraft to detect enemy forces and then launch missiles. Electronic warfare plays a large role in this battle.

DAMAGE CONTROL

When damage is inflicted on a ship, there is an equally desperate battle to repair it. Western navies have an edge in this respect. The ship that can recover from damage more quickly, can get back to the battle faster. See the section below on ship design.

Differences Between Russian and Western Techniques

Most Western nations have long and successful naval traditions. Russia has had a shorter and much less distinguished naval experience. Neither World War I nor World War II were particularly glorious chapters in Russian naval history. Since the 1950's, Russia has rewritten the book on naval practice. The Russians have been perceptive, imaginative, resourceful and desperate. Although the United States had taken the lead in nuclear attack submarine development, Russia has forged ahead in the even more critical areas of surface attack missiles and electronic warfare. Withal, Russians have clung to their traditional concept of putting all their power up front. This has meant many smaller, more heavily armed ships. By providing more targets, the attacker's job was made more difficult. The numerous ships were also able to defend each other against aircraft and missiles.

Belatedly, the West caught on to the potential of the Russian innovations. As we proceed through the 1980's, however, we find Russia imitating the West. Currently under construction in Russia are attack aircraft carriers. Previous Russian carriers were antisubmarine vessels. In addition, Russia has just put into service a 23,000-ton battle cruiser. This ship type has not been seen since World War II.

Both sides are becoming more like each other, but for the foreseeable future, fundamental differences will remain.

Ship design. Russia favors a larger number of smaller, somewhat cheaper and, ton for ton, less effective ships. These vessels appear to bristle with weapons compared with Western ships. Because of their design, work space and equipment access are less than those in Western ships. This makes it very difficult to get at anything that breaks down, but then, due to the shortage of skilled crews, Russian repairs await return to port. In the meantime the Russians depend on duplicate systems. For major missile systems, particularly suface to surface, there are often no reloads. A more serious problem is the less thorough vibration-proofing. Most electronic systems are easily put out of action by excess vibration, and combat operations produce a lot of it. The lack of vibration damping and onboard repair capability puts a lot of systems out of action which, on Western ships, would quickly be put right. Russian ships do tend to have more modern propulsion systems. They are often less compartmentalized, which leads to a higher incidence of catastrophic critical hits.

Command and control. Following long-standing Russian practice, operations are carefully planned and executed under centralized control. Little room is left for individual initiative or deviations from the plan. The command layout of Russian ships reflects this. Western ships use a *combat information center* (CIC). The CIC is a room, often in the bowels of the ship, where all sensor and fire-control information is centralized. The captain of the ship commands from the CIC while the executive officer, second in command, mans the bridge topside. Computers allow the ship commanders to grasp the entire battle. Individual ship data are sent to the task-force flagship, where the commander makes decisions affecting the entire task force.

The Russian commander issues more detailed orders before an action, thus leaving less latitude for individual initiative. The Russian ship captain normally commands from the bridge. The navigation, early warning and fire-control radar sections report to him, as do the sonar (antisubmarine warfare) and other sections. Unlike in CIC systems, all this information is combined only in the captain's head. This works fine if a previously prepared plan is carried out. Things begin to fall apart when unexpected events occur.

Concept of mission. As a continental power fighting oceanic powers, Russia's obvious mission is the denial of sea access to the oceanic powers. The U.S. mission is to maintain sea access. Merely presenting a potential danger to oceanic powers lessens the use of the oceans for moving goods. If, in addition, sufficient quantities of merchant shipping can be destroyed, then the economic and military power of the oceanic nations is diminished. This makes Russia relatively stronger. The attacker in this case is somewhat similar to a guerrilla fighter. He doesn't have to be everywhere at once. He can pick and choose his strikes.

Deployment. Only about 15 percent of Russian ships are at sea at any time versus 40 percent for Western navies. A long-standing tradition in the Russian armed forces is not to use equipment until a war breaks out. This guarantees that there will be somewhat more equipment available and that no one will know how to use it very well. There are practical reasons for this. There are not enough technicians to maintain heavily used equipment. The ships are not built for sustained comfortable cruising. They can build more ships with the money saved by keeping existing vessels in port most of the time. This Russian practice allows them to put more ships to sea when war comes, if they can make it that far and if they can do anything effective once they get there.

The Western practice, based on long experience, is to keep the crews at sea as much as possible. Practice, it has been found, makes the most effective crews. As the majority of crewmen are technicians, their skills can be maintained only through continual use. Even constant local cruises are no substitute for high seas experience.

Attack techniques. Ambush is the preferred technique of all navies. It has become more attainable with the growth of electronic warfare. Beyond this the major difference is that the Russians must stay out of the way of the more powerful Western task force while preparing their attacks. The Western forces will come looking for the Russian units.

In peacetime Russian ship locations are monitored much more accurately than Western ships. This is particularly true with submarines. United States submariners openly proclaim their ability to detect Russian subs at ten times the range of Russian sensors. The Russians attempt to balance this by maintaining small surface combat ships or submarines as "escorts" for all major Western task forces. These ships not only maintain location information for other Russian ships, but are expected to get in the first shot themselves. However, the Russian escort ships will most likely be destroyed whether they get off any missiles or not. Once the escorts and their up-to-date information are gone, the former location of the enemy task force becomes useless very rapidly. For example, in 6 hours of 30-knot steaming, a task force can travel 320 km in any direction. This search circle includes 320,000 square kilometers. Lacking surviving air-search capability, a Russian task force can search only 2000 square kilometers per hour. A U.S. aircraft carrier task force can search over 100,000 square kilometers per hour. Satellite and electronic surveillance can work if the satellites remain operational and/or the enemy task force has sloppy signal discipline. But these means are available to both sides, so it's anyone's guess who is likely to find who first.

When the Russians have located a task force, they attempt to launch a saturation attack of cruise missiles launched from aircraft, surface combat ships and submarines. First, they send the aircraft-launched cruise missiles against the carrier. Once the carrier is rendered unserviceable, any aircraft aloft will be down within a few hours. The surface combat ships and submarines will then have a better chance of closing with the remaining surface combat ships.

Thus the aircraft go in first. They do not have an easy time of it. A carrier task force has early-warning aircraft that can spot targets 700 km away. For this reason the Russians have one cruise missile (AS-6) that can be launched as far as 360 km from the target. It comes in on the deck at high speed, less than 8 minutes to the target. Guidance is a problem. An inertial-guidance system on a ship or plane can get the missile to 30 to 40 km from the target. At that point, terminal guidance homes in on radar emissions, large masses of metal, heat or images. All of these can be defeated if the target is well prepared.

To guide the cruise missiles to within 30 to 40 km of the target, the ships must be there already or the aircraft must go in behind the cruise missiles. Normally a large, slow, but electronically well-equipped, TU-95 aircraft will control air and ship operations for an attack on a task force. This aircraft is a lucrative target for carrier aircraft.

Against a noncarrier task force the Russians have a better chance. The basic attack strategy is the same except for more aggressive air observation and "over the horizon" guidance of cruise missiles from helicopters. The helicopters, flying low to avoid detection, can pinpoint the location of the enemy task force so that the guidance systems on the cruise missiles may be set. More risky is broadcasting midcourse corrections to the cruise missiles themselves, as enemy ECM could easily disrupt them.

Western navies have somewhat different combat techniques. These differences result primarily from their superior electronic and air power. Whether land- or carrier-based, Western navies will have more need for superior air power, as Russia's main naval bases are in out-of-the-way places (the Sea of Japan, east of northern Norway, the Black Sea, etc.). Western shipping routes are closer to Western naval bases.

An American aircraft carrier, with its 60 combat and 20-plus support aircraft, is capable of detecting enemy surface combat ships and submarines from hundreds of kilometers. Attack and support aircraft enable it to launch massive cruise-missile attacks without much risk. A typical attack would include 3 electronic-warfare aircraft and 12 attack aircraft carrying 24 cruise missiles plus torpedoes, guided bombs and radar homing missiles. They would encounter half a dozen Russian surface combat ships mounting about 30 surface-to-air missile launchers with 500 to 600 functioning missiles. The odds are against the defending ships. See Chapter 11 for more details.

Western submarines also have superior detection equipment that acquires targets at a greater distance and launches their weapons with greater accuracy. See Chapter 10 for further details.

The Sailor's Life

The majority of sailors are technicians who spend most of their time maintaining and operating equipment. They are constantly trained for damage control and combat operations. Because combat can only be practiced in peacetime, a sailor's life is divided between standing watch, manning equipment, maintaining the equipment and going to school to learn more about it. Combat, when it comes, falls into two forms. Passive combat consists of looking for the enemy and/or waiting to be found. Active combat is actual shooting and being shot at. In peacetime soldiers at sea get considerable practice at passive combat. In wartime they might spend a third to three quarters of their time in passive combat, particularly the submarines and antisubmarine-warfare personnel. Active combat occupies much less time, a few hours a week in a combat zone on the average, and when it comes, you wish it hadn't.

Mine Warfare

This is a subject most sailors wish would just go away. Mines are a deadly nuisance. The advantage in mine warfare is with the side that is able to go where it wants when its wants in order to place them. Western navies have a greater freedom of action to deliver mines. The Russians have a much larger stock of mines and far more mine-warfare ships. More Russian ships are equipped to lay mines. All of this evens things up a bit.

THE HISTORICAL EXPERIENCE

The first modern naval mines were used extensively during the Russo-Japanese war (1904–1905). They were contact mines, moored to the sea bottom and lurking just below the surface at the right tide. They exploded when a ship struck them. Sixteen ships were lost to about 2000 mines.

This war pointed out the importance of keeping track of your own mines. A number of ships were sunk by their own mines or while moving through previously charted minefields. During and after the war, several ships were sunk by free-floating mines as well as by moored ones that had broken free. Thereafter, more care was taken to reduce the number of loose mines floating about.

During World War I, most of the modern mine tactics were perfected. Defensive use was laying thousands of mines as barriers to enemy movement. Offensive use was attempting to place mines secretly in enemy shipping lanes. The latter was discovered only toward the end of the war, after most of the mines had been laid openly. Nevertheless, more than 1000 merchant ships and warships were sunk by 230,000 mines.

During World War II, mine warfare came of age. A total of 2665 ships were lost or damaged by 100,000 offensive mines. That's one ship for every 37 mines. Some 208,000 mines, used defensively, sank very few ships but inhibited enemy movement and tied up resources. British and German antimine forces in the North Sea totaled 99,000 men and 2400 ships and aircraft.

Using mines offensively achieved striking success. In the Pacific if the atom bomb had not ended Japan's resistance, naval mines would have. During a 10-week period, April–August, 1945, 12,000 mines were delivered by air. These accounted for 1,250,000 tons of enemy shipping (670 ships hit, 431 destroyed), or 18 mines for each ship hit. This was a situation in which the American forces had air superiority. They lost 15 aircraft, mostly due to operational failures. Fifteen hundred sorties were flown. Had these missions been flown against heavy resistance, the usual 2 percent to 5 percent loss per sortie rate would have been taken. As many as 80 aircraft could have been lost, plus escorting fighters. It still would have been a small price to pay considering the alternative: an amphibious attack with hundreds of thousands of losses.

By comparison, a submarine campaign was waged against Japanese shipping. This operation involved 100 subs and lasted throughout the war, 45 months. Sunk were 4,780,000 tons of enemy shipping. Enemy tonnage sunk per U.S. fatality was 3500 tons for mines and 560 tons for submarines. Just including the cost of lost enemy ships and subs, each ton lost to mines cost $6 ($500,000 per B-29), while each ton cost $55 for subs (at $5 million per boat). These data, classified as secret until the 1970's, indicated that mines might have been much more effective than torpedoes, even if many of them had been delivered by submarine.

The Germans waged a submarine minelaying campaign off the east coast of the United States between 1942 and 1944. Only 317 mines were used, which sank or damaged 11 ships, with a ratio of 29 mines per ship hit. In addition, 8 ports were closed for a total of 40 days. One port, Charleston, South Carolina, was closed for 16 days, tying up not only shipping but the thousands of men, ships and aircraft dealing with the situation. United States subs waged an even more successful mine campaign in the Pacific. For 658 mines used, 54 ships were sunk or damaged (1 ship per 12 mines). No subs were lost. Again, considerable Japanese resources were tied up in dealing with the mines. On the Palau atoll, the port was closed and was never reopened. Even surface ships got into the act. Three thousand mines were laid by destroyer. Only 12 ships were hit, but these mines also were a barrier, even though they were intended to be offensive.

In Korea the Russians provided 3000 mines, many of 1904 vintage, which were used at Wonsan Harbor. It took several weeks to clear the mines at a loss of about a dozen ships hit, half of which were destroyed.

During the Vietnam War, over 300,000 naval mines were used, primarily in rivers to interdict enemy shipping. The vast majority were not originally designed as mines, but as low drag bombs, with a parachute to soften their fall to earth and equipped with magnetic detectors in place of their normal fuses. In the water these devices functioned quite well as mines. Haiphong Harbor was actually mined with 11,000 of these "destructors," as the U.S. Air Force called them, and less than 100 conventional naval mines.

NAVAL MINE DESIGN

Today, a naval mine usually weighs from half a ton to a ton. As a rule of thumb, a submarine can carry two mines for every torpedo.

Naval mines come in several varieties. The first distinguishing characteristic is where they wait for a victim. *Free-floating* mines are dangerous to friend and enemy alike. They are also inefficient, as the flow of the water moves them away from wherever they are supposed to be. No longer used intentionally, this may happen also when moored mines break loose. Terrorists may also use them. *Moored* mines drop anchor and float either near the surface or several hundred feet down. The original moored mines were detonated by contact. Current mines are more sensitive and often don't need a mooring line for detonation, depending on water depth. *Bottom mines* just lie on the bottom. Two obvious advantages are the elimination of the mooring mechanism and the greater difficulty mine-clearing forces will have detecting an object just sitting there. *Mobile mines* are initially either moored or bottom mines. Once they detect a target, they move under their own power toward it. The most striking example is the U.S. CAPTOR mine. This is a moored mine equipped with a powerful acoustic sensor. Armed with an Mk 46 torpedo, it is intended for use against Russian nuclear submarines. It apparently can detect with a range of several kilometers.

The other distinguishing characteristic of mines is their target-detection systems. *Remote control:* Another sensor, or human observer, detects the target and detonates the mine. This is still used to arm or disarm minefields to allow friendly ships to pass. Some mines are radio controlled, others are wired. *Contact:* This is the most primitive form of detonation. A ship hits the mine, and it explodes. Some mines have used their mooring cable as a sensor, but this depended on the length of the cable and the power of the mine. Contact mines are generally obsolete because of the ease of clearing them. All you have to do is cut the cable and detonate them with a rifle or machine gun.

Influence: First came the magnetic-influence mine. A ship is a large hunk of metal that will "influence" a mechanism containing small magnets. This was the first bottom mine. Then came the acoustic mine. Ships make noise. A noise detector in the mine will sense this and detonate. Finally, there is the pressure mine, which senses the change in water pressure as a ship passes overhead. The influence sensors are often combined in the same mine to make sweeping more difficult. If the mine doesn't detect all three, or two out of three, it will not explode.

Microcomputers can now be put into mines to detect ships by type. If you only want to destroy submarines, the mine will be programmed accordingly. The mine can also be programmed to let the first ship or more pass by. The mine can also be programmed to activate, then deactivate and then activate again after a certain time. This will make clearing the mines even more nerve-racking than it already is. It will also allow the use of dummy mines—light, empty shells that just lie there. An additional feature of the programmable mine is its ability to self-destruct after a set period of time. This saves the hassle of clearing mines, particularly if they are difficult to clear.

MINE CLEARING

Mines, once laid, are indeed a deadly nuisance. Western navies have a greater capability to clear them, particularly the deadly pressure mines, but neither the West nor the Russians can clear the number of mines that are available.

Currently pressure mines are hunted with specialized sonar. Remote-controlled miniature submarines plant explosives to destroy them. It's a very slow and tedious procedure, but at least it's a solution. Under development are faster techniques.

Minesweepers obtained their name from the cable that swept along between two ships to cut the mooring cables. The sweepers had to be careful not to hit any mines themselves. This was possible, although casualties were not uncommon. With the appearance of magnetic mines, an electrified cable simulated a ship's magnetic field. The sweepers avoided detection by being made of wood or by electrically decreasing any magnetic field. To set off acoustic mines, a noisemaker was towed behind a very quiet sweeper. The United States has pioneered helicopter-towed minesweeping "sleds." They deploy much faster and are also a lot safer. The sleds can also be more rapidly modified and upgraded.

The best way to fight mines is to prevent the enemy from using them. Fortunately, most mines can be used only in more shallow coastal waters, which can be more easily patrolled. Unless you want to drop 10-ton mines, you must use bottom mines no deeper than 30 to 40 meters, moored mines or mobile mines that will rise to the surface before exploding. Mines are best used on the continental shelf. They are becoming more flexible as more accurate sensors are developed. The Western nations have the edge because of their lead in electronics. Unfortunately, the West is winning in the laboratory and losing on the battlefield. The Russians and their allies are alleged to have over 50,000 mines. Western navies apparently have fewer than a third that number. And U.S. techniques can work both ways. For example, the Russians obtained a number of bombs fused with sensors during the Vietnam War.

WHY AREN'T MINES USED MORE?

A reasonable question. The answer usually given is that mines don't belong to a union. In the navy neither aviators, submariners nor surface sailors will embrace this weapon. The air force doesn't mind using it, but the navy gets feisty when its territory is encroached upon. There is also the cowardly way in which the mine operates. After all, a mine has no pilot to fly or sail it. No gunner arms and fires a mine. It just sits there and waits. It's not the sort of weapon a warrior type would cleave to. Whoever bestirs himself to fall in love with this ugly duckling may indeed inherit the kingdom during a future war.

MINES IN THE NEXT WAR

Vietnam suggested how effective mines can be. Although some casualties were inflicted on American forces, it was obvious that the side with air and naval superiority was nearly immune to mines.

In a general, nonnuclear war, sea and air control will not be so one-sided. The number of areas critical to the West is small: the Persian Gulf, Djakarta and Singapore Strait, the east coast Japanese ports, the English Channel ports and the three largest ports on the east and west coasts of North America. Mine the majority of these areas and keep them supplied with fresh mines, and the Russians could paralyze the Western war and industrial effort. The Russian situation is more grim. Mine four areas—Murmansk, Vladivostok, the eastern Baltic, the Turkish straits—and the Russian fleet goes nowhere. Just mining the Murmansk and Vladivostok areas bottles up over two thirds of the Russian naval forces.

If the navies are blocked, there are potential unconventional ways to deliver the mines. Send the aircraft on one-way missions—not likely except in isolated cases. Equip cruise missiles with mines and launch them from aircraft—very expensive and limitedly effective due to the small number of missiles available. Secretly equip merchant ships with mines to be released at the outbreak of hostilities—rather outlandish but not impossible. Plant mines in peacetime that can be activated via an underwater signal—reliable underwater signals are still not perfected, although there may be ways around this.

Even with conventional methods, mainly submarines have drawbacks. Russian subs are likely to be cut off from their bases after a war begins. Most will likely be hunted down and destroyed before they can even attempt to return for reloads. If they avoid the NATO dragnet, they will have to negotiate the enemy minefields surrounding their bases.

The best chances for mining will present themselves during the early stages of the war. The greatest success will go to the side with the most effective mines and tactics. Thereafter, the mine war will likely be a much more closely fought affair. Should one side gain an advantage, it will be a crucial one.

Naval Forces

United States and Russian warships comprise nearly two thirds of all the world's combat shipping. The ships and their designs are exported to most of other navies worldwide.

CLASS TYPE is the code indicating the size and function of the ship. "N" means a nuclear-powered ship; "SS," submarine; "G," a ship using missiles as its primary armament; "B," a submarine whose main armament is ballistic missiles; "CV," an aircraft carrier; "H," a smaller carrier using primarily helicopters or other vertical takeoff aircraft; "C" (and "CA"), a cruiser-class ship (over 5000 tons displacement); "DD," a destroyer-class ship (3000 to 5000 tons displacement); "FF," a frigate (1000 to 3000 tons displacement); and "PC," seagoing boats of under 1000 tons displacement armed with missiles or torpedoes.

Ballistic missile boats, while possessing torpedo tubes and detection equipment, are not designed for ship-to-ship combat. Their primary functions are to avoid detection and to launch, when ordered, their ballistic missiles. Further naval combat would be redundant. These boats are generally a few thousand tons heavier than SSN's, although the most recent Trident (U.S.) and Typhoon (Russia) class SSBN's weigh over 12,000 tons.

Minesweepers are not heavily armed. They sweep, lay or hunt mines. Not included are small ships (under 1000 tons) armed only with light guns or depth

NAVAL FORCES

The traditional surface ship has become an auxiliary to the more powerful carriers and nuclear submarines.

United States

Class Type	Class Name	# in Class	Surf	Sub	Air	Prot.	Long	Weight	Power Ratio	Speed	Range	Men per Kton	EW Value	Guns	Air-craft	SAM	SSM	TT DCT	Class End Build
CVN	Nimitz	3	100	12	100	10	332	92	3.04	56	200	68	9	3	95	3	0	0	1988
CVN	Enterprise	1	85	12	90	10	336	89	3.15	61	200	62	8	3	84	2	0	0	1961
CV	Kitty Hawk	4	85	12	85	8	319	81	3.46	54	8	60	8	3	85	3	0	0	1968
CV	Forrestal	4	70	6	75	8	324	79	3.54	59	8	62	7	3	70	3	0	0	1959
CV	Midway	2	70	6	60	7	294	62	3.42	56	8	69	7	3	75	2	0	0	1947
SSN	Lipscomb	1	12	25	0	9	111	5.8	2.59	45	200	21	6	0	0			4	1974
SSN	Narwhal	1	10	20	0	8	96	4.5	3.78	54	200	24	6	0	0			4	1969
SSN	Los Angeles	27	20	40	0	10	110	6.1	5.74	72	200	21	9	0	0			4	1985
SSN	Sturgeon	37	16	25	0	9	89	3.6	4.17	63	200	30	7	0	0			4	1975
SSN	Thresher	13	12	20	0	7	90	3.7	4.05	57	200	28	6	0	0			4	1967
SSN	Tullibee	1	12	18	0	6	83	2.3	1.09	40	200	24	6	0	0			4	1960
SSN	Skipjack	5	10	15	0	5	77	3.1	4.84	59	200	30	5	0	0			6	1961
SS	Barbel	3	6	12	0	6	67	2.9	1.07	36	20	27	3	0	0			4	1959
SSN	SSBN Conversion	6	10	12	0	5	120	5.5	2.70	56	200	19	5	0	0			6	1962
SSN	Skate	4	9	12	0	5	82	2.3	2.87	47	200	38	4	0	0			4	1959
CG	Ticonderoga	2	30	25	20	3	172	9.1	9.89	58	14	35	8	4	4	4	16	4	1990
CGN	Virginia	4	12	18	8	2	178	10	6.00	58	200	47	7	2	2	4	4	4	1980
CGN	California	2	14	9	8	2	182	11.1	5.41	54	200	49	6	4	0	2	8	12	1975
CGN	Truxton	1	4	6	5	2	172	9.1	6.59	52	200	58	5	5	1	2	2	2	1967
CG	Belknap	9	4	12	5	2	167	7.9	10.76	57	14	53	5	5	1	2	2	8	1967
CG	Leahy	9	3	6	5	2	162	7.8	10.90	57	14	51	4	6	0	4	4	14	1964
CGN	Bainbridge	1	3	6	10	2	172	8.6	6.98	54	200	55	4	2	0	1	4	14	1962
CG	Spruance	30	6	30	2	3	172	7.8	10.26	59	11	38	7	1	2	1	4	14	1983
CG	Coontz	6	2	12	5	2	156	5.9	14.41	59	9	64	4	1	0	2	2	14	1960
CGN	Long Beach	1	3	12	4	4	220	17.1	4.68	54	200	68	4	4	1	4	4	14	1961

Type	Class	Ships	Surf	Sub	Air	Prot	Weight Long	Power	Ratio	Speed	Range	Men per Kton	EW Value	Guns	Air-craft	SAM	SSM	TT DCT	Class End Build
CG	Albany	1	1	12	5	4	205	18.2	6.59	58	9	67	3	2	1	4	0	14	1946
CG	Kidd	4	4	35	4	2	161	9.1	8.79	59	12	40	6	4	0	4	4	10	1982
DDG	Adams	23	2	12	4	1	133	4.5	15.56	54	12	79	4	2	0	2	2	14	1964
DDG	Hull	4	1	12	4	1	128	4.1	17.07	56	8	89	4	1	0	1	0	14	1959
DD	Sherman	6	1	12	0	1	128	3.9	17.95	59	8	75	2	3	0	0	0	6	1959
DD	Sherman ASW	8	1	20	0	1	128	4.2	16.67	59	8	72	2	2	0	0	0	14	1959
DD	Gearing	17	1	12	0	1	119	3.5	17.14	58	10	88	2	4	0	0	0	14	1946
DD	Carpenter	2	1	12	3	1	119	3.6	11.39	59	8	81	6	2	0	1	0	6	1949
DDG	Perry	24	2	24	5	1	136	3.4	10.29	52	8	46	4	1	2	1	1	14	1984
DDG	Brooke	6	1	18	0	1	126	3.9	8.97	48	8	73	5	1	1	1	0	12	1967
DDG	Knox	50	2	20	0	1	134	3.4	10.29	48	8	72	3	1	1	1	1	14	1974
DDG	Garcia	10	1	18	0	1	126	3.4	10.29	48	8	73	5	2	1	0	0	14	1965
DDG	Glover	1	2	18	0	1	126					73		1	1	0	0		1965
FFG	Bronstien	2	1	22	0	1	113	2.6	7.69	47	8	75	2	2	1	0	0	14	1963
PG	Missile Boats	6	2	0	0	1	44	.2	90.00	86	3	105	1	1	0	0	8	0	1982
	Minesweepers	25	0	0	0	1	52	.7	3.26	25	4	109	0	0	0	0	0	0	1956

Class Totals

	Ships	1000 Tons	% of Total	Avg Class	Surf	Sub	Air	Prot	Weight Long	Power	Ratio	Speed	Range	Men per Kton	EW Value	Guns	Air-craft	SAM	SSM	TT DCT	Class End Build
SSN	98	392	13	10	12	20	0	7	93	4	3	53	182	26	6	0	0	0	0	4	1967
SSBN	33	239	8	16	6	12	0	6	132	7	2	54	200	19	5	0	0	0	16	4	1971
CV	14	1129	38	3	82	10	82	9	281	81	3	57	85	64	8	3	82	3	3	0	1965
C	70	581	20	8	5	14	5	2	173	10	9	56	90	56	5	3	1	3	1	11	1968
D	153	588	20	13	1	17	2	1	126	4	13	53	47	75	3	2	1	1	1	13	1963
FFG	2	5	0	2	1	22	0	1	113	3	8	47	8	75	2	2	1	1	0	14	1963
PC	6	1	0	6	2	0	0	1	44	.2	90	86	3	105	1	1	0	0	0	0	1982
Total	376	2935	100																		

NAVAL FORCES

The traditional surface ship has become an auxiliary to the more powerful carriers and nuclear submarines.

Russian Navy Combat Ships

Class Type	Class Name	# in Class	Surf	Sub	Air	Prot.	Long	Weight	Power Ratio	Speed	Range	Men Per KTon	EW Value	Guns	Air-craft	SAM	SSM	TT DCT	Class End Build
CVH	Kiev	3	40	22	6	6	275	37	4.86	56	23	68	7	12	35	4	8	36	1983
SSGN	Oscar	2	20	8	0	4	145	12	5.00	54	200	12	6	0	0	0	24	6	1988
SSGN	Charlie II	7	12	7	0	4	103	4.3	5.12	54	200	21	5	0	0	0	8	6	1985
SSGN	Charlie	11	10	6	0	4	94	3.9	5.64	45	200	23	4	0	0	0	8	6	1974
SSGN	Papa	1	10	6	0	4	106	6	4.83	54	200	15	4	0	0	0	10	6	1972
SSGN	Echo II	29	6	4	0	2	117	4.8	4.58	40	200	21	3	0	0	0	8	6	1968
SSG	Juliet	16	4	4	0	2	87	2.8	2.50	31	15	29	3	0	0	0	4	10	1967
SSN	Alpha	8	10	16	0	8	79	2.8	8.57	72	200	21	2	0	0	0	0	8	1988
SSN	Victor II/III	10	10	18	0	6	102	5.5	5.45	54	200	16	3	0	0	0	0	8	1986
SSN	Echo I	3	5	10	0	5	110	4.6	6.52	54	200	20	2	0	0	0	0	10	1980
SSN	Victor I	16	4	10	0	4	94	4.3	6.98	58	200	21	2	0	0	0	0	8	1975
SSN	November	16	4	8	0	2	110	4.2	7.14	54	200	20	1	0	0	0	0	12	1962
SS	Tango	18	3	12	0	3	92	2.1	2.86	29	36	29	1	0	0	0	0	6	1985
SS	Foxtrot	45	2	10	0	2	92	2	3.00	32	36	38	1	0	0	0	0	10	1973
SS	Romeo	11	2	8	0	2	77	1.4	2.86	27	20	39	1	0	0	0	0	8	1961
SS	Whiskey	20	1	6	0	1	75	1.1	2.45	27	24	50	1	0	0	0	0	6	1958
CGN	Kirov	2	30	12	10	5	247	23	5.22	58	200	48	8	4	2	13	20	14	1988
CH	Moskva	2	3	20	3	4	190	18	5.56	54	22	44	6	6	18	4	0	26	1968
CG	Udaloy	3	3	12	6	2	150	5.7	21.05	59	12	53	6	5	1	4	2	32	1990
CG	Sovremenny	2	6	0	5	2	180	7.5	14.67	54	8	60	5	6	2	4	6	12	1990
CG	Kresta III	1	6	10	6	3	185	12	11.67	54	8	50	7	6	0	6	0	32	1990
CA	Sverdlov	12	2	0	1	4	210	17.5	6.29	54	15	57	2	40	0	0	0	0	1956
CG	Kara	7	3	6	4	2	174	9.8	12.24	56	14	56	4	8	1	8	0	54	1981
CG	Kresta II	10	3	5	3	2	159	7.5	13.33	59	10	53	3	8	1	4	4	54	1978
CG	Kresta I	4	4	3	2	2	156	7.5	13.33	59	10	53	2	8	1	4	4	46	1968
CG	Kynda	4	5	2	1	1	143	5.6	17.86	63	12	71	2	4	0	2	8	30	1965

Type	Class	N	Surf	Sub	Air	Prot	Long	Weight Ratio	Power	Speed	Range	Men Per MTon	EW Value	Guns	Air-craft	SAM	SSM	TT DCT	Class End Build
DDG	Kashin II	6	3	3	2	1	146	4.9	19.59	63	8	65	1	8	0	4	4	29	1978
DDG	Kashin I	13	1	2	1	1	144	4.7	20.43	63	8	60	1	4	0	4	0	41	1975
DDG	Kanin	8	0	0	2	1	139	4.7	17.87	61	8	74	1	16	0	2	0	46	1963
DD	Kotlin II	18	0	2	1	1	128	3.8	18.95	65	7	79	1	20	0	0	0	55	1960
DDG	Kotlin SAM	8	0	1	2	1	128	3.6	20.00	65	7	97	1	14	0	2	4	29	1960
DDG	Kildin II	4	0	1	3	1	127	3.8	18.95	63	7	79	1	20	0	4	0	36	1976
DD	Krivak	34	0	1	1	1	125	3.7	19.46	58	7	68	1	6	0	0	0	36	1986
DD	Skory	8	0	1	2	1	120	3.1	19.35	59	7	87	1	8	0	2	0	37	1954
FFG	Koni	2	0	1	1	1	96	2.3	18.26	58	2	48	1	8	0	1	0	24	1982
FFG	Tarantul	4	3	0	1	1	56	.8	14.00	63	5	63	1	3	0	1	4	0	1982
FFG	Pauk	4	0	2	1	1	56	.8	13.00	58	5	69	1	2	0	1	0	6	1986
FFG	Nanuchka	26	5	0	1	1	60	.9	22.22	58	6	67	1	2	0	2	6	0	1985
FFG	Grisha II	38	0	1	1	1	73	1.1	20.00	61	5	55	1	5	0	2	0	10	1986
FF	Mirka	18	0	2	0	1	82	1.1	38.18	61	9	91	1	4	0	0	0	12	1967
FF	Petya	46	0	1	0	1	82	1.2	21.67	54	9	83	1	11	0	0	0	14	1972
FF	Riga	30	0	2	0	1	92	1.3	10.77	50	4	135	1	4	0	0	0	34	1960
FF	Missile Boats	130	2	0	0	1	40	.2	50.00	63	2	175	1	4	0	0	4	0	1985
PC	Minesweepers	170	0	0	0	1	60	.5	12.00	36	3	110	0	4	0	0	0	0	1990

Class Totals

	Ships	1000 Tons	% of Total	Avg Class	Averages Surf	Sub	Air	Prot	Weight Long	Power Ratio		Speed	Range	Men Per MTon	EW Value	Guns	Air-craft	SAM	SSM	TT DCT	Class End Build
SSN	119	514	21	11	9	9	0	0	104	5	6	52	183	20	3	0	0	0	6	8	1977
SS	94	165	7	24	2	9	0	0	84	2	3	29	29	39	1	0	0	0	0	8	1969
SSBN	85	535	22	12	2	2	0	0	131	6	6	97	200	15	2	0	0	0	12	6	1976
CV	3	111	5	3	40	22	6	6	275	37	5	56	23	68	7	12	35	4	8	36	1983
C	47	532	22	5	7	8	4	3	179	11	12	57	31	55	5	10	3	5	4	30	1977
D	99	391	16	12	1	2	2	1	132	4	19	62	7	76	1	12	0	2	1	39	1969
F	168	190	8	21	1	1	1	1	75	1	20	58	6	76	1	5	0	1	1	13	1978
PC	130	26	1	130	2	0	0	0	40	.2	50	63	2	175	1	4	0	0	4	0	1985
Total	745	2465	100																		

charges. These are generally fit only for patrol work, as their antisubmarine equipment is largely ineffective against modern submarines. Russia still retains several hundred, although some are being disarmed.

These ship classifications are those used by most Western nations.

Navies have a tendency to play fast and loose with these categories. For one reason or another, an 8000-ton ship in one navy will be called a destroyer, while a smaller ship in another navy will be called a cruiser. The Russians tend to call most of theirs "antisubmarine ships," which is generally accurate.

CLASS NAME is the name of the lead ship of a class of generally identical ships. The ships of that class are referred to by the name of the lead ship; for example, a Spruance class ship.

IN CLASS is the number in that class in 1982.

COMBAT VALUES are numerical evaluations of the ship's combat capabilities against surface (SURF), submarine (SUB) and air (AIR) targets. These values take into account the quality and quantity of onboard weapons, equipment and crew as well as past performance.

PROT. is the protection value of that ship against attacks from enemy weapons. For submarines this includes the quietness of the boat and thus the difficulty of detection. The protection value suggests the number of large-caliber hits a surface ship would have to take to be destroyed.

LONG is the ship's length in meters.

WEIGHT is the ship's full-load displacement in thousands of tons. For submarines this is surface displacement.

POWER RATIO is horsepower per ton, an indicator of agility.

SPEED is the top speed in kilometers per hour.

RANGE is the unrefueled range in thousands of kilometers, at cruise speed of under 30 km per hour. For nuclear ships this is a minimum of 200,000 km, although some can go five times as far between refuelings.

MEN PER KTON is the number of crew per thousand tons of ship's weight. This indicates how labor-intensive the ship is.

EW VALUE is the effectiveness of the ship's electronics in general and its electronic-warfare capabilities in particular. The higher the number, the better.

GUNS is the number of gun systems the ship mounts. Multibarrel Gatling types count as one gun. Almost all guns are under 128mm and are used primarily for air defense.

AIRCRAFT is the number carried on board.

SAM is surface-to-air missile launchers carried; and SSM is surface-to-surface missile launchers carried. Many ships use launchers capable of firing SAM or SSM. Some SSM have torpedo warheads which, upon landing, release a homing torpedo.

TT DCT is the number of tubes for launching torpedoes (TT) or depth charges (DCT) against submarines. Torpedo tubes rarely exceed ten. The Russians still mount a large number of depth-charge throwers, a generally ineffective weapon against most submarines.

CLASS END BUILD is the year in which the last ship of that class was built. This indicates how up-to-date the class is, although ships over ten years old are often refitted with more modern weapons and equipment. Western navies are more apt to do this than the Russian Navy. Dates later than 1982 indicate that the class is still building.

CLASS TOTALS gives the average for each class, making it easier to compare one nation, and/or class, to another. For combat purposes one should not include SSBN's, which exist primarily to carry and launch strategic missiles, not fight other ships.

CLASS DIFFERENCES

While ship classes are a rough guide to different ships' capabilities, national policies give a more accurate appraisal. National policies come in three basic flavors: high seas Western, Russian and Third World coast protection.

The Western naval tradition stresses control of the high seas throughout the world. The U.S. Navy is currently the best example. The majority of U.S. ships are designed as multipurpose, high-seas, long-range vessels. The principal U.S. ship remains the large aircraft carrier. All other ships support the ability of these carriers to carry the war to the enemy homeland. A parallel mission is maintaining the security of the high seas for merchant vessels by sinking enemy subs or confining them in port. Those Western navies that no longer maintain carriers or major amphibious forces now concentrate on antisubmarine warfare.

The current Russian Navy developed out of a coastal-defense force. Although the Russians have some high-seas capability, this is secondary to their deliberate building of ships capable of destroying enemy carriers and submarines. The majority of Russian ships are primarily antisubmarine vessels.

Most small and poor nations have a navy to patrol and protect their coasts.

Over the past 300 years, ship classifications have constantly changed. Many terms are used interchangeably with different meanings. Until recently "capital ships" (originally battleships, then carriers and now SSN's) were for destroying enemy warships. "Cruisers" were for long-range scouting and maintaining control of areas free of enemy capital ships. Destroyers were "torpedo boat destroyers" at the turn of the century. The torpedo boats turned into submarines, and destroyers went after subs.

During the last few decades, another change in these classifications has taken place. Aircraft, including ship-to-ship missiles, have become a potent enemy. Ships now specialize in antiaircraft, antiship or antisubmarine activity. This is particularly true of Russian ships.

The oldest Russian ships still in service reflect this specialization. The Skory class destroyers and Whiskey class subs were antisubmarine and antishipping classes, respectively. The first big building program of the 1960's and 1970's produced a large number of ships with increasingly potent antisubmarine and antiair capability. Western navies were most impressed by the smaller increase in antiship firepower. The Russians also joined in the trend of blurring the difference between cruisers and destroyers. World War II cruisers were not antisubmarine ships. Russian cruisers and small carriers were their most capable antisub vessels.

Recognizing Western superiority in surface-ship strength, Russia developed classes of cruise-missile subs. These boats were, in effect, submerged aircraft carriers. Western superiority in detecting subs, and the poor design of the Russian boats, made the SSGN's a less than perfect solution. Russia is currently building its own large aircraft carrier. After a hiatus in the late 1970's, they are also building improved classes of surface ships to protect these carriers. It is unlikely that Russian surface warships would last long outside the cover of their own aircraft.

Western designs have kept pace with the Russian in air defense and antisubmarine warfare. Western surface-to-surface missiles, introduced a decade after the Russian models, appear to be superior. The ability to fire SSM's from a wide variety of launchers also gives the West a superiority of numbers. Russian designs appear to be having problems progressing beyond their first primitive, but effective, efforts. No nation is immune from the problems of developing complex weapons.

Western navies are also blurring the difference between destroyers and cruisers. Onboard ASW helicopters give Western cruisers a weapon against subs. More and more, cruisers are becoming little more than larger and more powerful destroyers.

The class of ships now called frigates are, in most cases, comparable to earlier destroyers. Their primary function is antisubmarine activity. The Russians prefer a large number of frigates, another manifestation of their belief in quantity.

WARSHIP DESIGN

Designing a warship means making compromises. What do you want to do and how do you want to do it, and how much do you have to spend?

There are three design classes of warships: aircraft carriers, submarines and surface combat ships. This last class includes supply and amphibious shipping. The surface combat ships, like any other surface vessel, carry a cargo of weapons. The aircraft carrier is distinctive because it carries an airfield plus storage and maintenance spaces for 50 to 100 aircraft. Submarines devote much of their weight, and expense, to hull and machinery modifications necessary for underwater operations.

As a model of warship design compromises, we can take the typical surface combat ships. The first consideration is weight. Metal costs money, as does the assembly labor. Normally, 40 percent to 45 percent of the ship's weight goes to the structure, hull and superstructure. Another 20 percent to 25 percent goes to the main power plant. About 15 percent to 30 percent goes to auxiliary machinery and equipment. Finally, about 6 percent to 14 percent goes to payload (weapons).

The structure normally amounts to 15 percent to 20 percent of the ship's cost. The main power plant adds another 10 percent to 15 percent and the auxiliary equipment 10 percent to 15 percent. Weapons and sensors amount to 50 percent to

65 percent. A merchant ship is much cheaper because it has no weapons, less auxiliary equipment and a less costly engine.

One way to save money is to build a tight ship, one with less internal space. But once you have saved this cost, something has to get squeezed out or squeezed together. While a typical Western ship has 20 percent more internal volume than a comparable Russian vessel, the Russians put more weapons on deck and allot less space to access, passageways and work areas, and internal bulkheads. This has serious repercussions in combat.

History is a harsh, no-nonsense teacher. Lessons may seem illogical to the uninitiated. Yet experience shows what works and what doesn't. Mere human logic has nothing to do with it. To some extent, experience is perishable. New weapons and technology subtly or drastically change conditions. Past wartime experience must be modified even though there is some risk. Not to modify past experiences is certain to lead to disaster. The Western naval warship design tradition is largely based on the extensive World War II experiences of the navies of those countries, particularly the American and British.

Many things have not changed since World War II. The crew's primary function is still to keep everything operational. This requires considerable storage for repair equipment as well as easy access to the ship's systems. For damage control, additional internal space is devoted to watertight compartments, additional pumps, redundant plumbing and power-control systems.

Unlike World War II ships, little attempt is made to armor modern ships for protection. Rather, more space, weight and expense are devoted to fire fighting and repair capability. The Russians slight all of this. This is how they obtain smaller, seemingly more heavily armed ships. Less internal space is allocated to access, bulkheads and repair facilities. Their practical reason is lack of skilled crews to operate on the same standard as Western navies. But the result is the ships are less capable of either maintaining themselves at sea or recovering from battle damage.

Therefore, a U.S. vessel allocates 12 percent of its space to access versus 8 percent for a Russian ship. Stores occupy up to 12 percent of space versus 2 percent to 4 percent on a Russian ship. Nearly twice as much space (2 percent to 3 percent) is devoted to ship control. Ironically, Western ships devote somewhat less space to personnel, because they tend to have smaller crews as a result of greater automation and more skilled manpower. Even so, less efficient layout gives Russian crews less useful space. All these elements are not readily apparent, yet in combat they become critical. Russian experience, largely land warfare experience transferred to the sea, maintains that all-out initial attacks are more important than ability to carry on a protracted conflict. It is a risky gamble, with the alternative certain defeat at sea.

Naval Weapons

This chart shows the principal weapons used by naval vessels. Except for missiles, no attempt is made to show every example of each weapon type. This is adequate because most other naval weapons have very similar characteristics and effects.

NAVAL WEAPONS

Missiles have given all ships the equivalent of a squadron of kamikaze aircraft. Every ship can now be an aircraft carrier.

Weapon	Made by	Range (km)	Weight (lbs)	Speed (mps)	Guid- ance	Impact Power	Launch from	Trpdo Tube?
SSN-2	Russia	45	5000	250	2	12	S	Yes
SSN-2C	Russia	50	3500	300	4	10	S	No
SSN-3	Russia	450	10400	450	3	45	S,U	No
SSN-7	Russia	60	7700	500	4	38	U	No
SSN-9	Russia	120	6600	250	5	16	S	No
SSN-10	Russia	50	4000	600	5	23	S	No
SSN-12	Russia	500	5000	800	4	36	U	No
SSN-14	Russia	15	3500	300	3	10	S	No
SSN-15	Russia	40	4000	400	3	23	U	Yes
SSN-19	Russia	500	4000	800	6	23	S	No
Harpoon	US	110	1100	280	8	14	A,S,U	Yes
Exocet	France	70	1870	300	7	27	A,S	No
Gabriel	Israel	35	1230	200	8	12	S	No
Penguin	Norway	25	1400	300	6	21	A,S	No
Otomat	Italy	180	1653	270	7	20	S	No
SUBROC	US	55	4000	400	4	15	U	Yes
ASROC	US	10	959	400	4	15	S	No
Torpedo Mk48	US	46	3500	25	8	20	U	Yes
Torpedo Mk46	US	8	565	25	6	8	S,A	Yes
Torpedo Mk37	US	12	1700	12	3	15	S,U	Yes
DTL Mortier	France	2.7	500	200	2	5	S,A	No
DTL Nelli	Sweden	3.6	500	200	2	5	S,A	No
Gun 76mm	Average	15	14	900	2	1	S	No
Gun 127mm	Average	23	70	800	2	5	S	No

WEAPON shows the name of the weapon. The weapons are divided into the following categories: missiles, torpedoes, depth-charge launchers and guns. Each category is discussed in greater detail below.

MADE BY gives the country of origin. Most of these weapons are used by many other nations.

RANGE is the effective range of the weapon in kilometers. Anything over 40 to 50 km is "over the (radar) horizon" and must be located by accurate sonar or aircraft.

WEIGHT is the launch weight of the weapon in pounds. The lighter it is, the more you can carry, or the smaller the ship that can carry it. With the exception of guns, these weapons are launched from a rather light apparatus, either a container or a rail.

SPEED is the average speed of the projectile in meters per second. The faster the weapon, the more difficult it is to evade or destroy.

GUIDANCE is an evaluation of the weapon's guidance system on a 1-to-9 scale. The higher the better. These evaluations also take into account the sensors of the launching ship. Without these sensors initially to locate the target, the missile's guidance system will be quite blind. Weapons that travel beyond the launching ship's sensor range suffer a penalty.

IMPACT POWER is the destructive power of the weapon, taking into account the quality of the guidance system and the power of a conventional warhead. Chemical or nuclear warheads will increase its destructive power considerably. Nuclear warheads will almost always destroy their targets, even with a near miss. Only when exploded underwater will they be less likely to assure destruction, as the destructive effect only extends a kilometer or so. Chemical weapons will not destroy a ship, but will make the crew uncomfortable, or dead, if they are not prepared.

Most weapons have an explosive warhead weighing between 100 and 500 pounds. Missile impact itself causes additional destruction. A 2-ton missile causes considerable damage even if its warhead does not explode.

LAUNCHED FROM indicates the type of vehicles the weapon is normally launched from. S = surface ship; U = submarine; A = aircraft. An aircraft-launched missile generally has greater range because it starts at a higher altitude. These weapons are covered in the chart on aircraft weapons.

TORPEDO TUBE indicates whether or not the weapon can be launched from a submarine torpedo tube. This is primarily a Western concept, which allows one weapon to be used easily on many different types of ships without much modification. The larger Russian missiles, when fired from submarines, generally require specially designed boats.

MISSILES

The Russians pioneered in naval missiles. Once the Western nations woke up to their potential, they quickly overwhelmed the Russians with superior technology. Despite this, the Russians have an impressive collection of naval missiles.

Their earliest missiles are still in use: the SSN-2 and SSN-3. The SSN-2 is still mounted on a number of small coastal missile boats. The SSN-3 is mounted on a number of elderly submarines. The SSN-2 was replaced in some boats by the SSN-2C. This class of missile has been completely replaced by the more modern SSN-9 design. The SSN-3 was replaced by the SSN-12, while subsequent missile subs carried the SSN-7. The Echo and Juliett class subs, which carried the SSN-3, were themselves superseded by the Charlie, Papa and Oscar class subs. Keep in mind the Russians tend not to retire weapons or ships until they literally fall to pieces. Many of the SSN-2 and SSN-3 missiles and the boats that carry them are at a grave disadvantage against more modern Western weapons.

The latest classes of Russian missiles are just barely keeping pace with Western developments. The SSN-14 is similar to the twenty-year-old U.S. ASROC. The SSN-15 is similar to the eighteen-year-old SUBROC. The SSN-19 is probably like the Exocet/Harpoon class missiles.

With 1500 missile launchers, the Russians don't lack for numbers. What is lacking is effective ship-to-ship attack capability. The majority of their missiles are either obsolescent or designed primarily for antisubmarine warfare. Nearly a third of the deployed Russian naval missiles are the older SSN-2 (including the improved 2C version). An additional 7 percent are the equally old SSN-3. Another 15 percent are the long-range SSN-12. An equally large number are the antisubmarine-warfare SSN-14. The SSN-14 illustrates the difficulty in identifying Russian missiles. When first seen, it was thought to be a ship-to-ship missile. More perceptive analysts studied the Russians' own pronouncements and eventually discovered that the SSN-14 was an antisubmarine-warfare weapon. Later it was discovered that some of the so-called SSN-14's were actually SSN-10's, ship-to-ship missiles. Incidentally, some still maintain that there is no SSN-10. The primary argument for the SSN-10 is that the SSN-14 can only be used on a ship with sufficiently powerful sensors to pick up a submarine target at the distance to which the missile can reach. Only the Russians know for sure.

Russian long-range missiles still require midcourse corrections from another ship or aircraft. This is due to unsophisticated sensor, guidance and terminal homing systems. To get around their shortcomings, they use more radio command guidance and multiple terminal-guidance systems. The terminal-guidance systems are particularly vulnerable to interference.

Part of peacetime espionage activities is to discover the characteristics of these guidance systems so that the most effective countermeasures can be developed. Lacking precise knowledge, it is possible to use general methods. When the weapons are actually employed, more precise knowledge will be obtained. If a missile is kept in use for too long, its secrets will become known and its initial wartime effect will be diminished. Russian missiles are a good example. During the 1967 Middle East War, these weapons were devastating. Six years later, these same weapons caused no damage.

Western missiles generally avoid many of the problems under which the Russians operate. The ship-to-ship missiles of the West are not dependent on command guidance. Unlike the Russian Navy, Western navies are more likely to control the air and seas, so they do not need to substitute long-range ship-to-ship missiles for air power.

Superior Western technology is nowhere more evident than in the fact that such small nations as Norway and Israel have produced quite effective ship-to-ship missiles. The Russians lack the ability to construct large numbers of complex machines to consistent quality, nor can they maintain them. For this reason Russian possession of the secrets of Western weapons does not always enable them to copy them.

The Western ship-to-ship missiles shown here are the principal ones. The Harpoon, in particular, is widely used, primarily because it can be easily launched from so many different vehicles: aircraft, torpedo tubes and ships of all sizes. The Exocet, the first Western ship-to-ship missile, has gone through a number of improvements. The Penguin was initially designed for coastal defense from a land base. Otomat is a longer-range ship-to-ship missile designed for ships operating without air support.

SUBROC and ASROC were designed for antisubmarine warfare. Their principle is simple—attach a rocket motor to a torpedo, fire it into the general area of the target submarine and let the homing torpedo do the rest. Some versions use a nuclear depth charge, which kills at ten times the radius of a conventional depth charge (30 meters). As homing torpedoes become more effective, the trend is to avoid nuclear weapons. With nukes, even at sea where you don't leave any holes in the water, there is still the possibility of nuclear escalation. SUBROC is launched from submarine torpedo tubes, ASROC from surface ships.

TORPEDOES

The appearance of these weapons has changed little over the years; improved propulsion and sensors have increased destructiveness enormously. The three models shown are all of U.S. manufacture. They are typical of most torpedoes.

At the top of the line is the Mk 48. This is an exceptionally capable weapon, and no other navy is likely to have anything like it. The Mk 48 is the current outer limit of torpedo technology.

The Mk 37 is more typical of the torpedoes found in all navies. The range can be increased, but this is only useful if a torpedo's sensors can detect the target. This problem is solved currently by a wire from the launching ship to the torpedo. The ship's sensors guide the torpedo close enough to the target so that its sensors can finish the job.

Against submarines that can travel as fast as a regular torpedo, you need a weapon that is extremely fast, accurate and quiet. Although torpedoes are still nominally useful against surface ships, in most cases the launching ship will not be able to get close enough to use them.

The Mk 46 is typical of lightweight, smaller torpedoes used by aircraft or rockets (SUBROC, ASROC, SSN-14/15, etc.). ·

A current problem is the ever stronger hulls of nuclear submarines. As subs are designed for plunging farther down into the depths, their hulls become correspondingly stronger. An additional problem is that torpedoes are not designed to go as deep as some modern submarines. There will always be a need for improvement.

DEPTH-CHARGE LAUNCHERS

Depth-charge launchers are an elderly but still effective means of destroying submarines, particularly diesel-electric boats. They are much less effective against nuclear subs. Their operation is quite simple. An explosive charge is propelled by a rocket to the general area of the sub. The charges are launched in groups to cover a pattern, and are set to explode at a preset depth and/or use sensors to detect the submarine. Obviously these weapons are highly dependent on the accuracy of the launching ship's sensors. The Russians, and less affluent navies, still use these weapons extensively.

GUNS

With the proliferation of missiles, not much space has been left on ships for guns. Those that remain are generally 3-inch (76mm) or 5-inch (127mm) weapons. Many smaller cannon (20mm to 40mm, for the most part) are still being used for air defense and, when equipped with sophisticated sensor/fire-control systems, for missile defense. See the chart of air-defense weapons.

Otherwise, guns can still be effective if you can get close enough to use them.

10

THE NAVY: RUN SILENT, RUN DEEP

SUBMARINE SERVICE requires considerable faith and self-confidence. You never see, hear or smell the enemy. Everything is done through instruments. Worse still, if you err and the enemy gains an advantage, there is no place to run. Your battlefield is a metal cylinder tapered at both ends. It is 200 to 600 feet long and 20 to 40 feet in diameter, but you can only move about in one third this volume. The space is crammed with equipment, weapons, supplies and fellow crewmen.

Yet submarines have become the premier naval weapon during the last twenty years. Why?

Hiding ability. A submarine, once submerged, cannot easily be detected, even by another submarine. For various reasons explained below, the sea is an excellent place to get lost in, and stay lost.

Nuclear power. The necessarily cramped design of a submarine limited the capabilities of its power plant until the introduction of nuclear power. Nuclear power enabled subs to stay submerged for as long as they wanted. The enormous resources of the nuclear reactor also powered life-support systems, weapon support and other equipment.

Improved sensors. With the introduction and steady development of more powerful sensors, submarines were no longer blind. Surface ships remained as easy to detect as ever.

Improved weapons. More accurate and longer-range torpedoes, as well as missiles, extended submarines' reach. No longer was it necessary to sight through a periscope before firing.

Modern Submarine Design

A submarine is a ship capable of cruising underwater. It is designed around a pressure hull, a steel tube strong enough to withstand water pressure at great depths (200 to 1000 meters). Within this pressure hull must be crowded all weapons, equipment and crew. Outside the pressure hull are tanks, which are flooded when the submarine submerges and emptied when it surfaces. Submarines differ in the following characteristics.

Size. By itself, size is a disadvantage. The bigger boats are easier to find. Submarines are only as large as their weapons and equipment require. Weights range from under 1000 tons to over 16,000 tons surface displacement. The heaviest submarines are nearly 600 feet long and 40 feet in diameter. The smallest are 180 feet long and 20 feet in diameter. Crew size ranges from 30 to 140 men.

The first modern submarines, introduced eighty years ago, had dual electric propulsion: diesel for surface cruising and battery powered for underwater movement. The diesel system required fuel and the batteries were heavy. Boats of this type rarely were larger than 2000 tons. Endurance was limited to 90 to 120 days. Actually the chief limitation was the crew's nerves. The diesel-electric boats were very cramped and uncomfortable. In 1943 the "snort" (snorkel) was introduced on a wide scale. When the boats were cruising at periscope depth, air could be brought in for the diesel engines. This made the boats more difficult to detect, but subjected the crew to bad air and surface turbulence. Nuclear power plants were somewhat heavier, but the boats became larger because power could more easily be increased; more effective weapons and sensors added; and crew comfort and endurance increased without the penalty of carrying more fuel. A nuclear-powered boat can cruise over 200,000 km before its minuscule amount of nuclear fuel has to be replaced.

Main Weapons

Originally torpedoes, mines and a deck gun were the main weapons. The gun was a practical recognition of the diesel-electric boat's status as a small surface ship that could submerge briefly to sneak up on its victims or evade a more powerful adversary. The introduction of better sensors and torpedoes has eliminated the deck gun. Mines are still carried whenever the tactical situation calls for them. "Torpedoes with brains" and missiles are the principal submarine weapons today. The enormous increase in the sensitivity of underwater sensors has made it possible to detect surface or submerged targets at ranges in excess of 100 km. At closer ranges the target can be located accurately enough to allow a wire-guided torpedo to be driven to the target. Wire control also allows the torpedo to relay its sensor findings to its operator and, through the submarine's attack computers, stay one jump ahead of any evasive actions by the target. The most sophisticated wire-guided torpedo is the U.S. Mk 48, with a range of 50 km. Non-wire-guided torpedoes are less capable and depend on an accurate bearing given by the submarine's sensors. The torpedo then homes in on the target using its acoustic sensor. Should the original direction prove too inaccurate, these torpedoes can also be programmed to run a search pattern until they either sense the target or run out of power. Range is typically 40 km or less.

Missiles fired from torpedo tubes, a Western innovation, are designed for long-range attack on surface or submarine targets. They are launched from the submarine's torpedo tube, pop out of the water, ignite their rocket motors and head for the general direction of the target. At this point the missile must depend on its own navigation equipment and sensors. If available, friendly ships or aircraft can provide additional guidance. Russian inability to miniaturize components has led them to develop special submarines for carrying larger tactical missiles. They operate on the same principles.

The larger ballistic missiles carried by submarines usually have fixed land targets. These are strategic weapons armed with nuclear warheads. Terminal homing sensors for these missiles are under research so that they can attack submarines and surface ships.

Sensors

By far the largest difference in performance among submarines is their sensor capability. To be able to detect enemy ships before discovery is the key to combat success and survival. American submariners openly proclaim their ability to detect Russian submarines at ten times the range at which U.S. submarines can be detected. There is ample opportunity in peacetime to test this claim. If true, it puts Russian submarines at a grave disadvantage.

SONAR INTERFERENCE

A sonar set consists of a directional sound-making apparatus, hydrophones to detect echoes and other sounds, and data-processing equipment. Depending on the power of the transmitter, the sensitivity of the hydrophones and the efficiency of the data processor, the effective range of the sonar varies from 1 to over 50 km. There are a number of other variables that affect the range. In deep water, varying water temperatures distort and misdirect signals. Water tends to form layers of different temperature through which sound travels at a different speed and direction. Some transmissions will even be deflected by a layer. The deeper the water, the greater the number of layers encountered. Each layer is a potential hiding place. Temperatures are read at different depths by dropping a cable with thermometers. This gives the sonar a constant profile of the local temperature layers. Another method is to lower the hydrophones and transmitter to different depths so that the sounds, which tend to travel through a layer, can be more accurately detected.

Ambient noise, that is, noise near the hydrophones that is caused by its own submarine, interferes with the accuracy of the data processing. The most effective way to filter it out is by towing the hydrophones and transmitter on a cable. The ambient noise of the sea itself—whales, schools of fish, etc.—is handled by signal processing.

The level of noise given off by the target itself also must be considered. For passive detection the louder the target is, the greater the detection range and the greater the difficulty of hiding within a different temperature layer.

Water salinity also affects the transmission of sound in water. The most accurate way to compensate is by collecting data on salinity and making adjustments when processing signals.

Signal processing. Sorting out all this noise into coherent information is a data-processing problem that has been solved with increasing efficiency by computers. The faster and more powerful the computer, the more quickly and accurately you will locate your target.

The simplest modern sonar puts a blip on a TV screen to show the contact. The more powerful your sonar-transmitting power is, the farther away you will detect targets, although the accuracy will be lower. The earliest method for solving this problem was to take the sonar information of two or more searching ships and triangulate to obtain a more accurate location. Where the contact lines intersected was where the enemy was likely to be. This method is still used and is a favored Russian technique. The Russians attempt to utilize as many searching ships as possible.

A much more accurate approach is to collect as much data as possible on temperature layers and salinity while suppressing ambient noise. The modern signal-processing computer maintains a large library of the sources of underwater sounds and conditions. A fast computer integrates all the sensor data and compares them with its library of sounds so quickly that not only is the blip accurate, it also has a name tag plus speed and course data. The signal library often has the unique sound of individual ships by type and class and frequently by name. Another function of a submarine equipped with a signal library is to collect tapes on all contacts made. These tapes are returned to a larger data-processing facility, where they are integrated into the existing library to keep it current. Submarines then receive current copies of the library on a spool of magnetic tape or a suitcase-sized magnetic disk.

ACTIVE VS. PASSIVE SONAR

Sonar that transmits sound, an active sensor, also announces their presence. Far more preferable is a passive sonar that simply listens and detects the location of the sound. However, passive sensors are totally dependent on the sound the target generates. A powerful passive sonar can detect a moving submarine generating a fair amount of noise over a distance three to five times as great as can the most powerful active sonar. Passive sonar requires greater signal-processing ability to squeeze more information out of less numerous and more subtle data.

The least capable passive sonar can pick up loud targets, ships moving at high speed, at ranges of over 300 km. More sophisticated rigs can triple that range. The least effective equipment can make contact with slow-moving submarines at 5 to 10 km at best. Superior equipment again will triple that range. A motionless diesel-electric submarine is virtually soundless. It can only be detected by moving close, which can enable the target's passive sensors to pick you up first if they are sufficiently powerful. Sensors are the key.

The most modern submarine, the U.S. Los Angeles class, using its towed array of very efficient active sensors, can detect submarines or surface targets over 100 km distant. The active sonars on sonobuoys (data transmitters that float while the hydrophone is lowered to a preselected depth; these passive devices have a range of 10 to 20 km) are effective to 2 km. "Dunking sonar" from helicopters is good to 8 km. Hull-mounted sonar in ships works up to 15 to 30 km. Towed and variable-depth sonar ("dunking sonar" on a ship) is good to 50 km. Bow-mounted sonar in U.S. submarines reads to 50 km. Lack of good maintenance or poorly trained operators can reduce the effectiveness of any sonar by five to ten times. Thus, detection ability can range from 1 to 100 km. When you add the differences in accuracy and promptness, the spread becomes even larger.

Silencing. This is what you do to make your submarine quiet and thus more difficult to detect by passive sensors. Water flowing over the hull, from currents or submarine movement, creates noise. A hull that resembles a fish allows the water to flow without noise-creating detours around protrusions. Many Russian submarines are poorly designed in this respect and produce additional noise. Diesel-electric submarines, because they must spend much of their time on the surface, are designed for surface cruising and thus have a less streamlined hull when traveling submerged. Engines make noise. Even the quiet diesel-electric motors generate a very small noise signature. This silence can be enhanced with soundproofing materials and vibration-damping mounts for propellers and other machinery. Nuclear submarines have an additional problem. Their reactors require a constant flow of cooling water, provided by pumps that can never be shut down. They can be silenced to a certain degree, but because of these pumps a stationary nuclear boat is always noisier than a diesel-electric submarine.

Machinery is noisy. You can avoid its noise through sound-proofing or by not using it. Also, any movement on the submarine creates noise. Crewmen wear rubber-soled shoes and practice "noise discipline" at all times, just so they won't get out of practice.

Active sensors transmit sound. If you use your active sonar, you are giving away your position.

ANTISUBMARINE WARFARE

It is not always possible to have a friendly submarine around when you are being harried by unfriendly ones. This problem is solved with a mixture of surface ships and aircraft with tactical sensors and torpedoes.

Strategic sensors. Only one worldwide system is in place: SOSUS (SOund SUrveillance System), a U.S.-built network of passive sensors located on the continental shelves of the North Atlantic (the CAESAR network) and the North Pacific (the COLOSSUS network), plus a few in the Indian Ocean. They listen to everything. The data are sent via cable to land stations, where the information is passed on, often via satellite link, to data-processing centers. The system works much like the passive sensors aboard submarines. Currently it is accurate enough to locate a submarine within a circle no larger than 180 km. That's a large area, but depending on the quality of the contact, the circle may be reduced by as much as ten times. The only drawback to the system is its inability to provide detailed coverage of deep ocean areas more than 500 km from the continental shelf. On the other hand, the Russians know this and congregate in these areas, making it easier to deploy the tactical forces, particularly nuclear submarines. A means of covering this gap has been developed: SURTASS (SURveillance Towed Array SyStem), a "sled" containing hydrophones towed by a tug in open ocean areas. Data are collected and sent, via satellite, to land stations. Once the decision is made to go after a submarine, then surface ships, submarines or aircraft can be on top of the target within hours.

In theory there are a number of potential methods for detecting submarines on a strategic scale. None of them have yet been put to practical use, although it's just a matter of time before they are. The most promising involves strategic reconnaissance with satellites to detect: heat (water must be constantly pumped through a submarine's nuclear reactor to keep it from overheating), magnetic disturbances (caused by the large metal object passing through the water) and water displacement (the bulge in the ocean surface caused by the submarines passing through the water).

Tactical sensors. Surface ships use the same sensors as submarines. In addition, they can have helicopters. In a class by themselves are the long-range antisubmarine aircraft.

In general, the helicopters extend the more powerful sensors and weapons on board a surface ship. Helicopters carry torpedoes or depth charges. Helicopters can hover to lower "dunking sonar" to various depths in order to search different temperature layers. In addition, the helicopters use sonobuoys, radar (for surfaced submarines or their periscopes) and a Magnetic Anomaly Detector (MAD) that senses disturbances in the magnetic field caused by a submarine's large metal hull passing through the water. Helicopters process some of their data on board, but because of their close proximity to surface ships, they often broadcast back this collected data.

Using the same equipment as helicopters, except for the "dunking sonar," aircraft have more extensive data-processing facilities and much longer range.

Tactics. Once a submarine's general location is determined, aircraft are usually the first on the scene. If the submarine is a nuclear boat, it will most likely be submerged when the aircraft show up. This means surface search radar will be useless.

At that point aircraft drop a pattern of sonobuoys. These will likely pick up the submarine, as it has no way of knowing an aircraft is in the area. Specialized sonobuoys that sense the water temperature at various depths are also used. One or more lines of sonobuoys are formed. The aircraft's computers process all these data. In addition, data links to the satellite communications system enables the aircraft to share data with SOSUS and other antisubmarine forces.

When the submarine passes through the sonobuoy line, the aircraft returns and, flying low (200 meters), attempts to use its MAD to get a precise fix on the submarine. It is effective only for a kilometer or so on either side of the aircraft's flight path. At the same time the MAD is being used, active sonobuoys are dropped. These can be turned on to operate just like normal sonar. They only have a range of a few kilometers and will announce their presence to an alert submarine crew.

Once the submarine has been located to within a kilometer or so, the aircraft drops a homing torpedo or two and/or depth charges to attempt to destroy the sub. Surprise is important, because a nuclear submarine can quickly increase its speed to 60 km per hour and dive very deeply. Although the aircraft will not have lost the submarine, it will have to repeat the dropping of sonobuoys and, of course, the submarine will now be on its guard. The sonobuoys can only operate up to eight hours, and even the largest aircraft cannot carry more than a hundred. Thus, it is important to use passive sensors as much as possible.

THE HUNTER BECOMES THE HUNTED

During World War II, only under rare circumstances did the submarine turn on the surface ship hunting it. Modern nuclear submarines are not nearly so timid. Being faster than surface ships confers one major advantage. Longer-range torpedoes and missiles, coupled with better sensors, make the submarine a potent adversary. Finally, because the surface ships cannot hide their noise as they move on the surface, they are easier to detect by underwater sensors than the other way around.

Like a lion stalking a herd of antelope, the submarine slowly patrols its sector until it detects fast-moving surface ships. If they are heading toward the submarine, an interception course is set. Speed is increased to ensure interception, but not if it would generate enough noise to give away the stalking submarine's position. Once close enough to use its long-range weapons, the sub attacks. If only torpedoes are used, the range will be 40 km or less. Cruise missiles have ranges of over 50 km. After the attack, the submarines will quite likely be located by the surviving surface ships. The best tactic is to dive deep and move slowly away; thus the motto: "Run silent, run deep."

Russian surface ships are more likely to receive this treatment. The U.S. SOSUS system provides enough significant tracking of Russian submarines to make a Russian ambush unlikely but not impossible or improbable.

Who Does What to Whom?

Russian deployment. Russia will first get as many submarines into the high seas as possible. This will be tricky as 85 percent to 90 percent of their subs are normally in port. Political tension will then escalate, since the deployment of Russian submarines is seen as an act of war. Russia and her allies could send as many as 200 submarines out to sea, where they would face an almost equal number of enemy submarines plus thousands of antisubmarine aircraft and helicopters, and the ever-watchful SOSUS, which ought to be their first target.

Once at sea the Russians plan to use all available reconnaissance and mass "wolf packs" of subs against key Western targets. The submarines can communicate with surface and land headquarters, but they risk detection. What happens when they meet a Western wolf pack is the stuff adventure movies are made of.

One must not forget Russian naval priorities. The protection of their ballistic-missile submarines (SSBN's) comes first. Therefore, the distribution of their SSBN's will dictate where the protective nuclear- and diesel-electric-submarine contingents will be. Twenty submarines could protect the 16 SSBN boats operating in the home waters of the northern fleet, Murmansk area. An additional 15 SSBN's, with shorter-range missiles, could head for the Atlantic. These could be escorted by 35 nuclear and 45 diesel-electric submarines. Five SSB's or SSBN's could be in the Mediterranean or in the Indian Ocean. These might be escorted by 10 nuclear and 20 diesel submarines.

In the Pacific there are 8 SSBN's that could operate under the cover of friendly air power. A further 10 SSB's or SSBN's could head for the high seas and be covered by 55 nuclear and diesel subs. Any submarines that are unoccupied would concentrate on the destruction of enemy ballistic-missile subs.

Next the Russians will go after enemy surface combat ships. Finally, they will try to destruct enemy shipping. However, there are never enough Russian submarines to go around.

NATO DEPLOYMENT

Western naval strategy takes advantage of the awkward locations of major Russian naval bases. Two bases, the northern (in the Murmansk area) and the Pacific (in the Vladivostok area), have over two thirds of the Russian ships and a disproportionate number of their most modern ones. For vessels to exit these bases, they must pass close by areas occupied by Western powers. The northern fleet must move past northern Norway and the British/Iceland/Greenland (BIG) gap. The BIG gap is the most difficult because it is covered by Western antisubmarine-warfare aircraft and is generally out of range of Russian planes. It is also covered with SOSUS.

Should the Russians manage to get most of their submarines to sea before hostilities commence, which could take two to three weeks, they could not prevent Western antisubmarine-warfare forces from shadowing them. After the initial shoot-out, the surviving Russian diesel-electric submarines would only last ninety days, the nuclear boats as long as food, ammunition and the crews' nerves lasted. At that point they would have to run the BIG gap again. It appears that the bulk of the naval war would be over in months, if not weeks.

It is not that nearly a hundred Russian submarines are an insignificant fighting force. Nor are their naval air force and surface fleet insignificant. But they are outnumbered and outclassed. Over a thousand long-range antisubmarine-warfare aircraft will harry their submarines on the high seas. SOSUS will constantly reveal their positions. Hundreds of Western surface ships and submarines will be all over the place. Russian forces can, and probably will, attack SOSUS installations and Western air bases, but inferior numbers and technology will take their toll quickly.

Western forces can put nearly as many submarines into the Atlantic as the Russians can. The West has a clear superiority in surface forces and control in the air, which it supplements with naval mines. The situation in the Pacific is not much different. The same applies to the other minor theaters (the Indian Ocean, the Mediterranean, the Baltic).

THE SHAPE OF THINGS TO COME

The Russian and Western navies are not unaware of the imbalance of power. The Russians realize that if they must go to war, they must do so in such a way to allow their naval forces to deploy. Being in an inferior position, they must be innovative.

One possible innovative solution is to destroy antisubmarine-warfare-aircraft bases in Britain, Iceland, Norway, Japan and elsewhere with submarine-launched missiles. A strong case can be made for ballistic missiles. Nuclear weapons could be used, but it could lead to escalation. Chemical warheads are less effective, but decrease the chances of nuclear escalation.

The Russians could try to destroy the SOSUS system by cutting the cables from the hydrophone arrays to the land stations. The arrays themselves can be destroyed by torpedoes, the land stations by missiles. It may also be possible to employ countermeasures like large-scale noisemakers against SOSUS and other sensors.

The apparent disparity between Western and Russian submarine capabilities is so great that even without Western naval air power and SOSUS it would be an unequal battle. To increase the chances of success, and probably not increase them enough, the Russians would have to use nuclear warheads on their naval missiles, torpedoes and depth charges. Nuclear warheads increase the kill radius underwater to a kilometer or more. Given the low accuracy of Russian sensors, this might not be enough. Western navies also have nuclear warheads.

It is possible too that despite all the evidence, Russian submarine forces are not so inferior. They do continually increase the technical quality of their subs, as shown by constant Western submarine tests at sea. The Russians may continue to close the effectiveness gap. But, short of war, there is no way of knowing precisely when the gap has been closed. Even this is not enough. When, or if, the Russians pull ahead, they will have a chance of success. At that point they could logically initiate hostilities. Before that time there is always the possibility of an irrational entry into a naval war. That, unfortunately, is how most wars begin.

11

THE NAVY:
IN THE AIR

EVEN BEFORE World War I it was recognized that aircraft would provide invaluable reconnaissance for the fleet. It wasn't until nearly a decade after World War I that a few navies realized aircraft were also a potent combat force. Experience in World War II confirmed this. Subsequently, nuclear submarines proved, in some respects, an even more potent weapon. But naval aircraft, either land- or carrier-based, still had substantial advantages, and they have been increased with the introduction of the cruise missile. The cruise missile is, for all practical purposes, a ship-based attack aircraft. Similarity to the Japanese *kamikaze* suicide aircraft of World War II is striking. There is the added grim reality that every cruise-missile-carrying ship becomes an aircraft carrier. Can you imagine what half a dozen modern cruise-missile ships could have done in World War II? Imagine what these ships can do right now.

Carrier-Task-Force Defense System

The range and flexibility of aircraft enable carrier task forces to put up a multizone defense extending to 700 km from the carrier and its escorts.

THE PRIMARY ZONE

The primary, or vital, zone extends 40 to 50 km from the carrier and is monitored by ship-borne sensors. This zone is defended primarily by electronic weapons, surface-to-air missiles and guns. Electronic jammers blind cruise-missile homing systems. "Blip" enhancers on smaller ships and low-flying helicopters make the cruise missile think that these are the carrier. Electronic pulses can even prematurely detonate a cruise-missile warhead. Nonelectronic devices include chaff, aluminum strips to create a sensor-proof cloud and flares, which create a more attractive target for heat-seeking homing devices.

THE MIDDLE ZONE

The middle zone is monitored by patrol aircraft from the carrier. They can spot surface warships 360 km away and other aircraft 700 km away. This zone is defended by aircraft and one or more detachments of surface warships and submarines. These detachments of one to four ships, if they can be maneuvered quickly into position, can use the same electronic-warfare techniques as ships in the primary zone. In addition, the submarines may be able to destroy the enemy ships with their own cruise missiles. The F-14's stand ready to go where needed with missiles of a maximum range of 200 km. Although U.S. carrier aircraft normally operate for 105 minutes in the air, they can be refueled in the air. Six to 12 F-14's, with 6 missiles each, would be quickly overwhelmed by a massive attack. They can be assisted by electronic-warfare aircraft from the carrier. Ideally the F-14's would go after the aircraft carrying the cruise missiles. Additional attack aircraft would also be sent up to attack enemy surface warships and submarines before cruise missiles could be launched.

THE OUTER ZONE

The outer zone extends to 700 km and beyond. Satellites, land-based P-3 patrol aircraft and SOSUS sensors supply monitoring support. Attack aircraft from the carrier go after any enemy aircraft, surface warships or submarines that have been detected. This is also the most likely zone in which to destroy the Russian airborne control aircraft, usually a four-engine bomber equipped with a heavy load of electronic equipment. This aircraft not only coordinates the Russian attack, but also attempts to provide midcourse guidance for cruise missiles fired over the horizon. Russian naval reconnaissance aircraft attempt to get a precise fix on the task-force location so that the cruise-missile guidance systems can be given the most accurate target data. Because of the heavy use of electronic warfare, it is important that the cruise missile get within terminal homing range of the task forces without additional radio signals that could be jammed.

This outer zone is quite possibly the most important. It presents the best opportunities for destroying cruise-missile carriers before they can launch. Each surface warship or submarine carries up to ten cruise-missile launchers. The ability of carriers to spot these threats so far away, and launch effective attacks, demonstrates vividly their advantage. The fact that all these aircraft come from one carrier also demonstrates their vulnerability.

ANTISUBMARINE WARFARE

Both on land and at sea, helicopters provide the backbone of antisubmarine warfare. Aircraft like the P-3 have the range that helicopters lack, but they are often too fast. Helicopters can hang around and pursue submarines with more precise diligence. Helicopters can also be carried by small ships, 3000 tons or so. Many navies have smaller carriers with nothing but helicopters.

ORGANIZATION OF CARRIER AIRCRAFT

A modern attack aircraft carrier is assigned an air wing of up to 90 aircraft. Among current U.S. carriers, six can handle 90 aircraft, four, 85, and the remainder about 75. There are 12 carrier air wings in the U.S. Navy, plus 2 active-reserve wings. At least 1 carrier, more likely up to 3, is in dry dock for refit at any one time. Each air wing contains 48 to 60 or more combat aircraft and two dozen support aircraft.

Only seven American wings have 24 F-14's, the remainder have F-4's. The larger wings have 24 A-7's and 12 A-6's. These are the primary combat aircraft. The F-14's or F-4's are fighters, the A-7's and A-6's are bombers. The A-6's in particular are superb all-weather aircraft. The A-7's also have all-weather capability but not as extensive as the A-6. The support aircraft consist of three reconnaissance aircraft (RF-4) for finding distant targets and taking pictures. The antisubmarine contingent consists of ten S-3A fixed-wing aircraft and eight helicopters (SH-3). The S-3A's in particular have enormous capabilities; they can stay in the air for more than six hours at a time while carrying torpedoes, sonobuoys, harpoons and other weapons. Other specialist aircraft include four electronic-warfare aircraft (EA-6B) for countermeasures. The four search aircraft (E-2C) maintain patrols that extend the carrier's vision out to 700 km. The four tankers (KA-6D) refuel the search and other aircraft to keep them aloft for extended periods. Other specialized aircraft are carried if space allows. No other nation has attack carriers quite in the same class as the United States.

The U.S. Navy has 11 helicopter carriers for amphibious operations; they contain 16 to 20 helicopters each. Russia has 5 helicopter carriers. A number of other navies are also building helicopter carriers.

Normally carriers do not keep aircraft in the air for very long periods. Although American carriers have tankers, they can only service a few aircraft each per flight. Keeping CAP (combat air patrol) and early warning aircraft up for more than twenty-four hours severely strains aircraft and pilots. Carriers are dependent on strategic warning reconnaissance from satellites, SOSUS and electronic monitoring.

Land-based Naval Aircraft Operations

All navies use land-based aircraft, primarily for patrolling, secondarily for ship destruction. Western navies have over 1000 four-engine, long-range patrol aircraft. These are primarily P-3's, which can also carry cruise missiles. The U.S. Air Force has trained some of its B-52 crews to launch cruise missiles, patrol, and drop naval mines (CAPTOR's).

As was the case in World War II, getting within range of land-based interceptors will surely lead to high losses. Russian naval bases are well defended by interceptors. They will not be easy to destroy.

Over 3000 fixed-wing and more than 2000 helicopter naval patrol aircraft operate from land bases worldwide. The majority are equipped primarily for hunting submarines. Naturally many weapons that can be used against submarines can also be used against surface warships. Many antisubmarine aircraft are also equipped with air-to-surface weapons with sufficient range to allow the lumbering patrol aircraft to get out of harm's way after launching.

Although restricted to land bases, if sufficient bases are available the patrol aircraft are as flexible as carriers. They are a fast-moving reserve, able to pinpoint and attack ships far more rapidly than other vessels. So why have ships at all?

Patrol aircraft, although many can stay aloft for over twelve hours at a time, have limited endurance. A ship can remain on station for months at a time. Depending on the endurance of its ground maintenance crew, a patrol aircraft can fly one sortie a day for a number of weeks. After that, things start to fall apart. Most patrol aircraft also have minimal defense against air attack.

PATROL-AIRCRAFT MISSIONS

Surface ship search. This is the simplest form of patrol and is usually performed with radar-equipped aircraft. Depending on the sophistication of the radar, an aircraft can cover an area up to 360 km wide while traveling 400 km per hour (the U.S. E-2's). Most patrol-aircraft search radars have more modest ranges, less than 100 km for helicopters, less than 200 km for fixed-wing aircraft. Human observers can see at best 20 to 30 km during clear weather. These searches are performed either offensively or defensively. That is, the aircraft can either establish a barrier of observation at sea to prevent undetected intrusion of enemy ships, or scour an area to fix the location of enemy ships so that they can be attacked with more massive forces.

Antisubmarine search. This is performed by fixed-wing and helicopter aircraft. Because most submarines submerge when being searched for, aircraft will not be able to cover as much area. Helicopters are much more useful because they can hang around.

Strike. In this case the patrol aircraft are armed for attack, usually with missiles. Most task forces have too strong an anti-aircraft defense to make close-in attacks effective. Most Russian patrol aircraft are actually bombers equipped with better radar and trained to travel long distances over water while carrying air-to-surface missiles.

Cruise-Missile Warfare: The Probable Outcome

During World War II, American task forces in the Pacific encountered massive *kamikaze* attacks by Japanese aircraft crewed by hastily trained suicide pilots. The Okinawa campaign saw 1900 aircraft attacking over a period of 100 days, March–June 1945. Under attack were 587 ships, of which 320 were warships. Each attack averaged 150 aircraft, one as many as 350. Between the defending aircraft and ships' guns, only 7 percent of the aircraft scored hits. Eighteen percent of the ships were hit severely enough to be sunk or put out of action. Although the *kamikazes* were unexpected, the defending fleet had the usual strong sense of self-preservation. Because the American ships were equipped with radar and effective air defense, there was no reason why this new weapon should have been decisive. The *kamikazes* were, in effect, cruise missiles, and they performed at near-maximum effectiveness due to their element of surprise.

Twenty-two years later, electronic pilots were developed to replace the human ones, and Israel lost a destroyer to Egyptian cruise missiles. Over the next five years the cruise missile was used in combat a number of times, severely damaging at least a dozen warships. During the 1973 Middle East War, Israel demonstrated that the cruise missile could be stopped cold with a combination of electronic warfare, gunfire and evasive maneuvers. Over 50 Arab cruise missiles were fired without scoring a single hit. Fifty percent of the Israeli cruise missiles scored hits, even if another cruise missile had just disintegrated the target. It would appear that an unsuspecting target has a 90 percent chance of being hit by a modern cruise missile. Various defensive measures—guns, missiles, electronics, maneuvers—can bring this hit probability down to 0 percent.

The crucial question is, what would be the percentage of hits by Russian attackers on defending Western forces? Based on historical experience, between 0 percent and 10 percent. Considering the previously demonstrated ineffectiveness of Russian antiaircraft and cruise missiles, 1 percent to 3 percent is more likely. Western cruise missiles will probably hit three to ten times that rate (3 percent to 30 percent).

What, then, is the expected outcome of combat involving cruise missiles, aircraft and surface-to-air missiles? Consider the resources of the opposing forces. The Russians have 70 oceangoing surface warships mounting 260 cruise-missile launchers. In addition, 69 submarines carry 460 cruise-missile launchers. Finally, 380 naval aircraft can carry up to 500 cruise missiles. That's a potential total of 1220 cruise missiles. The U.S. fleet has a maximum of 13 carrier task forces. Other allies can muster an additional dozen or so noncarrier task forces. At most, the Russians could send 300 to 400 cruise missiles after one task force. This number is low because of the separation of Russian fleets, other task-force threats and the at least 20 percent of ships and planes out for maintenance. With 400 cruise missiles coming in with a 3 percent hit probability, the West would sustain a dozen hits. A U.S. carrier can take up to four hits and still function. (This statistic is continually tested in peacetime as carrier landings turn into crashes.) Smaller ships can often sustain no more than one hit. Therefore, a dozen hits, with most directed at the carrier, would take a task force out of action. Perhaps half of the ten ships would be sunk or dead in the water, and some of the remainder would be mobile but defenseless.

But the attacker would also take losses. An attack of such magnitude would commit almost all the usable strength of the Russian northern (Murmansk) fleet. With other Western task forces in the area, the attacking ships and aircraft could expect to take up to 50 percent losses. Trading 12 percent of Russian naval power for less—3 percent to 4 percent of Western naval forces—is a losing exchange rate.

Nothing is certain in warfare, but the trends are not in Russia's favor. They indicate that the Russian fleet would go to war with optimism as one of their primary weapons.

Naval Patrol Aircraft in Service

This chart shows the distribution of strategic and naval patrol aircraft among the world's nations. As with most major weapon systems, there is considerable concentration of the most powerful naval patrol aircraft in a few nations. The United States alone has 40 percent of the total. The top three nations possess over 70 percent of the world's naval patrol aircraft.

There is somewhat less concentration of aircraft types. The top three types comprise only 43 percent of all aircraft. However, when each is adjusted to equal the most capable naval patrol aircraft, the P-3, the top three types comprise 55 percent of all naval patrol aircraft capability. More on how this adjustment process works will be found below.

It should not be surprising that those nations which are most dependent on seaborne trade possess over 80 percent of the world's naval patrol aircraft capability. The primary reason for their investment is the over 200 Russian submarines they expect will be unleashed on their merchant shipping. Whether or not the equivalent of 1200 P-3 aircraft will defeat them from this potential menace remains to be seen.

The first column gives the nation possessing the naval patrol aircraft indicated in the columns to the right. The second column gives the total naval patrol aircraft each nation has. The third column gives the percentage of the world total of naval patrol aircraft which that nation possesses. The subsequent columns show quantity of each type of aircraft possessed by each nation.

At the top of each of the subsequent columns is the name of the aircraft. Below the name is the total number in use. Next is the percentage of all naval patrol aircraft which that type comprises. Below that is an interesting exercise in comparing apples and cabbages.

PATROL VALUE is an evaluation of each aircraft. The chart on naval-patrol-aircraft characteristics shows that not all are equal in capability. Most are equipped for antisubmarine-warfare operations. The remainder either have, or can be equipped to have, attack capability against surface shipping. All are search aircraft. In an attempt to show the general qualitative difference between all these types, a value has been assigned to each. The highest value ("100") has been assigned to the P-3, generally considered the most capable all-around naval patrol aircraft, in spite of that fact that it is primarily designed for antisubmarine warfare.

P-3 EQUIVALENTS gives the number of each type of aircraft in P-3 equivalents. Since none of the other aircraft were given as high a value, the number of all, except for the P-3, are reduced in proportion to their rating. If an aircraft was rated 50, then its quantity in P-3 equivalents was reduced by half.

% OF TOTAL gives the percentage of the total P-3 equivalents each type now comprises. As imperfect as this method is, it does suggest relative capabilities. All naval patrol aircraft are not created equal.

AIRCRAFT SPECIALTIES

The following aircraft are primarily for antisubmarine-warfare (country of origin shown in parentheses); P-3 (US), S-3 (US), S-2 (US), P-2 (US), M-12 (Russia), 1150 (France), Albatross (US), IL-38 (Russia), 1050 (France). The remaining aircraft are primarily equipped for surface attack. The All Other category is primarily surface attack aircraft.

Naval Patrol-Aircraft Characteristics

This chart gives the characteristics and capabilities of over 75 percent of the naval patrol aircraft in service today.

Without reconnaissance, a navy, no matter how powerful its ships, will be blind, and thus will be picked to pieces by its adversaries. Capable recon aircraft are the classic force multiplier. That is, the ships, aircraft or ground forces that the recon aircraft support become much more destructive. Being able to see and attack the enemy before he can return the favor is a decisive advantage.

Soon after the introduction of aircraft seventy years ago, navies were the first to recognize the advantages of reconnaissance from the air. It soon became obvious that the enormous size of the oceans required a large aircraft. It was not enough to fly, one had to stay in the air for a long time to cover all that water. In addition to having a longer range, large aircraft could carry better sensors and weapons.

When airborne radar was introduced during World War II, the age of modern

NAVAL PATROL AIRCRAFT IN SERVICE

The quantity and quality of these aircraft allow them to rule the oceans to a greater degree than any ships.

	Total	% of total	P-3 Orion	Tu-16 Badger	A-6E	S-3A	S-2E	P-2H Neptune	EA-6 A/B	E2B/C	M-12 Mail	1150 Atlantic	Tu-142 Bear	Tu-22M Backfire	IL-38 May	TR-1 (U-2)	SR-71	E-3A AWACS	All Other
Total	3107		604	425	320	180	174	125	104	94	90	82	75	50	50	20	10	26	678
% of Total		100	19	14	10	6	6	4	3	3	3	3	2	1.61	1.61	0.64	0.32	0.84	22
Patrol Value			100	35	22	45	33	55	32	86	42	62	54	35	55	65	80	100	22
P-3 Equivalents >	1515		604	149	70	81	57	69	33	81	38	51	41	18	28	13	8	26	149
% of Total			40	10	5	5	4	5	2	5	2	3	3	1.16	1.82	0.86	0.53	1.72	10
US	1303	41.94	400		320	180			104	94						15	10	26	154
Russia	760	24.46		425							90		75	50	50				70
Japan	176	5.66	45				23	90											18
West Germany	131	4.22										19							112
France	123	3.96						12				35							76
Britain	113	3.64	85																28
Canada	55	1.77	18				15												22
Australia	48	1.54	19				19												10
Taiwan	35	1.13					27												8
Netherlands	33	1.06	13					13				7							
Brazil	28	0.90					16												12
South Africa	25	0.80																	25
India	25	0.80																	25
Sweden	24	0.77																	24
Indonesia	23	0.74																	23

			P-3	Tu-16	A-6E	S-3A	S-2E	P-2H	EA-6 A/B	E2B/C	M-12 Mail	1150	Tu-142 Bear	Tu-22M May	IL-38	TR-1 (U-2)	SR-71	E-3A AWACS	All Other
Turkey	20	0.64					20												
South Korea	20	0.64					20												
Italy	18	0.58										18							
Argentina	16	0.51					6												10
Spain	16	0.51	6																
Thailand	14	0.45					10												4
Chile	14	0.45																	14
Mexico	11	0.35																	11
Peru	11	0.35					9												2
Venezuela	10	0.32					6												4
Greece	8	0.26																	8
Poland	8	0.26																	8
Norway	7	0.23	7																
Iran	6	0.19	6																
Portugal	5	0.16	5																
Pakistan	5	0.16										3							2
New Zealand	5	0.16	5																
Philippines	5	0.16																	5
Israel	3	0.10					3												
Uruguay	3	0.10																	3

NAVAL PATROL-AIRCRAFT CHARACTERISTICS

Compared with ships, these aircraft are faster, better armed and more likely to attack without being attacked first.

	P-3 Orion	Tu-16 Badger	A-6E	S-3A	S-2E	P-2H Neptune	EA-6 A/B	E2B/C	M-12 Mail	1150 Atlantic	Tu-142 Bear	Tu-22M Backfire	IL-38 May	TR-1 (U-2)	SR-71	E-3A AWACS
Total 3107	604	425	320	180	174	125	104	94	90	82	75	50	50	20	10	26
% of total 100	19	14	10	6	6	4	3	3	3	3	2	1.61	1.61	0.64	0.32	0.84
Capability Ratings																
Surface Search	6	3	3	4	4	5	5	8	1	5	4	3	4	8	9	10
Submarine Search	10	0	0	8	6	8	0	0	4	7	0	0	6	0	0	2
Air Search	3	3	4	3	2	3	5	8	1	3	4	3	3	4	6	10
Surface Attack	8	7	9	6	5	5	4	0	3	3	7	8	3	0	0	0
Submarine Attack	10	0	0	8	6	8	0	0	4	6	0	0	6	0	0	0
Air Attack	0	0	4	0	0	0	5	0	0	0	0	0	0	0	6	0
Manufactured by	US	Rus	US	US	US	US	US	US	Rus	Fr.	Rus	Rus	Rus	US	US	US
Cruise Speed (km/hour)	608	900	765	680	440	400	774	500	550	550	800	900	645	650	3186	800
Range (kilometers)	7600	5700	5500	5500	2100	4500	3800	4000	4000	6400	12500	8000	7200	4800	4800	8000
Max Air Time (hrs)	13	6	7	8	5	11	5	8	7	12	16	9	11	7+	2	10
In-flight Refueling?	No	Yes	Yes	Yes	No	No	Yes	No	No	No	Yes	Yes	No	No	Yes	Yes
Max Weight (tons)	64	72	27.4	24.8	13.2	34	29.4	23.4	29.4	42.5	188	130	60.5	13	77	147
Crew	12	6	2	4	4	7	5	5	5	14	10	7	12	1	2	17
Length (meters)	35.6	34.8	16.6	16.3	13.3	28	18.2	17.6	30.2	31.7	49.5	42.5	40.2	19.2	32.7	46.6
Disposable Weapons (tons)	9	4.5	6.8	1.4	3	3	1	0	3	3	11.3	7.5	5	0	2	0

maritime air patrols arrived. Equipped with long-range sensors capable of operating in all weather, these aircraft could find and destroy submarines and light surface ships. It was no longer necessary to possess a navy in order to control the seas.

This search capability was eventually extended to land operations. Soon computer-controlled radars became sensitive enough to pick out air targets flying below the search aircraft without being confused by all the clutter on the ground. This was the critical breakthrough, one which the Russians have not yet achieved. These aircraft, the E-2 and the E-3, are the only means of spotting low-flying planes and helicopters. In addition, enemy aircraft can be spotted as they take off and form up for their missions.

The development of exotic metal and engine technologies in the West also led the United States to develop high-altitude reconnaissance aircraft. Operating at 30,000-meter altitudes, the SR-71 and the TR-1 can avoid most enemy counter-measures.

CAPABILITY RATINGS are the various capabilities of the aircraft expressed from 0 to 10. A 0 indicates no capability in that area, usually because the aircraft is not equipped for it. A 10 indicates the best capability available. Where a certain capability does not have a 10 on this chart, this indicates that some other type of aircraft possesses that rating.

Improvements in aircraft weapons and equipment can increase an aircraft's rating by a point or two (or less if the "new, improved version" is less reliable, as is often the case). With the increasing use of microcomputers, such improvements are becoming more frequent and reliable.

Keep in mind that the skill and training of the aircraft crews can modify these capability ratings drastically. A rating can be reduced by half or more due to crew failings.

SURFACE SEARCH is the ability to detect objects on the land or water. Generally, this means radar search. Other sensors detect heat, engine exhaust (for diesel-electric submarines), electronic transmissions, etc. Visual search is also used, but is limited by the need for clear daytime weather.

SUBMARINE SEARCH is the ability to detect submerged submarines through MAD (magnetic anomaly detector) and sonobuoys. See Chapter 10 for more details.

AIR SEARCH is the ability to detect aircraft, especially those flying close to the ground. This is done primarily with radar.

SURFACE ATTACK is the ability to attack surface targets. The most effective weapon is the air-to-surface missile. Other weapons include torpedoes, rockets, bombs and cannon. Also taken into account in this evaluation is the quality of the aircraft's fire-control system.

SUBMARINE ATTACK is the ability to attack submerged submarines with homing torpedoes or depth charges. Therefore, submarine attack ability is highly dependent on submarine search ability.

AIR ATTACK is the ability to attack, not just defend against, other aircraft. This is a rare quality in patrol aircraft. Their large size and design best equip them for long, slow patrolling. The prompt, violent maneuvers associated with air combat are not possible with most patrol aircraft.

MANUFACTURED BY. Primary nation of manufacture.

CRUISE SPEED (in kilometers per hour) is the most fuel-efficient flying speed (for maximum time in the air). Often the aircraft must fly slower to use certain sensors or weapons. Because the TR-1 is a glider with a jet engine, it can shut off its engine and glide if the situation permits.

RANGE (in kilometers) is the maximum distance the aircraft can fly in one trip without inflight refueling.

MAX AIR TIME (in hours) is the maximum flying time at cruising speed. For antisubmarine work, this will often be 10 percent to 20 percent less.

IN-FLIGHT REFUELING? If the aircraft can be refueled in flight, it can greatly increase its range. The limit to range then becomes crew endurance. If two crews are carried (possible in the largest aircraft), an aircraft can stay up for about twenty-four hours.

MAX WEIGHT (tons) is the aircraft's maximum takeoff weight. It indicates its size.

CREW represents the usual number of crewmen carried. This number will sometimes vary with the mission. Generally, the crew is divided into two sections, flight (to operate the aircraft) and operations (to take care of the sensors and weapons).

LENGTH (in meters) is the length of the aircraft. This is another indicator of the aircraft size.

DISPOSABLE WEAPONS (tons) is the amount of weapons that can be dropped or launched against the enemy. It indicates the aircraft's destructive potential. Keep in mind the importance of quality in fire-control and sensor systems.

THE AIRCRAFT

The aircraft are arranged in order of numbers in service. At the top of each column is the aircraft designation. Next comes the number in service, followed by the percentage this number represents of all naval patrol aircraft in service.

The *P-3 Orion* (US) is the most powerful naval patrol aircraft currently in service. It is primarily an antisubmarine aircraft, but can also perform reconnaissance missions.

The *Tu-16* (Russia) was originally designed as a bomber, but has since been used on a large scale as a naval patrol aircraft. Many also serve as tankers for inflight refueling.

The *A-6E* (US) usually operates from a carrier and is lavishly equipped with sensors. Aircraft such as this are borderline patrol aircraft; they are basically attack aircraft with a generous allocation of sensors. Nearly all carrier aircraft are used for patrol. Only those with sufficiently effective sensors to enable them to cover large areas are included.

The *S-3A* (US) is a carrier-based naval patrol (antisubmarine) aircraft.

The *S-2E* (US) was the predecessor of the S-3 and is still used by many nations from land bases.

The *P-2H* (US) was the predecessor of the P-3 and is still widely used.

The *E-6* (US) is an electronic-warfare version of the A-6. Because of its extensive sensor equipment, it falls into the category of a patrol aircraft.

The *E-2* (US) is a long-range air-surface reconnaissance version of the S-2.

The *M-12* (Russia) is an amphibious naval patrol aircraft.

The *1150* (France) is similar to, but smaller than, the P-2.

The *Tu-142* (Russia) is a long-range bomber used as a naval patrol aircraft. The bomber version is called the Tu-95.

The *Tu-22M* (Russia) is a long-range bomber used as an attack aircraft as well as a naval patrol aircraft.

The *IL-38* (Russia) is similar to the P-2.

The *TR-1* (US) is the latest version of the venerable U-2. Although capable of staying in the air for twelve hours, the TR (tactical reconnaissance) 1 is now equipped to fly at 27,000 meters altitude adjacent to enemy territory in order to locate radars.

The *SR-71* (US) is the premier high-altitude strategic recon aircraft in the world. Cruising at 30,000 meters altitude at speeds of up to 900 meters a second, the SR-71 can fly anywhere with near impunity. Protected by speed, altitude and electronic devices, the SR-71 collects information with a wide range of sensors and delivers the data to the user in less than two hours.

The *E-3A* is a more powerful version of the E-2. Essentially it is a Boeing 707 crammed with electronic gear. It can track over 1000 aircraft at once and control over 100 friendly combat aircraft. It is also capable of tracking land vehicles and ships. Russia is still trying to develop an effective version. Their first attempt (the MOSS) was not a success. Their next version will use the IL-76, a much heavier aircraft.

Naval Helicopters in Service

This chart shows the distribution of the world's naval helicopters. Two facts are apparent immediately: Three navies possess two thirds of the world's naval helicopters and six types of helicopters account for over 70 percent of those in service. If the relative capabilities are taken into account, the three top naval powers own nearly 80 percent of the world's naval helicopter capability. The six

NAVAL HELICOPTERS IN SERVICE

Operating from land or small ships, these craft give even the poorest nation a degree of sea control.

	Total	% of Total	Sea King	CH-46	UH-1+	LAMPS II	Sea Stallion	Ka-25	Lynx	Wessex	Alouette III	Wasp	AH-1	Mi-4	Bell Jet-Ranger	Mi-8/14	H500D	LAMPS III	Super Frelon	Other
Total	2846	100	568	431	339	270	248	176	156	134	94	93	86	56	44	41	24	20	20	46
% of Total			20	15	12	9	9	6	5	5	3	3	3	2	2	1	1	1	1	2
US	1387	48.74	225	400	150	270	242						80					20		
Britain	332	11.67	83						57	130		62								
Russia	197	6.92						162								35				
Italy	121	4.25	42		79															
France	100	3.51							40		40								20	
Japan	89	3.13	78	11																
Brazil	53	1.86	5						9			8			18					13
China	50	1.76												50						
Spain	47	1.65	11		18								6				12			
Sweden	46	1.62	15		16			5							10					
India	40	1.41		20							20									
Netherlands	34	1.19							24			10								
West Germany	33	1.16	21						12											
Canada	32	1.12	32																	
Iran	32	1.12	20		6		6													

Country	Total	Ratio								
Peru	30	1.05		12			2		10	
Taiwan	22	0.77		10					6	
Turkey	19	0.67		19					12	
Australia	16	0.56	6	4		4			2	6
Chile	16	0.56				6			4	
Norway	16	0.56	10		6					
Denmark	16	0.56			8	8				
Argentina	9	0.32			0	6				
Poland	12	0.42	3							12
Indonesia	9	0.32				3				6
South Africa	11	0.39						11		
Greece	11	0.39		7		4				
Venezuela	10	0.35		10						
Syria	9	0.32			9	3				
Belgium	8	0.28	5							
Thailand	8	0.28		8						2
Bulgaria	8	0.28							6	
Pakistan	6	0.21	6							
Egypt	6	0.21							2	
Philippines	5	0.18	6							5
Mexico	2	0.07							2	
New Zealand	2	0.07							2	
Uruguay	2	0.07								2

most numerous types actually account for nearly 90 percent of total. Other nations usually have a naval helicopter capability that is more form than substance.

The first column of the chart gives the user nation. The second column gives the number of naval helicopters. The third column gives the percentage of all the world's naval helicopters controlled by that nation. The fourth and subsequent columns give the number of naval helicopters of a particular type controlled by that nation.

Across the top of the chart are the principal types listed according to their number. Beneath that is the percentage that type forms of all naval helicopters.

THE HELICOPTERS

Sea King is a U.S. design manufactured by many other Western nations (Italy, Britain, Japan, etc.). Primarily used for search.

CH-46 is a U.S. Marine Corps transport helicopter.

UH-1 + is the many variations of this U.S. design. Widely used for both search and transport.

Ka-25 is the standard Russian antisubmarine helicopter.

Lamps II and *III* are antisubmarine helicopters operating from ships. The III model is a naval version of the U.S. Army UH-64. The III model will gradually replace the II version by the end of the decade.

Sea Stallion is a U.S. Marine Corps transport helicopter.

Lynx is a British-French antisubmarine helicopter.

Wessex is an older antisubmarine type being replaced by Lynx and others.

Alouette III is primarily a search-antisubmarine type.

Wasp is an older antisubmarine type being replaced.

AH-1 is a gunship used by the U.S. Marine Corps.

Mi-4 is primarily used for search.

Bell Jet Ranger is primarily used for search.

Mi-8/14 is primarily a search helicopter.

H500D is primarily used for search.

Super Frelon is similar in design and capabilities to the Sea King.

HOW TO DETERMINE A NATION'S NAVAL HELICOPTER CAPABILITY

Refer to the helicopter characteristics chart for the capabilities of the most effective individual craft. Those nations possessing a larger number of the more capable helicopters obviously possess the greater capability.

Naval helicopters are used for search missions and for transport from sea to land. Except for the transport helicopters of the U.S. Marine Corps, most of the world's naval helicopters are used primarily for search. Often flying from ships at sea, the objects of these searches are either surface ships or submarines. In peacetime many search helicopters patrol and rescue.

When going after surface ships, the purpose is not just to find the enemy ship, but often to keep it in sight to help guide missiles from the launching ship. This type of mission allows ships to fire missiles accurately over the horizon. The helicopter remains out of antiaircraft range of the target ships. Against a navy with aircraft carriers, this tactic is more risky.

Against submarines, helicopters are extremely effective weapons. Antisubmarine helicopters are the most capable, expensive and numerous. The antisubmarine equipment is quite expensive, more than doubling the cost of the helicopter. It consists primarily of electronic sensors (sonar, MAD, sonobuoys and associated communications equipment). It is possible to convert most helicopters to antisubmarine models simply by adding sensors and lightweight torpedoes or depth charges. Most of these weapons weigh in at less than 600 pounds each. A helicopter does not possess great lifting power.

The adaptability of helicopters through the installation of specialized equipment provides a wide variation in the capabilities of those of the same type. This is particularly true of the UH-1 types. The " + " is added to their designation on the chart to emphasize this point. Many of these UH-1 helicopters are made under license in Italy. Most of the U.S. versions are transports.

Helicopter Characteristics

This chart covers over 95 percent of the types and over 98 percent of the numbers of helicopters used by the world's combat forces.

NATION is the country that originally manufactured the helicopter. Many are subsequently manufactured or assembled in other countries.

TYPE is the official designation of the helicopter. For Russian helicopters, the NATO code name is also given.

LAND FORCES

UH-1H is the latest version of the original U.S. tactical troop-transport helicopter. Over 3000 of the older models are still in use. They have lower lift capacity and reliability. Still in limited production.

OH-58 is the most current U.S. observation helicopter. Still in production.

Mi-8 was, until the arrival of the Mi-24, the standard combat helicopter in Russian service. Still the most widely used machine and still in production.

Mi-2 was originally a Russian design. It is now manufactured only in Poland and is the standard observation helicopter in Russian-equipped armies.

500D Defender is a further development of the OH-6. It is sold to civilian and non-U.S. military users. It is particularly effective with antitank missiles. Still in production.

AH-1S is an upgrade of the AH-1G, which, among other improvements, can use TOW missiles.

AH-1G is the original U.S. gunship. It is not equipped to fire antitank missiles. It is no longer in production and is gradually being phased out or rebuilt to AH-1S standards.

OH-6 is an older observation helicopter that is still in use. Gradually being phased out of service.

HELICOPTER CHARACTERISTICS

Land craft are used primarily for killing tanks and moving troops and equipment around the battlefield. Naval helicopters are larger, better armed and more heavily equipped than many fixed-wing aircraft.

LAND

Nation	Type	Number Deployed	Ground Attack	Anti-tank	Maneuverability	Fire Control	Protection	Combat Load	Weight	HP:Wt Ratio	Radius	Max Speed	Crew	Passengers
US	UH-1H	4500	1	0	9	1	4	1.5	2.4	359	160	200	1	14
US	OH-58 Kiowa	2200	2	0	8	3	3	.4	.7	382	160	222	2	2
Russia	Mi-8 Hip	1660	3	1	4	2	2	4	6.7	318	150	250	3	24
Poland	Mi-2	1000	2	1	7	1	3	.8	2.4	281	140	200	2	6
US	500D Defender	1000	5	6	8	4	4	.5	.6	382	170	280	1	6
US	AH-1S	900	8	7	9	7	7	1.6	2.9	400	180	225	2	0
US	AH-1G	800	7	3	9	7	7	1.5	2.7	333	200	275	2	0
US	OH-6 Loach	800	4	0	7	4	5	.3	.6	280	200	240	2	2
US	CH-47 Chinook	800	0	0	3	0	5	12.7	9.7	335	180	300	3	44
Russia	Mi-4 Hound	700	2	0	3	1	2	1.7	5.1	250	120	210	3	14
Russia	Mi-6 Hook	500	0	0	1	0	1	12	27.2	281	210	300	5	65
US	CH-46D	431	1	0	6	1	3	6.8	4.5	331	120	260	2	25
Russia	Mi-24 Hind-A	340	5	5	5	3	5	2.1	6	370	120	260	4	8
US	AS-64 (AAH)	200	10	10	10	10	10	1.2	4.7	519	200	300	2	0
US	UH-60A	200	2	0	7	1	7	3.6	4.8	371	200	290	3	11

Nation	Type													
Russia	Mi-24 Hind-D	310	6	5	4	3	5	1.3	7	361	120	260	2	8
WGer	BO-105	227	0	6	6	5	3	.7	1.2	442	210	270	1	4
US	CH-53D	200	1	0	3	1	3	3.6	13.4	462	130	300	3	38
UK	Lynx	181	4	7	8	6	6	.8	3.1	462	210	220	2	10
FR	SA-342 Gazelle	150	2	6	7	5	5	.4	.9	454	220	260	2	3
US	CH-53E	49	1	0	3	1	4	14.5	14.9	447	130	350	3	56
FR	SA-330 Puma	24	4	6	6	5	5	3.2	3.8	450	570	270	2	20
Averages		781	3	3	6	3	5	3	6	376	186	261	2	16

NAVAL

Nation	Type	Deployed	Surface Search	Sub Surface Search	Surface Attack	Sub Attack	Combat Load	Weight	HP:Wt Ratio	Range	Max Speed	Crew	Passengers
US	SH-3 Sea King	571	7	7	3	8	1.5	5.1	424	320	250	4	12
Russia	Ka-25	284	4	5	3	5	1.5	4.8	314	220	180	4	12
US	SH-2F LAMPS II	270	6	5	3	6	1.8	3.2	540	220	260	3	1
UK	Wessex	144	4	5	3	4	1.6	2.8	364	250	200	2	12
US	SH-60B LAMPS III	20	7	7	4	9	.2	6.2	412	200	240	3	11
UK	Alouette III	97	3	4	2	3	.8	1.1	458	200	220	1	6
UK	Wasp	96	2	4	2	4	.7	1.2	374	150	180	1	5
France	Super Frelon	20	7	7	5	8	5	6.8	394	270	275	2	27
	Averages	188	5	6	3	6	2	4	374	229	226	3	11

CH-47 is the primary U.S. Army heavy transport helicopter. Still in production.

Mi-4 is an older design now being phased out of service. Still used for transport. No longer in production.

Mi-6 is the standard Russian heavy lift helicopter. Still in production.

CH-46D is a heavy-duty U.S. Navy-Marine Corps machine similar to the U.S. Army CH-47. It is no longer in production.

Mi-24 A/D is the first Russian helicopter designed specifically for combat use. Like the U.S. UH-1 series, the Mi-24 appears in two versions: combat transport and gunship. Although a large and unwieldy machine, it is heavily armed. In production, about 200 a year.

AH-64 is the latest U.S. gunship. It incorporates all the Army's past experience in this area. Over 500 are on order.

UH-60A is the latest U.S. troop-transport helicopter. Small numbers will be equipped for electronic warfare. Several thousand will eventually be built.

BO-105 is similar in concept to the 500D, that is, a light observation helicopter upgraded to gunship status through the addition of modern, lightweight weapons (antitank missiles, primarily). The antitank version is called the PAH-1. In production.

CH-53D is the U.S. Marine Corps' primary amphibious landing helicopter. Operates from special helicopter carriers. No longer in production.

Lynx is a British multipurpose (land-naval) helicopter. In production.

SA-342 is a French multipurpose helicopter that is used for both land and naval operations. In production.

CH-53E is the latest version of the U.S. Marine Corps' heavy lift helicopter. The main improvement has been the installation of a more powerful engine unit. In production.

SA-330 Puma is a recent French design used for combat and tactical transport. In production.

NAVAL HELICOPTERS

SH-3 Sea King was first introduced in 1959 and is still produced by U.S. and other licensed manufacturers. Over 900 naval versions built.

Ka-25 is the standard Russian shipborne antisubmarine/search helicopter. Still in production.

SH-2F is the first of the LAMPS (light airborne multipurpose systems) helicopters designed to operate from surface ships. The F version is an upgrade of the earlier D model. No longer in production (upgrade work ends in 1982).

Wessex is no longer produced.

SH-60B is a Navy version of the U.S. Army's UH-60. It will be designated LAMPS III, for use on ships. Over 200 will be produced.

Alouette III is used for both land and naval operations. Still in production.

Wasp is an older helicopter that is being replaced by the Lynx. No longer in production.

Super Frelon is the standard French heavy transport naval helicopter. Also used for antisubmarine work.

NUMBER DEPLOYED is the number of helicopters of each type in use by combat forces worldwide. Numbers are not 100 percent accurate because of reporting errors, unexpected accidents or war losses, etc.

GROUND-FORCE HELICOPTERS—RELATIVE CAPABILITIES

GROUND ATTACK is the ability to deliver attacks accurately on ground targets. This is dependent more on the quality of fire-control equipment than on the amount of ordnance. Most helicopters can have rocket pods hung on them, or machine guns. The more specialized helicopters have heavier automatic weapons (20mm to 30mm cannon) built right into them. These often destroy most armored vehicles.

ANTITANK represents the heavier cannon and rockets and, most of all, guided missiles. Not counted here is the ability to drop antitank mines. Western helicopters in particular can do this. These small (3- to 4-pound) mines either blow off the tank's track or blast a hole in its underbelly with a shaped charge. The helicopters must drop them in front of the vehicle's path without coming close enough to be fired at.

MANEUVERABILITY is a combination of horsepower-to-weight ratio (see below) and the basic "controllability" of the helicopter. A high value indicates a helicopter that is able to travel rapidly while close to the ground. Pilot skill being equal, the helicopters with the more effective flying instruments (all-weather radar, etc.) and flight controls (including the shape, size and weight of the helicopter) will be much less vulnerable to detection and much more capable of effective attacks.

FIRE CONTROL. This is the effectiveness of fire-control equipment and has a direct bearing on combat capability.

PROTECTION is the helicopter's ability to avoid and survive being hit. This is a combination of small size, maneuverability, detection-prevention features and armor protection. The armor is not usually metal but a composite of lightweight plastic, ceramic and metal. The rotor blades are often made of such materials to survive fragment or shell hits. Detection prevention is achieved by small size and, more important, by design features that degrade infrared (heat) or radar detection, like special paints and engine-exhaust deflectors. Electronic counter-measures (radar warning devices and jammers) also increase protection.

NAVAL HELICOPTERS—RELATIVE CAPABILITIES

SURFACE SEARCH is the ability to detect objects on the water surface and is primarily a function of the helicopter's search radar. Without radar you can still use visual observation, but this is limited to clear weather daylight search.

SUB SEARCH is the ability to detect submerged submarines. The primary equipment for this is "dunking sonar"—sonobuoys and data links to the mother ship, where many of these data are processed with the more powerful ship computers. The ability to hover and lower a sonar sensor into the water allows the helicopter to detect a submarine without detection. This makes the submarine more vulnerable to a surprise attack by torpedoes or depth charges from the helicopter.

SURFACE ATTACK is the ability to attack surface targets. Most helicopters are used only for antisubmarine operations. Some are equipped with surface attack missiles. In addition, antisubmarine torpedoes can also be used against surface ships. The problem is in getting close enough without being shot down. Most helicopter-carried torpedoes have ranges not exceeding 5 km.

SUBMARINE ATTACK is the ability to attack submerged submarines. This is done with torpedoes and depth charges. Success is largely a function of weapons quality, detection ability and effectiveness of the target's deception equipment.

CAPABILITIES IN COMMON

COMBAT LOAD is given in metric tons. This is the maximum load carried of weapons, equipment, passengers, etc. For transport helicopters this often indicates a load suspended from a sling underneath the helicopter. Normally a higher load can be carried this way. To obtain longer range, less combat load is carried (one half to one third less).

WEIGHT is the empty weight of the helicopter. Before it can function you must add crew, fuel and combat load.

HP:WT is the horsepower-to-weight ratio, that is, the horsepower per ton of helicopter weight. The higher this ratio is, the more maneuverable the helicopter. This is particularly important for low-flying operations. The helicopter's weight is calculated by adding its empty weight to its combat load. The fuel and combat load varies during a mission. Fuel is burned, ammunition expended and other cargo unloaded. This computation of weight gives a useful average for the purpose of calculating the ratio.

RADIUS (RANGE) (km) represents the normal operating radius. The ferry range (going from one point to another without returning) is three times the radius. Normally one third of the helicopter's total range (radius time 3) is reserved for maneuvering, combat, getting lost, etc. In well-planned operations some of this can be regained, thus increasing the radius to 50 percent. More fuel may also be carried to increase radius at the expense of combat load. This is not done with most helicopters. This also requires additional weight of additional tanks and fueling hoses.

MAX SPEED is the maximum level speed and is given in kilometers per hour. A helicopter is not built for speed. Its normal, most fuel-efficient speed is 85 percent to 90 percent of maximum speed. Just hovering, not moving at all, uses fuel, although time in the air can be increased to 50 percent by doing mainly this. Most helicopter missions do not exceed three hours.

CREW is the normal number of people required to keep the helicopter operating. One is the pilot. If the helicopter is an attack aircraft, another crew member is required to operate the weapon systems. On more specialized helicopters (antisubmarine, for example) more crew members will be required to operate the additional sensors and weapons.

PASSENGERS is the maximum number of people that can be accommodated in the helicopter. When passengers are carried, they will count as part of the combat load (nine to twelve per ton, depending on how much equipment each man carries).

CAPABILITIES

The most widely used land-combat helicopter currently in service, the UH-1 series, is a design first introduced in the early 1960's. The most widely used naval helicopter, Sea King, first saw service in 1959. Most other helicopters in service are also approaching the twenty-year mark. Age is much less a factor in a helicopter's effectiveness than are the efficiency of its equipment and weapons. Basically a helicopter is nothing more than a platform that can hover, land and take off practically anywhere, and carry a load of men or equipment. Equipment is more a factor in its effectiveness than speed, range or carrying capacity.

Aircraft and other complex machines require such extensive maintenance that they become somewhat immortal. Useful lives of over twenty years are not uncommon. This tends to make the year of introduction somewhat irrelevant. The airframe is built to withstand high stress over long periods. As helicopters are less affected by drag because of their low speeds, their basic shape will be little changed by more efficient designs. Engines wear out and are replaced. More powerful engines of more recent design are often installed in older helicopters. This will increase range or combat load. Finally, and most important, the helicopter's weapons and equipment will show the most remarkable improvements over the years. Antitank guided missiles were only perfected in the late 1960's and were promptly installed on helicopters built ten or more years before. Fire control, sensors and electronic-warfare equipment have shown similar increases in quality. This improved equipment is routinely installed in 10- to 20-year-old helicopters, thus making them more effective than those that have been recently manufactured without such equipment.

Pilot skill is also a key element in the effectiveness of a helicopter, because it is generally more difficult to fly than fixed-wing aircraft. Modern tactics frequently call for flying it a few meters from the ground or tree tops. Given the basic instability of these machines and weather conditions close to the ground, this type of flying is even more difficult, fatiguing and dangerous.

12

THE NAVY:
ON THE GROUND

"POWER PROJECTION" is the ability to project military power thousands of miles in order to get others to bend to your political will. Naval forces, like air forces, cannot occupy the ground, unless the fleet is capable of putting ground forces ashore. Most major navies have marine infantry forces. These were originally organized as close-combat specialists when ships still rammed each other. During the last one hundred years, their function has changed to on-board military police and landing parties.

Being a maritime power, the United States has always had a marine corps. This corps was fortunate to be staffed by an unbroken succession of able, innovative and persevering officers. Before World War II, the USMC developed amphibious tactics, techniques and doctrine for an expected war in the Pacific against an increasingly militaristic Japan. The war came and the marines went. Amphibious warfare would never be the same.

In Europe the British Royal Marines and the British Navy also developed many specialized ships and techniques for amphibious warfare. The result was the ability to move hundreds of thousands of men, vehicles, aircraft and ships thousands of miles; land, opposed, on an enemy shore; and always win. Most of the amphibious and landing operations were undertaken by army troops.

After 1945 only the USMC maintained a significant amphibious force. Of the 400,000 officially designated marine troops in the world, nearly half of them are in the USMC. Moreover, only the USMC maintains its own air force in addition to the U.S. Navy's specialized landing ships and equipment.

Capabilities

Although all marine forces are amphibious, they fulfill three kinds of functions.

Raiding. In effect, commando troops for missions requiring above-average combat-skill and training levels. Although all marine troops are elite, when used as raiders they are even more so. At times these operations are not even amphibious. They are often conducted by parachuting aircraft or helicopters.

Spearheading amphibious operations for army troops. In this case the marines handle the difficult task of establishing the beachhead, normally against enemy opposition. Functioning much like combat engineers, they use a wide variety of specialized skills, so that the army troops can come ashore with a minimum of trouble. At that point the marines can be withdrawn. Often they have to stick around. The marines are also superb infantry.

Performing independent amphibious operations where conventional ground forces are either not available or are incapable of getting there quickly enough. Some likely areas for USMC operations are Europe, northern Norway, the Azores Islands and North Germany. Japan or Korea may be in great need of a marine division in time of war.

Tactics and Techniques of an Amphibious Landing

PREPARATIONS

The fleet approaches to 40 km off the enemy shore. Friendly naval forces have established control of the local sea areas, swept mines and cleared underwater beach obstacles. Naval aircraft—land-based if within 600 km of the beach—have established air superiority. Intensive air, electronic, naval, and, if possible, landing-patrol reconnaissance is imperative. Undetected water or beach defenses can easily wreck the entire operation.

All this takes place within five days of the actual landing. Even though detailed plans may be drawn up far in advance, it is necessary to double-check all information. Planning is a highly perishable item. It must be constantly refreshed to be effective.

Surprise, although not always complete, must be achieved as much as possible to prevent the enemy from reinforcing the landing area to an unbreachable level. The MAF (marine amphibious force) can steam 800 km per day. The longer it is at sea, the more likely it will be spotted by the enemy. This not only puts the force in danger of enemy attack, but also gets the enemy to thinking hard about where the landing will be. The most favored solution is to assemble the convoy as close as possible to the target area yet still out of range of enemy air power. One or two days' steaming would normally be suitable.

Before departing, there will be a frantic period of planning and organizing. Time will most likely be critical. Ships and the units they are to carry will arrive in a random sequence, but will have to be loaded properly for the landing. Once on the way the MAF will hold onboard exercises to shake down. If possible, a rehearsal landing will be held. Because a division rarely practices together, this rehearsal can prevent costly errors in the presence of the enemy. The two or three dozen ships of the carrier task force apply their firepower to the defender, paying particular attention to the transportation system and any reinforcing units moving up. By now, if the enemy intelligence system is functioning at all, the defender is aware of an amphibious task force in the area. If so, his reserve air, ground and naval forces have been alerted.

In theory, only a few hours after the MAF steams into their holding pattern 40 km off the enemy coast, the marines will be ashore.

The minesweepers have cleared a series of 1-km-wide lanes to the landing area. The landing area, 4 km wide or more and 4 km deep, is marked with two primary control ships, one usually containing the primary control officer and his staff. The 4-km depth of the landing area keeps the larger ships out of small-arms range but not artillery range. The larger ships carrying the landing craft take about an hour to enter the landing area plus another half-hour to launch their landing craft. These amphibious boats line up at the line of departure, 3 to 3.5 km from shore, from which the wave-guide commander, on the primary control ship, directs them to their landing beaches.

THE BEACH ASSAULT

Each regiment lands assault elements of two battalions on its own landing beach. Each battalion lands the assault elements of two companies plus some combat engineers. The success of the entire MAF of 54,000 men hinges on the ability of these two dozen LVTP-7's to get ashore in one piece.

This first wave takes about 25 minutes to hit the shore. At 2-minute intervals the next three waves arrive. The second wave contains the other rifle and weapons companies of each battalion. The third wave contains a tank company (in LCU's) and battalion support units (in LVT's and LCM's). The fourth wave contains more support units. In 8 minutes over 3000 men and over 150 armored vehicles and artillery pieces are ashore. In support, 4 km offshore, are 6 destroyers and 2 cruisers. Overhead, up to 400 navy and marine aircraft are on call.

THE HELICOPTER ASSAULT

Ten to 30 km inland are the marine battalions that were landed by helicopter before the landing force even hit the beach. The third marine regiment in the division goes in by helicopter. Reconnaissance aircraft constantly patrol the area, while helicopter gunships and attack aircraft clean out helicopter approach and retirement lanes. Sixty-five helicopters form the first wave: an infantry battalion and a battery of artillery. Once the landing zone is cleared of enemy resistance, the remainder of the regiment is brought in. Ideally you land where the enemy is not. Sometimes more than one landing zone is established.

CONSOLIDATION

Once the beaches have been cleared and linkup has been achieved with the helicopter-landed marines inland, combat proceeds until a tenable beachhead is taken, an area large enough to hold an airfield and supply dumps. An area 30 km deep and 50 km wide will usually do. At this point, two or three days after the initial landing, army troops should be coming ashore. The carrier task force will have already withdrawn. Within a week or two the marines also withdraw. Supply will continue to come over the beach until a port is taken or engineers build a temporary one. Landing craft and marine helicopters can bring supplies but only at great cost. Every time an amphibious ship runs up on a beach it suffers wear and tear. Helicopters require many man-hours of maintenance for each hour flown. Sustained operations, even without enemy fire, wear down the machines and their maintenance crews. Losses resulting from wear and tear are missed just as much as those caused by enemy fire.

HOW TO STOP AN AMPHIBIOUS LANDING

From the enemy's point of view, the situation is grim. Within the space of a few hours, over 20,000 combative marines have been landed and are carving out a beachhead 30 km deep and 50 km wide. Normally an amphibious landing will be made in the vicinity of an enemy port and/or airfield. A landing is made only because a port isn't open. Once a port is captured and rapidly repaired, whole armies can be brought ashore quickly.

The closer the attacker lands to the port, the stronger the initial resistance is likely to be. But the farther away from the port, the longer the march and the greater the opportunity for the enemy to organize an effective resistance. Usually the number of available landing areas is limited. This makes the enemy defense easier. The defender can gamble on which beach the attacker will choose. The attacker depends on his reconnaissance to be where the defender is not.

The enemy, given sufficient resources and time, can prepare numerous passive defenses: naval and land minefields, beach obstructions, roadblocks, etc. Active measures include naval and air forces. The marines must neutralize, if not destroy, these defenses before the MAF can land. If not completely destroyed—and this may not be possible—at least in time to achieve surprise, enemy air and naval forces can still make desperate attacks. Preparing for them takes air power away from ground support.

The Russians, likely defenders against amphibious operations, plan to use chemical weapons delivered by unstoppable ballistic missiles. Chemical weapons can be dealt with. Nuclear weapons are another matter. Current amphibious doctrine spreads ships out to mitigate the effects of nuclear weapons. But enough nukes will cripple the MAF. The marines can also use nuclear weapons. See the chapter on chemical and nuclear weapons.

DIFFERENT APPROACHES

The United States' amphibious doctrine is most often quoted because the USMC is its foremost practitioner. Many Western marine forces follow USMC practice out of convenience if not always conviction. Many Western marine forces were trained by

USMC advisors and are organized along USMC lines. The only major amphibious forces other than the USMC and its allies are the Russian and Chinese.

Only in the last twenty years has Russia developed an amphibious capability. It is built around five amphibious brigades. These brigades spearhead larger ground forces of conventional infantry units landed from LST's and conventional transports. Russian marines are currently like combat engineers; they assist in breaching this unusually difficult enemy defense. The use of follow-up waves of less well-trained army troops in amphibious warfare is risky. Past experience actually shows it to be very risky. The Russians make greater use of land-based aviation. Potential areas of activity are Norway, West Germany, Turkey and the Sea of Japan. All are within reach of Russian land-based aviation. The Russian amphibious forces are much less capable than an MAF of breaching shore defenses or expanding a beachhead against significant resistance. But the mere presence of such a capability means that it must be defended against.

Chinese marines are actually army troops trained for landing on Taiwan. A collection of aging ex-LST's and some more recent small landing craft give the Chinese a slight capability. Most are being used for commercial purposes much of the time. The Chinese marines are numerous but less well trained than Russian or U.S. marines. All depends on their getting across the water without interference from enemy air or naval forces.

MURPHY'S LAW

More so than any other type of operation, amphibious operations are very susceptible to things going wrong, with the worst possible results. Speed is usually imperative both in planning and execution. It is likely that marines will have to undertake operations with fewer amphibious ships than they would like, due to ships being lost or not available in time. Thus, a marine's greatest asset is an ability to improvise effectively. It is not enough to possess uncommon courage in the face of the enemy. Equally important is the ability to make do in the face of your own side's shortcomings.

Things will go wrong, often very wrong. The USMC is accustomed to taking heavy losses, much heavier than those of normal combat, in order to pay for these mistakes. It's easier to avoid, or improvise around, the mistakes in the first place.

THE WORLD'S AMPHIBIOUS FORCES

The amphibious force of any significance belongs to the United States. Everyone else's is a distant second.

Rank	Nation	Amphib ious Ability	Troops	Units Available	Dsplcmt Tons Amphib. Ships	Troop Lift	Qual ity Rating
1	USA	1000	189000	3 Divisions, 3 Air Wng	942000	85636	106.2
2	Russia	137	22000	5 Brigades	196000	17818	70
3	Taiwan	92	38000	2 Divisions	122000	11091	75
4	China	63	38000	3 Divisions	180000	16364	35
5	Great Britain	58	7500	6 Commandoes, 4 Rgmnt	61000	5545	95
6	France	43	4500	1 Brigade	51000	4636	85
7	South Korea	39	20000	1 Division, 2 Brig	49000	4455	80
8	East Germany	34	2100	1 Regiment	48000	4364	70
9	Philippines	32	7000	6 Battalions	91000	8273	35
10	Greece	32	2000	1 Brigade	59000	5364	54
11	Turkey	22	2400	1 Brigade	44000	4000	50
12	Spain	21	11000	4 Brigades, 2 Btlns	32000	2909	65
13	Poland	15	7000	1 Division, 1 Brig	23000	2091	65
14	Italy	14	1700	2 Battalions	21000	1909	65
15	Thailand	10	8000	1 Brigade	25000	2273	40
16	Indonesia	7	5000	1 Brigade	37000	3364	20
17	Argentina	7	10000	12 Battalions, 6 Comp	17000	1545	40
18	Yugoslavia	6	15000	1 Brigade	11000	1000	55
19	Venezuela	6	4000	3 Battalions	14000	1273	40
20	Brazil	5	12000	1 Division	13000	1182	40
21	Chile	5	5000	1 Brigade	13000	1182	40
22	Iran	2	1800	3 Battalions	6000	520	30
23	Netherlands	2	2900	2 Battalions, 1 Comp	3000	273	65
24	Portugal	0	3200	1 Regiment	1000	91	45
	Totals	1652	419100		2059 (x1000)	187182	

The World's Amphibious Forces

The twenty-four nations covered represent over 98 percent of the world's ability to launch amphibious operations. Keep in mind that just because a force has the capability to perform certain types of operations, unless trained to do so, it won't. If it attempts to do so, it will do badly. Many nations have a small amphibious capability that is used primarily for local transport or police functions. These forces do not amount to much militarily and are not included in the chart. The U.S. Marine Corps is unique in that its mandate is to launch amphibious operations anywhere in the world, at any time, under any conditions. As a component of the U.S. Navy, the USMC is technically available primarily to establish naval bases. As a practical matter, it is ready to go anywhere and do just about anything. All other marine forces have far more limited missions. At best, these other forces can conduct raids or short-range landings against light opposition.

AMPHIBIOUS ABILITY. This rating represents the overall quantity and quality of the nation's amphibious forces. It was calculated by multiplying the DIS-PLACEMENT TONS AMPHIBIOUS SHIPS by its QUALITY RATING.

TROOPS represents the number of men organized for amphibious operations. Not all of these belong to the navy or even to an organization called "marines." The primary criterion is training and ability to engage in amphibious operations.

UNITS AVAILABLE gives the designations of the marine units referred to under TROOPS. Div = Division; Comdo = Commando (a force with a strength of 100 to 400 or more organized and trained especially for raids); Btln = Battalion.

DISPLACEMENT TONS AMPHIB. SHIPS gives the displacement tonnage at full load of that nation's seagoing amphibious ships. See amphibious shipping chart for more detail.

TROOP LIFT is the number of troops that the nation's amphibious ships can lift for an amphibious operation. This was calculated by dividing the total amphibious tonnage by the average number of tons normally needed to carry out an operation. You can get by with less by using a greater percentage of conventional ships. For example, passenger vessels can be used if you are willing to take the additional time to transfer the troops to the landing craft. You will also have fewer landing craft, as these are normally carried on, and are considered a part of, the amphibious ships. Overcrowding can also increase capacity, but this only works for short voyages of less than a day. Otherwise your troops are not going to be in very good shape to fight. Without overcrowding or using nonamphibious ships, you can increase your lift capacity by simply shuttling back and forth to pick up more amphibious troops, which can be loaded in a few hours. This depends on the closeness of your base to the landing operation.

Keep in mind that the amphibious ships are also responsible for supplying the amphibious forces—up to a hundred or more pounds per man per day. (See the chapter on logistics.) The first day you will be lucky to get one day's supply ashore because of the need to get the force itself ashore. Fortunately, the troops carry with them one to three days' supplies. After that you can build up two to five days or more of supplies per day. You may need to build a reserve supply for up to thirty days, depending on how many additional combat and combat support troops you bring ashore and the loss of landing craft to enemy action or wear and tear.

QUALITY RATING. This quality rating reflects the size and quality of that nation's marine forces as well as its overall ability to conduct amphibious warfare. There is no formula for this qualitative rating. An examination of each nation's abilities and resources should not vary from my rating. Each nation's abilities and past performance speak for themselves.

Amphibious Shipping

This chart shows the capabilities of amphibious ships used by the United States and Russia (USSR), the two major amphibious powers. Most other nations use amphibious ships manufactured by the United States or Russia. Other nations build their own, either to the same designs or very similar ones. A further source

AMPHIBIOUS SHIPPING

Amphibious ships are small, numerous and essential for any landings.

Type	Class Name	Number in Class	Last One Built	Displace- Ment (tons)	Length (meters)
LCC	Blue Ridge	2	1970	19000	189
LHA	Tarawa	5	1980	29300	250
LPH	Iwo Jima	7	1970	18000	180
LPD	Austin	12	1971	17000	173
LPD	Raleigh	2	1964	13900	159
LSD	Anchorage	5	1972	13700	163
LSD	Thomaston	8	1957	11300	155
LST	Newport	20	1972	8300	159
LST	Suffolk	3	1959	8000	135
LCU	Landing Craft	60	1976	390	41
LCU	Landing Craft	31	1957	347	36
LCU	Landing Craft	23	1945	310	37
Landing Craft (carried on landing ships, above)					
LCM(8)	Mod 1			107	22
LCM(8)	Mod 2			130	22
LCM(6)	Mod 2			62	17
LVT				26	
LVTP-7	APC	940	Building	13	11
Helo	CH-46F	162	1966	10	
	CH-53D	126	1979	19	
	UH-1N	96	1965	4	
Russia					
LPD	Ivan Rogov	1	1982	13000	158
LST	Ropucha	13	1978	4400	110
LST	Polnocny	76	Building	1100	77
	MP-4	12	1959	780	56
	Vydra	28	1972	600	55
ACV	Aist	10	Building	220	46
Landing Craft (carried on landing ships, above)					
ACV	Gus	40	Building	27	21

of amphibious shipping is specialized peacetime shipping. Examples are the air-cushion vehicles used on the English Channel, ferries, and RO-RO ships (see the chapter on naval shipping).

THE TYPES

LCC. Command and control ship with extensive communications and other command facilities to enable it to handle all the control functions for a major amphibious operation.

LHA. Helicopter assault ships. Internal loading dock for landing craft. Best all-around amphibious assault ships in existence.

LPH. Helicopter assault ships without loading dock for landing craft. Extensive medical facilities. Because it lacks loading dock and storage space, it cannot carry heavy equipment (APC's, heavy trucks, etc.). Excellent for independent raids.

Crew	Passengers	Cargo	Landing Craft & Aircraft Carried	AA Defense
720	700		5 LCVP	2 G, 2 M
730	2000		4 LCU, 2 LCM, 30 Helo	9 G, 2 M
530	2100		24 Helo	4 G, 2 M
490	930	3900	6 Helo	4 G
490	930	2000	6 Helo	8 G
400	380		20 LCVP, 4 Helo	6 G
400	350		3 LCU, 2 Helo	6 G
225	390	500	Helo Platform	4 G
185	575		2 LCM	6 G
12	400	180	or 4 Tanks, or cargo only	1 G
12	400	180	or 4 Tanks, or cargo only	2 G
12	80	140	or 3 Tanks, or cargo only	2 G
4	80	65	Cargo or passengers	
4	80	65	Cargo or passengers	
4	80	34	Cargo or passengers	
2		5		
3	25	5	Cargo or passengers	
2	20	2.9	Cargo or passengers	
3	38	3.6	Cargo or passengers	
2	7	1	Cargo or passengers	
400	1000		40 tanks 3 ACV	22 G, 2 M
80		600	2 tanks	4 G
40		350	or 8 tanks	6 G
50		550		4 G
40		550		
12		200	or 2 tanks	4 G
6	60	30	Passengers or cargo	

LPD, LSD. Landing ships with internal loading docks for landing craft. LPD also have a helicopter platform for launching air assaults. LPD does not carry its own helicopters.

LST. Landing ship, tank. Originally designed during World War II, it lands tanks and other vehicles directly onto beach. Prone to damage as a result. Some have helicopter platforms or landing craft for offshore delivery of cargo.

LCU. Landing craft, utility. Carried on larger amphibious ships, although large enough for limited seagoing movement. Many smaller nations use LCU's by themselves. Runs up on beach to discharge vehicles or cargo.

LCM. Landing craft, mechanized. This is the standard landing craft for carrying men and cargo.

LVT. Landing vehicle, tracked. This is a truly amphibious landing craft that is tracked for moving off the beach. Usually for carrying cargo.

LVTP-7. This is actually an amphibious personnel carrier (APC) for carrying combat troops from the ships right into combat. See the chart in Chapter 3 for more detail.

ACV. Air-cushion vehicle. Floats on cushion of air and can travel over water or land at speeds of up to 70 kilometers per hour.

HELO. Helicopters. Various types carried on amphibious ships. Used for carrying cargo, passengers and light vehicles. Their weight is full-load weight.

CLASS NAME is the name given to the lead ship of a series. It is a convenient way to identify a ship type.

NUMBER IN CLASS is the number in use as of 1982.

LAST ONE BUILT is the year in which the last ship of that class was put in service. The term "building" indicates that, as of 1982, ships of this class were still being built.

DISPLACEMENT TONS is the weight the ship displaces in the water under full load.

LENGTH in meters gives you an idea of the ship's relative size. Amphibious ships tend to be broad. Those that run up on a beach (LST's, landing craft) have a shallow draft.

CREW is the number of personnel needed to run the ship in wartime. In peacetime some navies run these ships with 20 percent or more fewer men.

PASSENGERS are usually troops. This category includes provision for sleeping, eating and little else.

CARGO carried, in metric tons.

LANDING CRAFT & AIRCRAFT CARRIED are for getting the passengers and/or cargo to the land.

AA is antiaircraft defenses—G = gun, M = missile launcher. For the most part, amphibious ships depend on the protection of other combat ships.

THE ORGANIZATION OF A MARINE AMPHIBIOUS FORCE

The 60- to 65-ship MAF (marine amphibious force) consists of a marine division and a marine air wing (MAW).

One MAW has 110 aircraft, including 36 F-4N fighters; 16 A-4N's; 20 AV-8A's; 12 A-6E's attack aircraft; 10 OV-10's; 8 RF-4B's; 8 EA-6A's for reconnaissance and electronic warfare; 141 helicopters, including 54 CH-46D's; 42 CH-43E's; 6 CH-53D's; 21 UH-1N's for transport; 18 AH-1J's for attack; 1 SAM battalion; 17,000 men. The MAW may operate from carriers initially, but will shift to land operation as soon as a field is captured or prepared.

The 18,000-man marine division consists of 9 infantry battalions, each with 1041 men, 32 ATGM's, 8 81mm mortars, 30 machine guns; 1 tank battalion (70 tanks, 72 ATGM's); 4 artillery battalions (72 155mm towed howitzers); 1 amphibious tractor battalion (208 LVTP-7's, each carrying 25 marines or 5 tons of supplies). Plus the usual support units.

The MAW requires 28 ships (480,000 displacement tons). The marine division uses 13 ships for each of its two landing regiments (160,000 displacement tons, 3 LPD's, 5 LSD's, 5 LST's) and 7 ships for the helicopter assault regiment (172,000 displacement tons, 2 LHA's, 5 LPH's). Another 15 ships (250,000 displacement tons) are used for the support units.

All these ships are "combat loaded." That is, men and equipment (especially equipment) are spread among many ships so that the loss of one craft will not result in the loss of a specific item, like the entire engineer battalion, all the tanks, all the artillery, or all the land mines. All equipment goes into the ships so that it may be quickly unloaded in the order needed.

Total: 54,000 men, up to 330 armored vehicles, 251 aircraft, 1.22 million tons of shipping (60 to 65 ships).

Part 4
THE HUMAN FACTORS

13

GETTING PSYCHED: WHY SOLDIERS FIGHT

FIGHTING ABILITIES VARY starkly among armies. Even more remarkable is the fact that anyone fights at all. Both phenomena have much to do with the illusions men are capable of creating, and then believing.

It Won't Happen to Me

In the first place, most men do not enter combat thinking they will be killed or injured. In warfare during this century, the odds of serving in the infantry during combat and being uninjured have been less than one in three. If potential recruits knew their chances, it would be much more difficult to get anyone into the infantry.

Indeed, given a choice, many would volunteer for any other branch of the armed forces to avoid the infantry. Most other branches are no more dangerous than civilian life. Even the armor and artillery branches offer a better-than-even chance of seeing a war's end uninjured.

Indoctrination

Those selected for the infantry are usually subjected to an age-old indoctrination stressing the following points.

Pride. The recruit is told that the infantry is the premier branch of the armed forces, the most noble calling, and the most respected and patriotic service one can render one's country. This is true, if getting killed or injured for one's fellow citizens is recognized as the highest form of patriotism. The pride taken in the dangerous business of infantry fighting is reinforced by the respect given to combat veterans. Like those of many other bad experiences, the memories lose their hard edges over time. Hearing the veterans' stories, the potential recruits tend to fixate on the glory instead of on the more common terrors. This is human nature, and it is drawn on generously to get the troops into the fighting without losing them to panic. A recent example was the comments heard from combat troops on a possible invasion of Iran in 1979–1980. The young troops were quite eager. The older, combat-experienced NCO's and officers were also ready to go if less enthusiastic. No one who's been shot at retains the enthusiasm of the uninitiated.

Effective preparation. The soldier is told he will be given the best equipment, training and leadership to enable him to carry out his duties as effectively and safely as possible. This is only rarely true. Infantry recruits tend to be young and often below average in education and general intelligence. If their armed forces have a winning tradition, they tend to believe that their training is adequate. This is particularly true during peacetime, when there is not enough embarrassing contrary evidence, dead bodies and maimed veterans, to contradict the official line.

A nation without a military tradition, or one most noted for defeat, will have problems from the start. There will be a feeling of inferiority among the troops. This is not an auspicious way to enter into combat. On the positive side, their reluctance is more realistic than optimism, which can lead to rashly aggressive actions in combat. The opening stages of World War I were infamous for this. Hundreds of thousands of troops were needlessly killed charging into machine-gun and artillery fire. Raw, unthinking courage is no match for firepower.

Friends. A vastly underestimated influence in combat performance is the "primary group." This is nothing more than the smallest unit of soldiers, five to forty men, organized for mutual support. Not all armies see to it that effective primary groups are formed. The primary group must be well trained and led and, most important, must know and trust one another personally and professionally. The troops must believe in their own skills and the abilities of their leaders. The members of the group must serve together for at least a few months before entering combat. New members must then not be introduced until the unit is taken out of combat. The members cannot get to know strangers while being shot at, especially if they are supposed to depend on them in life-and-death situations.

The transformation from green troops to battle-hardened ones is nothing more than the creation of these groups. Just getting men successfully in and out of combat does not form primary groups; leaders make it happen. The German Army, although it lost World Wars I and II, was quite efficient at the unit level. The Germans stressed rigorous selection of leaders. In the Germany Army, NCO's were given longer and more thorough training—six months—than were many junior officers in the U.S. Army—the "ninety-day wonders." Throughout World War II, when a German and a U.S. unit fought, the Germans, man for man, inflicted a larger number of casualties. Even when they were the attacker or were vastly outnumbered, the Germans gave better than they got. Studies after the war demonstrated that this was largely due to their more efficient formation of primary groups. During World War II, some U.S. division commanders set up their own training camps for new and experienced troops. Such units, particularly the 79th Infantry Division, became noticeably more effective for it.

Fight or else. The soldier is told that if he doesn't fight, he will face severe penalties. The closer the fighting, the more this technique is applied. It comes in various flavors. Western nations apply much social pressure to get out there and be shot at. This is a powerful motivator in all armies. Many armies, the Russian in particular, use even more powerful motivators, the most famous being officers shooting reluctant warriors on the spot. During the Russian invasion of Czechoslovakia, over a hundred Russian soldiers were reported shot by their officers for disciplinary reasons. Other measures include posting a line of military police behind the advancing units to arrest any reluctant troops found moving in the wrong direction. This approach was used by the U.S. Army in Korea. A U.S. innovation in Vietnam was landing troops by helicopter inside enemy-held territory. They could fight their way out or be killed by the enemy. This unofficial policy was very effective. The troops, however, were not fooled; some would refuse to board the helicopters.

What Works

Superior motivation, leadership and training have consistently proved the formula that produces victorious armies. Leaders who are willing to get out front and get shot at, and killed, are the best. Training that draws from experience, not pet theories, produces the most competent troops. Equipment that works most of the time, and does what needs to be done, is the most effective. Men will start fighting for any number of reasons, but will continue fighting, and do so successfully, only if they have confidence in their leaders, equipment, training and themselves.

CONVICTION

Conviction comes when the soldier believes he should be fighting. Too often, men are simply put into uniform and called a military force. Conviction is a powerful motive and includes what is usually termed "morale." As Napoleon said, "The moral is to the physical as three is to one."

There are three motivators: loyalty, personal gain, and desire for adventure, a common defect among the young. The most common form of loyalty is patriotism. Patriotism can be in the name of a nation, region, ethnic group, family, organization or group of friends. Often, patriotism is to several of these groups. Patriotism tends to be a group endeavor. If enough individuals are so motivated, they will inspire each other as well as the less patriotic members of the group. As wars throughout history have shown, patriotism propels people into situations of almost certain death or injury. The opening stages of a war between patriotic groups are always bloodier than the later stages.

As the fighting grinds on, the convictions begin to weaken. Taken away from the good things one is fighting for, the soldiers justify the fight by the prospect of victory and the end of combat. Eventually the losing side senses defeat, and the less determined individuals shrink from danger. This defeatism spreads until the losing side's armed forces fall apart. For this reason, most wars or battles are not fought to the death.

Some groups have such a high degree of conviction that they will continue until all are dead or incapacitated. Recently in Afghanistan, multiple loyalties—to tribe, ethnic group and nation, in that order—coupled with a tradition of warlike behavior produced a frighteningly high conviction. This conviction is the main reason the Afghans have never been subdued. Being the poorest nation in Asia has not substantially changed this intense conviction, nor has invasion by one of the most powerful armies in the world.

Other motivators, such as personal gain and sense of adventure, are much less common. A sense of adventure combined with patriotism often does get large numbers of young men onto the battlefield. At that point the adventure quickly dissipates for most. Because the bloody horrors of warfare are not likely to encourage anyone to go to war, adventure and duty are stressed during recruitment and training. Once under fire, most soldiers fight well enough because it seems a reasonable thing to do in order to survive.

Mercenaries are adventurous people who can be tempted into dangerous situations by the prospect of some monetary gain. For a true mercenary, money is not the overwhelming factor. Adventure and political convictions also play a large part. Much of the French Foreign Legion consists of true mercenaries.

Some poverty-stricken people fight for economic reasons. Current examples can be found in many parts of the world, particularly Asia and Africa. The Persian Gulf nation of Oman is defended by a force composed mostly of Baluchi tribesmen with a long tradition of such employment. Gurkha mercenaries still serve in the British Army.

MAGIC BULLETS

In predicting the superiority of troops, weapon effectiveness is much overrated. Superior weapons in the hands of superior troops will generally produce victory, but the same weapons in the hands of poorly led, unmotivated, incompetent troops will just as surely produce defeat.

Weapon performance is more capable of measurement in peacetime than motivation, leadership and competence. Therefore, troops throughout history have tended to rely more on technical superiority than the more slippery factors. The Russian Army in 1941 was one of the most lavishly equipped in history. Poorly led and trained, the Russians melted before the onslaught of the Germans, who were not only outnumbered but also equipped with inferior weapons.

With more wealth and technology available today than ever before, there is a tendency to rely even more on technology. This preference for hardware over human values results in magnificently equipped armies manned by the incompetent and led by the mesmerized. There are numerous examples of superior weapons in the hands of untrained or poorly led troops. When Iran and Iraq went to war in 1980, they used 1970's weapons with 1917 tactics. The highly sophisticated aircraft were unable to use their expensive sensors to find each other. Visual contact and primitive dogfight combat ensued. The numerous expensive missile systems proved much less dangerous than the traditional sixty-year-old cannon and machine guns. On the ground the armies lined up like the infantry masses of 1917 and ineffectually hurled artillery fire at each other. When combat did occur, it was of the old line-up-and-charge variety.

Israel has shown what superior leadership and training can do. During the 1967 war, they had generally less modern and sophisticated weapons than their opponents. That war ended in six days with an Israeli victory. And the Israelis were doing most of the attacking.

Many Serve, But Few Fight

It is important to remember that in modern warfare, the vast majority of troops do not engage in combat. In the U.S. Army in World War II, no more than 25 percent of those who served ever came under enemy fire. The infantrymen, the real fighters, comprise less than 10 percent of armed-forces strength. In the air force and navy, only combat aircraft pilots expose themselves to anything near the level of suffering and loss of the infantry. In the Russian Army the infantry comprises no more than 20 percent of total strength. Generally, for every infantryman there is one other combat soldier exposed to combat, although at a lesser degree of suffering than the infantry. Modern warfare's greater quantities of firepower may change this. However, if the past is any guide, the infantry will continue to gather unto itself the lion's share of combat misery and suffering.

14

LEADERSHIP

MANY OF THE most incompetent wartime leaders have been highly regarded peacetime military leaders. This has been common throughout history, and there is a reason. You can prepare for war, but you can't actually practice the real thing. No one has ever developed a satisfactory, mutually agreed upon method for determining an army's wartime potential short of an actual war. Therefore, peacetime military leaders spend most of their efforts convincing everyone that their services are adequate for wartime needs. Peacetime military commanders expend a lot of energy convincing themselves that what they are doing is the proper preparation for war.

Hardware is often viewed in a different light than the software of training, experience and general effectiveness of the troops themselves. Hardware you can see and feel. The troops? The goal is often to have the troops smartly turned out. Never mind that the most effective ones often look like a bunch of bandits. Perfectly aligned and garbed formations of soldiers are easier to perceive than their ability efficiently to inflict mayhem upon the enemy.

Nations with a long and systematically preserved military tradition develop better wartime leaders than those without it. Military tradition means a history of a war every generation so that the hard lessons haven't been forgotten. A long tradition means warfare going back a few generations, so that the right ideas become accepted as eternal truth and not as just passing solutions to unique situations. Systematically preserved means armed forces that maintain their traditions and personnel from war to war so that the lessons can be passed on.

The Germans have a military tradition, even though they haven't won a war in over a hundred years. So have Britain, France, Russia and some other nations. The United States has this status with its navy, to a lesser extent with its air force and to an even lesser extent with its army. This is a common pattern. In peacetime, air forces and navies do much the same as they would do in wartime. Just moving all their machines around comes very close to wartime operating conditions. Armies attempt to do the same thing, but it is just too expensive to move large masses of army troops around in peacetime.

Human nature being what it is, most leaders, military or otherwise, seek the easy way out. Unless feedback corrects ineffective actions, the wrong procedures become standard. On ships and in aircraft, mistakes are painfully apparent. Pilots and sailors tend to get buried with their mistakes. Therefore, air force and navy leaders are forced to get to know their subordinates' strengths and weaknesses. Army leaders must rock the boat to gain the same knowledge. Making a commotion is dangerous in any large organization. Most military leaders are cowards when faced with peacetime administrative danger, while the same men would be fearless in the face of wartime danger. It's a case of the pen, indeed, being mightier than the sword.

Even a leader who either knows from experience or intuitively feels that a subordinate is ineffective, cannot usually take any action. On the face of it, an ineffectual leader would be arbitrarily dismissed. Unless the commander has effective judgment and is backed up by the military tradition, he will not be able to remove ineffectual subordinates, particularly those who appear adequate under peacetime conditions.

Even in the most experienced armies a large proportion, sometimes even 50 percent or more, of leaders will prove incompetent in wartime. Many will get themselves killed or captured. How quickly these inadequate leaders can be replaced once the shooting starts is a key ingredient for success.

Meaningful reform of the military in peacetime is no easy task. It can be carried out only by a truly exceptional leader or after a particularly traumatic national defeat. It hardly ever occurs in a nation that has won its previous wars. A rare example of just such an effective reform is the current Russian Navy under Admiral Gorshkov. But, then, one can make a case for the Russian Navy having "lost" during World War II.

Scientific Leadership

Before the revolution the Russian armies were not much worse than any other European nation's. The most significant defect was that the highest commands in the Russian armed forces were given out on the basis of nobility and not merit. Fairly well-led troops were often squandered.

After the revolution three major changes were made. First, commands were given out on the basis of merit. Second, a scientific approach was embraced for all training and command. Third, the most technically advanced weapons and techniques were sought.

All of this affected leadership profoundly. Initiative, resourcefulness and imagination in combat were to be replaced with scientific planning and precision. During World War II, it was modified by a very large dose of combat experience. But combat experience is highly perishable, and the doctrine of scientific leadership, not so perishable.

Today the Russian Army is much like its pre-World War II predecessor. It looks splendidly equipped and scientifically led, but neither the splendid equipment nor the scientific doctrine and leadership have been tested. Whenever their armed forces have been tested, as during the Afghanistan invasion or the Arab-Israeli wars, the Soviet approach has been found wanting. Even the noncombat experience of Czechoslovakia in 1968 exposed glaring leadership and control problems.

The Russians have a strength in that if their training works, their troops will perform their duties according to set procedures. This will give the commanders a degree of control and knowledge of who is doing what, even if what the troops are doing is ineffective.

Western armies, on the other hand, often place too much responsibility on the individual leaders and much less on the system. This is fine if the leaders are up to the demands. But the only substitute for combat experience is some form of system. The Germans were good at developing both system and leadership. The German Army entered World War II with a tactical doctrine based upon a careful analysis of Germany's World War I experience. This doctrine was put into a manual that was regarded as a bible by combat officers. The procedures found in "Tante Friede" ("Aunt Friede," the nickname for the manual) got most officers through enough combat to give them experience. At that point they could use their resourcefulness and imagination to stay one jump ahead of the enemy.

Operating Within the Cycle

Combat is generally a series of intentional or unintentional ambushes. Self-preservation and the obvious appeal of hitting the other fellow when he can't fight back produce this syndrome. Successful armies use the ambush concept in all phases of operations, even in the attack. To do this you must practice another skill: operating within the other fellow's cycle.

We all have limits. In sports, athletes outperform their less skillful opponents by preempting the latter's actions. A superior fighter, for example, doesn't just parry his opponent's punch, he gets one of his own in. A superior fighter operates within his opponent's ability to respond. Skillful soldiers operate the same way. Even if a defender seems to have every advantage, a more skillful attacker can uncover his weaknesses. An attacker's superior ability to use weapons, artillery and cover negate the defender's advantages and allow the attacker to succeed with minimal losses.

Leadership makes this work. The old saying "there are no bad troops, only bad officers" has been proved too often in history to be ignored.

Push-Button Leadership

Less than a hundred years ago, effective leadership was conducted without radio, telephone and other electronic tools. Today effective operations are impossible without electronic aids. It is possible vastly to increase a leader's effectiveness through electronics, not just in terms of simple communications, but also by processing and analyzing large masses of information. This area of leadership is called C^3I (command, communications, control, intelligence).

At higher levels, where much of the information is compiled, the electronically assisted leader has a clearer picture of the enemy, not to mention his own forces. A leader with less natural talent for accurately perceiving a murky situation, guessing accurately and sorting out conflicting information manually will find himself better equipped with a host of functioning electronic aids.

If the electronic tools are not working, the results can be devastating. This is the origin of electronic warfare. Electronic warfare (EW) is directed primarily against communications, but much computational equipment must receive and send data. Exclusive reliance on electronic tools, without adequate backup systems and procedures, can prove fatal in wartime.

Another characteristic of superior leadership is the appreciation of this problem and the implementation of safeguards. Effective safeguards can be nothing more than exercising your forces under accurately simulated EW conditions. The forces had better be prepared to operate without many of their electronic leadership aids.

Creating Leaders

There are a number of basic approaches to selecting and training military leaders.

The Gentleman Officer is the oldest approach. Get an overeducated fellow with a lust for blood and adventure, and give him a pretty uniform and some authority. The United States inherited this terrible system from the British, "Burgoynes revenge," so to speak. It should be noted that the British Army's long combat experience produced habits that negated many of the bad effects of this system. For example, British NCO's tend to be superb. The officers are expected, if nothing else, to provide a good example for the troops. If this means nothing more than standing up and leading the attack, and being quickly killed, the officer has done his duty.

The Aspirant is a more effective system, usually employed during all major wars. It is a "trial-by-experience" method of selection. Germany and Israel, an odd pair, have both used this method. Potential officers are selected from the recruits and systematically put into higher and higher positions of responsibility. The theory is that a man who can't command a squad isn't going to do much better with a platoon, company, battalion or army. The officer aspirant is also given technical and scholarly education, if he isn't already a university graduate. But the cut is made at each level of command. At worst, you will end up with good small-unit commanders. Troops and subordinate leaders respond well to this system, as they know that the senior leaders have done it all. The gentleman officer approach leads to a credibility gap between the troops and the leaders appointed to lead them.

Trial by examination. This is a pervasive curse of our overeducated culture. Academics are obsessed by the thought that you can evaluate a person's abilities through written examination. It is true that people who excel at examinations are often bright, but what does that have to do with leadership? Particularly leadership in combat? Very little. The examination process does select leaders who at least have book knowledge of their profession. Russia uses this method extensively as do, to a much lesser extent, most other industrialized nations. Its major drawback is that it tends to depend on the examination process to rate other qualities. In particular, promotions are often dependent on examination performance. This path of least resistance becomes a fatal flaw when the shooting starts.

Each nation selects its military leaders on the basis of its military traditions, experience and perceived needs. Nations that have been defeated are the most prone to change their systems. Similarly, nations that have not been defeated, or have not fought for a long time, will not change. This is why obviously inefficient systems have persisted. It is unlikely that change will occur any other way. In the chapter on armed forces of the world, charts give each nation a force multiplier value. This reflects the quality of that nation's military leadership. It also suggests ways to improve leadership.

It is easier to create superior leaders if your raw material is superior to begin with. This happens only if the military profession attracts superior candidates. Social attitudes toward the military profession in peacetime vary from nation to nation. Generally, societies look with considerable disdain on peacetime officers. The military absorb considerable sums of money in return for an intangible, and often debatable, degree of national security. There is no way, short of war, for the military to prove that they are doing their job adequately. To obtain additional funds from unenthusiastic taxpayers, the military tend to downplay their abilities relative to their potential enemies. The average citizen wonders where all the money is going when the armed forces always protest that they are not up to doing their job without seemingly endless additional funds.

In nations bordered by obviously belligerent and historically hostile forces, the average citizen regards the military in a different light. Service in the military is seen as a high calling, a true public service that can attract superior manpower. The military gain further credibility through the presence of so many superior leaders. This is particularly true if the nation fills the ranks by universal conscription. Israel, Switzerland and West Germany are examples of this type. Germany has a long tradition of threatened borders and a need for strong and effective military forces. Even the trauma of World War II has not eliminated this tradition.

Another method of attracting superior talent is by less competition from the civilian economy. In less developed nations there are often not enough suitable jobs for young, talented, ambitious and educated youngsters. If the armed forces aren't a police force for the party in power, the result is a highly professional, well-led, although often poorly equipped, military. Examples abound: India, China, South Korea and a number of African countries. As opportunities increase in the civilian economy, the armed forces find themselves less able to attract the best recruits. This is happening in mainland China.

THE OBVIOUS EFFECTS OF SUPERIOR LEADERSHIP

Military institutions have a difficult time cultivating effective wartime leadership in peacetime. The ability to create such superior leadership is largely the result of military traditions, social attitudes toward the military and (a rare occurrence in itself) a truly outstanding leader at the head of the armed forces. Most of the time military leadership is mediocre. There are many good and compelling reasons for this, but the leaders are still mediocre, and we all suffer accordingly in time of war.

Throughout history, successful theorists have noted that the most successful military leaders are those who can achieve their goals with minimum combat and destruction. Mutual bloodbaths are generally the result of bad leadership on both sides. Good leaders provide quick, relatively bloodless victories.

Examples are the Israeli victories during their wars with the Arabs. The initial Egyptian success in 1973 was a result of first-rate Egyptian leadership. The American counteroffensive against the North Koreans (1950), the initial successes of the Chinese in Korea (1950), the early German victories in World War II, and many other examples show the lifesaving effects of superior leadership.

Superior leadership need not be earthshaking. As long as one side's leaders are demonstrably better than the other's, rapid victory will usually result. If a war breaks out and does not come to a rapid conclusion, you can be fairly certain that neither side has a marked leadership advantage. If one side is much larger than the other, you can assume that the smaller force, by avoiding rapid defeat, has superior leaders.

15

INTELLIGENCE

INTELLIGENCE MEANS getting information about the enemy and, even more important, keeping information from your opponent. There are three distinct layers of intelligence.

Strategic. This covers everything the enemy is capable of doing. Information obtained for this level is simplified and generalized; otherwise the sheer mass would be unwieldy. The nation's senior decision makers use strategic intelligence. Every form of information goes into this pot, including satellite reconnaissance. Data collected at the operational and tactical levels are passed up to fill out the strategic picture.

Operational. This level gets into more detail as it deals with smaller areas: a continent like Europe, or the Pacific Ocean. It is often further divided by activity: land, naval, air operations, economic, political, etc. Generals and admirals use operational intelligence. The means of gathering intelligence are the same as for strategic intelligence, but the data for the area in question are studied in more detail. Also, faster analysis is required. Strategic intelligence must be updated monthly, at the operational level, weekly or daily.

Tactical. This is battlefield intelligence. The more detail, the better, and applied as quickly as possible, minute by minute in some cases, otherwise a few times a day. Most units down to the battalion, or a naval task force's major ship, assign people to collect data. They analyze as much as they can, pass the information on as quickly as possible and receive and pass along analysis performed at other levels.

In ground operations patrolling for information is a major activity. A multitude of sensors are also used: radars, listening devices, aircraft, etc. Generally, the longer a unit is in contact with an enemy unit, the more will be known about the enemy. A good rule of thumb is 10 percent to 20 percent of the enemy's situation will be revealed each day. After about ten days, you will know the enemy's strength, dispositions, capabilities. This assumes two units with equal intelligence ability. A unit that has a low capability for intelligence gathering will stay in the dark longer. A unit more adept at counterintelligence will keep its particulars to itself longer.

Western armed forces in particular have introduced much automation to their intelligence gathering and analysis. A great deal of information is collected by machines and passed on to computers for analysis without any human intervention. This increases the speed of intelligence gathering three or more times. Armies still using card indexes are at a grave disadvantage.

The human factor is still decisive. Unfortunately, intelligence has not always been one of the most sought-after assignments in peacetime. In some armies the intelligence branch is a dumping ground for marginal officers. This attitude varies from nation to nation. Its effects are not usually felt until after the shooting starts.

Recent Developments

Things have become considerably more complex since World War II, much more complicated than most people realize.

At the beginning of this century, intelligence had not changed for thousands of years. Spies, diplomats and diligent trivia seekers collected strategic intelligence, often of dubious value. Much the same crew collected operational intelligence, although combat units often got involved. Tactical intelligence was usually as fresh and accurate as the enemy fire coming in your direction.

Intelligence began to change during World War I. Aircraft reconnaissance furnished a good look at enemy territories with something short of a large group of hard-riding horsemen. Radio became widely used during 1914–19. Radio transmissions, which could be easily overheard by the enemy, also gave a shot in the arm to the ancient craft of cryptography, the making and breaking of codes for transmitting messages.

World War II mainly refined World War I developments. Air photography and photoanalysis became much more effective. The biggest breakthrough was on the Allied side with the very successful breaking of Japanese and German codes.

The Data Explosion

Today we are up to our ears in machines. So many data are being collected that it is a real problem to pick the most valuable information out of the avalanche.

ELECTRONIC RECONNAISSANCE

Originally this was just listening in on enemy radio transmissions. Now intelligence listens in on just about every type of electronic transmission as well as some that aren't exactly electronic—infrared images and magnetic disturbances. At the strategic level, space satellites scan the entire earth. At the tactical level, infantrymen use small radar sets and sensor systems. Aircraft and ground monitoring stations also collect transmissions. Aircraft are particularly effective, especially helicopters. A helicopter hovering at an altitude of 1000 meters, 25 km behind the firing line, can monitor transmissions for hundreds of kilometers. Any transmitter broadcasting continuously for more than 30 seconds can be located within 1 or 2 km. More precise location can be obtained by photo reconnaissance.

PHOTORECONNAISSANCE

Again, satellites and aircraft do much of this work. Aircraft still carry a variety of cameras and are responsible for instant analysis. Unconventional photographs can record images that the human eye cannot see: heat, radiation, magnetic fields and as many other elements as imaginative scientists can catch. A U.S. satellite camera, 128 km away, can distinguish images on the ground one foot in diameter. It can distinguish between uniforms and civilian clothes. Photoreconnaissance will eventually be merged with electronic reconnaissance as photography becomes digital, using an electronic sensor in place of a lens.

SPIES, INFORMANTS AND PRISONERS

"Human intelligence" is still potentially the most valuable, but because people are so much more difficult to deal with and interpret than photography and electronic data, they are often submerged under the flood of so-called precise information. Human intelligence is invaluable for determining the reasons behind the physical evidence. Such data get a bad reputation in peacetime because of the difficulty of agreeing on interpretation. A wrong interpretation makes the information worse than useless; in its absence one could at least apply some logic to the situation.

In wartime the large number of prisoners can be used to check new data constantly against other sources. The truth usually comes out of such a large mosaic.

Analysis

Currently the cutting edge of intelligence work is not in collecting but in analysis. There is still the age-old problem of trying to determine what the enemy intends to do. People do not think alike, people from different cultures even less so.

It is relatively easy to count rifles, tanks, ships or missiles. It is more difficult to answer questions like: What is a particular piece of equipment really capable of? What does the enemy believe his equipment is capable of? What is the enemy equipment capable of in relation to your own equipment? (Their tanks versus our antitank weapons as opposed to our tanks versus our own antitank weapons.) What does the enemy intend to do with his equipment, given what he thinks it can do? (The Russians have a large army in Eastern Europe. Do they believe it is powerful enough to attack Western Europe or simply to defend Eastern Europe?) What could the enemy do with his equipment if he decided to change his doctrine? How aware is the enemy of his options?

Current examples of insufficient analysis abound. During the Vietnam War, our pilots discovered that the air combat techniques developed since the Korean War were not effective. The United States had a considerable amount of intelligence on Russian aircraft, tactics and pilot training. Despite this knowledge, American pilots initially went into combat trained to fight American pilots in American aircraft. Intelligence had not driven home that the Russians, who supplied and trained the North Vietnamese Air Force, used different aircraft and tactics than the Americans. It took a few years to adapt and achieve superiority.

During the 1973 Arab-Israeli War, the Israeli Air Force ignored intelligence data on the upgrading of the Arab air-defense systems. In particular, Israeli commanders thought they could counter new radar-controlled cannon and missiles with pilot skill, which was cheaper than buying expensive electronic countermeasures (ECM) equipment from the Americans. Heavy Israeli air losses taught them a hard lesson. The information was not lacking; the application was.

Not just the air force gets caught up in faulty analysis. During the Russians' invasion of Afghanistan in 1980, they were first thought to have pulled off a masterful land operation. Soon afterward, first-person reports began to surface indicating that they had misinterpreted their intelligence data on the Afghans. Overoptimism, a common problem with intelligence gathering, caused the Russians to go charging into numerous combat situations in which they found themselves more disadvantaged than their analysis had predicted. They are still trying to analyze, or fight, their way out of Afghanistan.

The abortive American raid to free the diplomatic hostages in Iran is just one more example of plenty of data ill used.

When one nation has half a dozen or more intelligence agencies, it is even harder to determine who is right. It's often not just a matter of different analysts coming to different conclusions. Each intelligence group represents a different interest. Each branch of the armed forces, foreign office, national intelligence group and perhaps an internal intelligence department has its own institutional requirements when interpreting information. In the United States, for example, not only does each of the armed forces have its own intelligence group, but there is also a Department of Defense intelligence agency. On top of that, the Secretary of Defense has his own personal group of analysts. In addition, of course, there are the CIA, the State Department, the National Reconnaissance Office and many more. Who's where on the playing field? It depends on which team you belong to and whose game you are in.

GOOD ANALYSIS IS ONLY HALF THE JOB

The mass of data, billions of words, collected by a major nation's intelligence agencies is too enormous for any individual to handle. Even the summaries are millions of words. The problem is compounded by the dynamics of secrecy; much information cannot even be looked at by all analysts.

The reason analysts exist is to handle these data. The analyst client, the decision maker, is usually not involved in intelligence work. Normally this person is a political leader, senior civil servant or military commander.

An analyst's job consists of much more than simply going through masses of information and deriving conclusions. First, he must find the data. A familiar phrase among analysts is "I didn't know that!"

Additional data can be expected to jump, or be driven, out of the woodwork at any time. Analyst output is always tentative, which disturbs most decision makers. What disturbs them even more is a conclusion that conflicts with what they would like to see. The analyst also must take this into account.

A good intelligence analyst makes many enemies. A good analyst has the misfortune to spot new developments before others do. This is especially vexing in peacetime, when there is no clear-cut way to prove yourself right. A good analyst survives in such a situation by becoming very persuasive.

The analyst must also defend his conclusions well. Analysis has three phases:

Phase one is accurately determining what the client wants and finding the raw data.

Phase two is doing the analysis, including perceiving the decision maker's biases.

Phase three is following up his analysis with an effective rebuttal of decision-maker objections. There is also the possibility that the decision maker has valid objections. These must be answered effectively also.

A competent analyst is part detective, part analyst and part politician. The analyst is up for election every time the work goes out the door.

THE GREAT GAME

Intelligence is more than collecting and analyzing data. It also presents an opportunity to deceive the enemy by feeding false data or manipulating events to encourage erroneous conclusions. This activity is sometimes called counterintelligence. But it involves more than thwarting enemy intelligence activity. It is often called the great game, and indeed it is.

Camouflage. Physically concealing troops and weapons from the enemy with camouflage nets, natural cover and special paints that resist infrared or radar detection. Equipment can also be designed to be more difficult to see, either physically or electronically. The U.S. stealth aircraft is a good example.

Electronic deception. Turning the enemy's collection of electronic emissions against him. Units can change their pattern of using electronic equipment. The most obvious example is radio silence. It is simple and it still works. This can apply to any transmitting electronic equipment. A small number of troops can operate a large number of transmitters to simulate an actual unit. For example, a real naval task force could move in one direction while maintaining radio silence. A smaller group of ships could go in another direction, making electronic noises simulating the task force. This technique has been used successfully for the past forty years on land, at sea and in the air.

Plants. Planting other types of evidence through agents to jibe with other deceptions.

The great game is presenting the enemy with all the elements of a puzzle which, once assembled, will say what you want said, not what actually is said.

With all the loose ends, deceptions, preconceived ideas and general murkiness of intelligence operations, you must be careful not to get sidetracked yourself. When playing the great game, you are exploiting the nature of intelligence analysts.

Much lip service is paid to the concept of national, and other, differences in the perception of events. But most analysts and decision makers are unable to get successfully into the mind-set of their potential adversaries. Scattered individuals possess this perception. Group mentality and party lines work against them in democracies and totalitarian countries.

In wartime perceptions are brought into line with reality very quickly. This is an expensive way to obtain wisdom. It also does not help prevent the war. Excellent examples abound. After the fact, the seizure of hostages in Iran was fairly predictable. After the fact was too late. This fracas may yet get us involved in a war.

Vietnam was anti-Chinese and pro-American in the late 1940's. It is still anti-Chinese, and may someday again be pro-American if the intervening unpleasantness can be forgotten.

Since World War II, Russia has been antiwar, paranoid and beset by internal problems. Think about your reaction to that statement and then try to look at things like a Russian.

16

THE PRIMARY LAW OF WARFARE: MURPHY'S

MURPHY'S LAW is "Everything that can go wrong will, at the worst possible moment." Military affairs are particularly subject to this rule. Combat and the endless preparations for it are loaded with unanticipated troubles. This is an acute problem during the opening stages of a conflict, when all the differences in weapons characteristics, tactics, doctrine and quality become concrete. Once the conflict settles down to the steady grind of mutual destruction, it is possible to get a fix on many of the interactions. More precise planning is then possible. Before that occurs, key factors are largely unknown.

Weapon Effectiveness

The weapons used in the last war will perform predictably, unless they have been improved. Just before World War II the U.S. Navy introduced a new torpedo. Unlike previous models that simply hit a ship and exploded, it ran under the target and exploded there with greater effect. Testing demonstrated this. Soon reports about dud torpedoes began coming back from U.S. submarines in the Pacific. For more than a year navy ordnance insisted that the torpedo was all right. Then more tests were conducted. Sure enough, the torpedo didn't work under all conditions. Peacetime testing had not revealed that it was sensitive to different water temperatures.

During the 1973 Arab-Israeli War, the U.S. TOW antitank missile often became unstable when fired across the Suez Canal. As salt water has different conductive properties than fresh water, the maneuver commands sent along the trailing guidance wire were scrambled and so was the direction in which the missile was going. The problem has since been fixed. In a larger war such an incident would have damaged greatly the troops' confidence in the weapon.

Also during the 1973 war it was discovered that the hydraulic fluid of the US M-60 tanks was too readily ignited by a hit. Minor, or at least nonfatal, damage was turned into a major problem. The fluid has since been replaced with a less flammable type. Speaking of hydraulic fluid, Russian tanks use alcohol that can be consumed. If the troops lack sufficient vodka, many tanks may be inoperable due to lack of fluid.

At the start of World War I, two types of artillery ammunition were used: high explosive and shrapnel. The shrapnel was considered more effective. The shrapnel projectile was designed like a shotgun shell. Its time fuse exploded it in the air in front of the enemy troops, showering them with "lethal" lead balls. The shrapnel shell was more expensive than the high-explosive shell, and most armies bought small quantities. The Germans, however, equipped their artillery with 50 percent shrapnel shell. Years after the war an accident during a weapons test peppered a number of technicians with shrapnel balls. They survived. Further investigation revealed that the effectiveness test for shrapnel was in error. Although during tests the shrapnel balls had penetrated wooden planks calculated to equal the resistance of human flesh, actual human beings proved more resistant. This had never been noted officially, and millions of very humane shrapnel shell had been used. This was not the first case of the persistent use of ineffective weapons. Bayonets and cavalry sabers continued to be used during the American Civil War, even though military surgeons kept noting the paucity of wounds from them.

Throughout its service life, the Russian MiG-21 fighter suffered a number of fatal shortcomings only discovered during combat. Two were the side effects of the violent maneuvering expected during combat. The gunsight was thrown out of alignment, leaving the pilot with little beyond "the force" to aim his weapons accurately. Luke Skywalker could handle it; most MiG pilots headed for home, if the second fatal flaw of the MiG didn't do them in. When more than half the fuel was gone and the aircraft was being violently maneuvered, fuel could no longer get to the engine and the power failed. Very embarrassing.

During the 1970's, a peculiar problem of the Russian T-62 main battle tank was finally fixed. Combat stress threw the tracks off the road wheels, immobilizing the tank. A Czech civilian technician finally came up with a mechanical modification. The Russians, however, have discovered a number of other disconcerting flaws in the T-62's and have been retiring them much more rapidly than usual.

To show that superior Western technology is not immune, we have the case of the harmless missiles. During the early 1960's, the warhead of the Polaris ballistic missiles would not detonate. The error was not detected for a while. When it was, the problem proved immune to numerous solutions. Meanwhile, the missiles might as well have carried rocks in their warheads.

Weapons have historically been debugged in combat. It has proven difficult, nearly impossible, thoroughly to perfect them in peacetime. The principal reason is the impossibility of simulating troops using the weapons under the stresses of combat.

Another increasingly more common problem is the cost of testing. Antitank missiles costing up to $10,000 each and anti-aircraft versions costing over $2 million make extensive live fire training risky for the budget-conscious commander. The old axiom "The more you sweat in peace, the less you bleed in war" also covers fiscal exertion.

Another reason for the inability to perfect weapons effectively in peacetime is unrealistic training by unimaginative and unenergetic leaders. Efficiently re-creating this chaos is the stage director's art, and military leaders are not selected for their theatrical abilities. Those armies that have more successfully gone from peacetime theory to wartime practice have done so on the backs of effective training exercises.

National doctrine and standard operating procedures play significant roles. In the Russian armed forces, weapons and equipment are used as little as possible to preserve them for war. No armed force is immune to this idea. Naturally this discourages the use of weapons and equipment with wartime fervor and severity. The dangers of waiting until the shooting starts to discover equipment incapable of hard use is a lesson that is never really learned.

Tactical Principles

The same weapons are often used quite differently by nations.

This creates problems in cooperating with allies and recognizing the different tactics of opponents. Chauvinism, inertia and sundry other factors lead armies to view potential enemies as clones of themselves.

The results are interesting history and ugly incidents. On the strategic level, during the opening stages of World War I, the French were so obsessed with the recapture of Alsace-Lorraine that they nearly lost Paris, and the war, as the Germans marched through Belgium. On the tactical level the French had not examined the German defensive tactics. Sending masses of French infantry up against German machine guns was a tactical error of the highest order. Prior to the war the French had deluded themselves that the fervor of the infantry would overcome the bullets. World War I was full of tacticians unwilling to perceive the differences between their own tactics and those of their opponents.

World War II was equally embarrassing in its repetition of many of the perception errors of World War I. The Germans had developed a new set of tactics, *blitzkrieg,* based on motorized units and armored vehicles. Germany's enemies had the same, and, in some cases, superior equipment. German tactics were developed from the writings of British theorists. The Germans made no secret of their tactics. Still, *blitzkrieg* was an unpleasant surprise to their opponents. Granted, much of the success was due to superior training and leadership. Yet without their dynamic tactics, and their enemies lack of same, Germany's early success would not have been nearly as complete.

After World War II tactical blindness continued unabated. In Korea Chinese infantry tactics smothered enemy units unaccustomed to opponents running up and down hills without benefit of large logistical tails. Chinese tactics were less successful during the static fighting that followed. United Nations morale might not have survived more such surprises. The 1973 Middle East War held surprises for all parties. Israel did not anticipate Egypt blasting its way across the Suez Canal and then just digging in. Israeli disdain for Egyptian tactical competence led to one very embarrassing incident in which well-prepared Egyptian defenses destroyed an Israeli armored brigade.

In Syria another illuminating example unfolded. Syrian armored units, well drilled in Russian tactics, found themselves defeated by a moving ambush. Outnumbered Israeli tanks refused to stand and be overrun. Leapfrogging backward, Israeli tanks destroyed the advancing Syrian divisions piecemeal. Russia immediately began revising its tactics, and the Israelis designed a tank with a larger ammunition capacity.

Tactical myopia is frighteningly persistent.

Lurching Forward

Planning is further hampered by the contending "unions." Between and within each branch of the armed forces, sundry factions battle over limited resources. The U.S. Navy has five major groups: surface ships, the marines, aviation, submarines and strategic missiles. The U.S. Defense Department is probably the worst example of office politics. The U.S. defense budget contains over 5000 line items. There are over 1500 different programs, each with an interest group pulling for it. To confuse the issue totally, over 130 different accounting systems are used by the U.S. military. Numbers cannot usually be compared reliably.

Pragmatic military men soon adopt the attitude that they'll attempt to get what they think they need. Get what they can. Do what they can with what they've got. Hope they guessed right.

This reality does not sit well with many military planners. Over the past thirty years a debate has gone on between the historians and technocrats. Before World War II, history, not science, was invoked during planning debates. All the new gadgets, many quite effective, gave the U.S. technocrats the upper hand after World War II. Since then, experience has shown that warfare is ill suited to scientific solutions. History-minded planners got a bad reputation from their frequent resistance to new technology. During the past thirty years, however, the historians have become more comfortable with technology. The technocrats, while not becoming much more concerned with history, have lost some ground to the historians, especially in the United States. Russia has always worshipped historical experience. European armies were also less mesmerized by technology, but the United States had forced the military technology approach on many of its allies.

The study and analysis of historical experience have returned to favor. A synergy may grow between historical experience and rapidly changing technology. Although planning would still lurch forward into the unknown, there would be less lurching forward *from* the unknown.

17

WHO WINS

WARS ARE EASY to start, expensive to continue and difficult to stop. Wars usually begin because one or both sides feel victory is assured. The wars are then continued for national and personal pride. The war ends when one or both sides are devastated, demoralized or, most rarely, suddenly enlightened by the absurdity of it all.

Starting Wars

Armies are almost always raised for defense. Once available, a powerful, or seemingly powerful, military force can tempt a nation to deal with others more adventurously and aggressively. Nations can start wars by accident—playing with fire—or by design—or "grab what we want and sue for peace."

The illusion of power is not easily given up in the face of defeat. Political leaders have constituents who absorb the losses in lives and property and demand revenge. They are usually aware of the true situation, but remain optimistic to avoid replacement by less defeatist leaders. Military commanders prefer simple objectives. Their jobs are difficult enough without the fine print.

Wars start because hope triumphs over experience. In all cases, the defender will be extremely resistant. Even after defeat a loser will not forgive, forget or cease plotting retribution.

Illusions survive less well on the battlefield. On the modern battlefield victory and defeat are less distinct than in the past. Until World War I, most battles were over in a day. The winner and loser were fairly obvious. Starting in World War I, situations became more ambiguous. Armies faced each other continually. Periods of increased activity were called battles. The results were reported by press release. One hardly ever lost a battle. When one side or the other was obviously getting mauled, the troops would decide for themselves who had won or lost. Even so, anyone with a modicum of optimism could maintain a reasonable belief in continuing victory. Without knowledge of the big picture, a locally bad situation could be explained away as an exception to the favorable prospects everywhere else.

Perceptions of Victory in the Foxhole

Putting the troops in situations where they have to fight their way out is a variation on the old saying that "a hero is a coward that got cornered." Whether accidental or intentional, this situation is typical of modern combat. Individuals and small groups fight to survive. More frequently, they avoid fighting to survive. A combination of discipline, convincing leadership, fear of reprisal, self-delusion and peer pressure get soldiers into fighting situations. Once there, combat reflexes enable one side to prevail. Soldiers recognize that the war will not end until someone wins. The individual soldier cannot be completely informed of the current overall picture. Therefore, they attempt to survive one day at a time. Victory is surviving the next assignment. The more dangerous the activity, the more each minor part of that activity becomes an occasion for victory or defeat. On a patrol or fire fight, each action that endangers the soldier becomes a win-or-lose situation. Crossing a street possibly covered by enemy fire, sticking his head around a corner, firing a weapon and a multitude of other actions constitute the hundreds of little battles in which the soldier participates. The individual soldier's concept of victory is very short-range. He doesn't have much choice.

Victory in the Middle

At the platoon, company and battalion levels, individuals are no longer the most important element. Although the leader is exposed to many of the random dangers of combat, his primary concern is the contribution of his unit to a larger operation. The commander also has personal conditions for victory. The unit is important because it is his means of influencing the war's conduct. The individuals in that unit are less important. At this level warfare begins to become impersonal. Below this level survival is more important than victory; and above this level, victory is the ultimate goal that crushes men beneath it.

Historians play down what middle commanders do with most of their time. The combat situation, although usually obvious in hindsight, is obscured at the time by the unreliability of incompetent subordinates. Issuing orders for the obvious solution is a futile exercise if those orders cannot be competently carried out.

Commanders at this level are truly middlemen. They are not privy to all the highest-level decisions, nor do they participate in the daily routine of the fighting troops. Yet they are directly charged, by the highest military command, to carry out the nation's military policy. Commanders at this level fight a two-front war. They fight the enemy, if somewhat abstractly. They fight their superiors and peers for scarce resources.

When you are commanding 5000 to 50,000 troops, you never face the enemy. You do face your own subordinates. If you can't win with them, you cannot defeat the adversary.

Winning at the Top

Combat-experienced commanders of mechanized armies can be counted on one hand. Most are Israelis and Egyptians. The severe dearth of experience is especially critical, as most potential army combat commanders will be operating in the urbanized, forested, often rainy and foggy terrain of Central Europe, while the most recent mechanized experience has been in the deserts of the Middle East.

Generals and admirals have fairly substantial egos. The system encourages them. A person in this situation tends to believe he knows what he's doing. Who is going to contradict him? Not some civilian who was elected to be the commander in chief!

Competent military leaders at this level often prefer to avoid war. Generals like to retire with honor. Having a command during a victorious war is nice, but not worth the risk of being a defeated general. History is unkind to losers, particularly to losing generals. Effective leaders are usually aware of their limitations. They also know that it is easier to defend than to attack. Let some other miscalculating egomaniac start something.

High military command is a political activity. Generals learn along the way the importance of good public relations. Strong-willed men with considerable self-confidence must contend with similar individuals for limited resources. This does not make them venal or any less public-spirited. Civil wars and coups result when this political system breaks down.

Winning in the News

Secrecy is a goal most governments pursue at all times. During a war, secrecy is a veritable article of faith. The temptation to manipulate the news during a war is usually overwhelming. The farther away the slaughter, the more optimism replaces reality. Reality is often nonexistent at the highest decision-making levels. This is especially true when you are losing the war.

News of what is happening at the fighting front comes down from on high. Some of the most fantastic fiction ever written appears in a nation's news media the day before surrender. Correspondents at the front see only a small portion of what is going on. Optimism can easily prevail over an unpleasant reality, especially as "defeatist" journalists will be replaced.

War is such a discouraging process that media manipulation is often the margin of victory. People are making many sacrifices and without encouragement, defeat will soon appear preferable to continued fighting.

The winning side often calls upon the loser to surrender, sometimes on very favorable terms. A government recognizes that its war effort will not survive long after negotiations are announced. Thus, news of a possible settlement or negotiations is kept secret or disguised.

National Victories

If the military high command is part of the government and not subordinate to it, the conditions of victory are clearer. Military-controlled governments are also more inclined to get involved in wars of aggression. Civilian governments usually debate and haggle too much to pull off a major aggressive war.

Yet competent professional military leaders generally avoid war. They know the risks only too well. They also know that most nations will put their differences aside in order to defeat an aggressive power. Most nations prefer that things remain as they are. In war there is more to lose than to gain.

Even victory is often an illusion. Victory often begets yet another war. The Franco-Prussian War of 1870 is an excellent example. Its primary purpose was to unite Germany. France was decisively defeated. Unfortunately, Germany seized the partially German, but traditionally French, provinces of Alsace-Lorraine. The loss of the provinces made the defeat impossible to forget. This set the stage for World War I, which in turn defeated and dismembered Germany. The result was an even more destructive war in 1939. The result of *that* war, the partition of Germany and Russian hegemony over Eastern Europe, is the most likely cause of yet another destructive war to come.

Both Russia and the United States see themselves as surrounded and threatened by the other. Anything that goes wrong, or is simply perceived as wrong, is somehow blamed on the other superpower. Because both nations are so powerful, they do not dare take each other on directly. Indirectly there is a lot more activity. Afghanistan holds no vital interest for the United States, so Russia can invade it directly. Iran, being a neighbor of Russia, cannot easily be invaded by America. If Iraq attacks Iran, that's another story. Often, a nation's vital interests are wherever one says they are. Stake out too many spheres of influence and eventually one will be violated.

Large wars frequently start between two strong nations and then escalate. World War II began like that, especially the war in the Pacific. Cutting off Japan's oil was a decisive move by the United States. Japan's only choice was surrender or war. This is all the more illuminating when you remember that most Japanese military leaders recognized that a war with America could not be won.

In Afghanistan the United States could support the Afghans against the Russians by cooperating with one of Afghanistan's neighbors, Pakistan, Iran or China. Any of these nations would then be subject to Russian reprisals. Things could get sticky.

Eastern Europe is another potential sinkhole. Russia is very touchy about losing control here. In the future, still more wars may arise out of misconceptions of victory and defeat.

18

WHAT ARMED FORCES DO IN PEACETIME

ARMED FORCES do not usually fight much during peacetime. Depending on the nation, they may put down a rebellion or civil disorder. Perhaps they will have a minor foreign adventure. If the country is wealthy, the combat troops will spend their time maintaining equipment, learning individual skills and exercising in large units, much in that order. Noncombat troops follow a workaday routine little different from that of a civilian job. Less wealthy armies spend a lot of time farming and road building.

The Daily Routine

Generally unmarried troops live in barracks. Depending on the nation's wealth, these range from crudely heated barns with no plumbing to a Western college-style dormitory. Married men, and sometimes unmarried troops in wealthier armies, usually live with their families, either in army housing on a military base or in civilian accommodations off base. The quality of housing, as of most other amenities, increases with rank. Officers usually live apart from the troops. Usually when troops serve outside their native country, they are not allowed to take their families with them. Everyone becomes unmarried as far as living conditions are concerned.

The day usually begins at an early hour, five or six A.M. Armies tend to be "early to bed and early to rise." Units usually go "on parade" (British) or stand "in formation" (U.S.) at least once a day so that announcements can be made and assignments given. The daily routine is usually set by a long-established schedule. Meals, taken in large dining halls, are at fixed times.

Western armies tend to require five or six eight-hour workdays a week. Russian-style armies often schedule every waking hour for six to six and a half days a week. The more restrictive the schedule, the more it is abused.

In the West most days are nine-to-five. The married troops, who are not regimented by barracks life, go to work in the morning and return in the evening. This schedule is disrupted by "alerts," primarily in combat units, when the entire unit must turn out as if for combat. Other disruptions for combat units are night training and field exercises. Noncombat units often work overtime and, if part of a division, also participate in some major field exercises.

The armies of less wealthy nations do fewer large unit exercises, which are expensive, and more economically productive work like raising their own food or working in factories part-time. Otherwise they attempt to operate just like those of more wealthy countries.

Armed-forces' routine varies from nation to nation. The U.S. armed forces are probably the most easygoing, with a routine as much like civilian life as possible. The Russian system is at the other extreme. When you are in the Russian Army, there's no mistaking where you are.

Keeping Score

All facilities and equipment are inspected at regular intervals: daily, weekly and, most dreaded of all, annually and semiannually. The last two inspections are often conducted by the high command's inspector general, and a failing unit can damage its commander's career.

A more difficult type of inspection attempts to evaluate the unit's combat capability. Evaluating individuals' ability to handle weapons is relatively simple compared with evaluating larger units. Some armies take a technical approach. Judges score a large number of activities: How long did a unit take to road march 22 km? Were the vehicles properly spaced while moving? Was an adequate aircraft watch maintained during the march? Were the route and destination properly scouted for suitability and freedom from enemy ambushes and mines? Were the vehicles quickly and adequately camouflaged upon reaching the destination?

The lists sometimes cover hundreds of items. Theoretically such systems should come close to evaluating a unit's capabilities. Unfortunately, weak leadership often colors the evaluations to show what the high command wants to see, particularly if promotions and reputations are at stake. This distortion starts at the lowest command levels, where officers must evaluate their own troops. The honest evaluator stands alone in exposing his share of incompetents and poor performers. Conformist evaluators make everyone look good, pleasing most superior commanders, who have enough problems already. It's an example of Gresham's Law: "The bad drives out the good." This problem can be held in check only by the highest standards of professionalism. Since such standards are largely mythical, nothing short of an armed enemy will overcome the delusions. The only partial cure is good leadership at the top. Hard-nosed behavior is generally considered slightly bizarre in peacetime and is only recognized as correct when the shooting starts.

Sometimes being forced to operate on a tighter budget will bring forth more effective leadership. More often, less affluent armies keep up appearances and just get by, particularly if there is no external threat or strong military tradition.

Forces That Create Peacetime Myths

SHORT MEMORIES

Armies remember no more of the past than their oldest members. History presents an endless cycle of armies sinking into a peacetime routine that prepares them less for war than for the establishment of another bureaucracy. War comes and the bureaucracy is transformed into a fighting organization through a bloody and expensive process. The war ends and the veterans, as long as they remain, maintain a sense of what must be done. The veterans age and depart, and the cycle begins again.

Peacetime armed forces have a generation gap. The age range of the constant influx of new recruits, comprising 50 percent to 85 percent of most armed forces, is usually eighteen to twenty-one years old. The establishment is the 15 percent to 50 percent of the troops, who are called regulars, lifers or some other term of endearment by the younger men. Most spend twenty or more years in the service. They have different attitudes than the younger troops.

Such peacetime armies will initially fight each other in tragi-comic fashion. A mutual bloodbath results until one side relearns the lessons of warfare before the other.

Soldiers are not stupid. Most senior officers are well read and knowledgeable about current military affairs. Few are serious students of history and endlessly reinvent, often ineffectively, solutions to age-old problems. Like most people, they have to take the world as they find it and make the most of a bad situation. Those few officers that perceive solutions that are effective but politically or institutionally unacceptable are forced to go along and get along, or get out. Examples of unpopular, but militarily necessary, policies are: live-fire exercises (deaths and injuries result), conscription (high-quality volunteer forces are too small and expensive), high-performance standards (careers are hurt). The list is long.

APPARENT VERSUS REAL STRENGTH

Even the wealthiest countries are torn between spending money to buy more equipment and spending to maintain and use what they have. Size and actual power are not the same thing. Size is more visible than power. Buy more tanks, put more men in uniform. These you can see and count. Training is expensive, especially firing those expensive weapons frequently. This expense cannot be seen and its results cannot be counted in peacetime. But this expense will pay off in wartime.

Thus, countries tend to have a lot of troops and equipment that are not being used together. They train as cheaply as possible.

NO OPPORTUNITY TO TEST CONFLICTING CONCEPTS OF WEAPONS, EQUIPMENT OR DOCTRINE

Without a real war to settle the arguments, solutions are found elsewhere. The U.S. armed forces have led the way in this area lately, practicing against clones of accurately equipped and trained Russian combat units. This is unusual; political expedience usually decides which weapons are needed.

The military should be the instrument of the state's political policy. But the preparation of the military for war is a technical matter. Warfare is conducted with bullets, not ballots, but ballots outvote bullets in peacetime. It should come as no surprise that decisions on weapons, equipment and doctrine made in peacetime appear idiotic when the shooting starts.

THE LOSS OF GOOD LEADERS

People join the military because of patriotism, adventure, a desire to render public service, careerism, a need to accomplish something. People of vastly differing abilities join. Too many of the best leave in peacetime. The "warrior leader" and the "political leader" are two distinct types. The warrior is uncompromising, striving always for unambiguous results. A warrior searching for trial by combat leaves the peacetime military for the "real" world. Battles conducted with balance sheets and market shares are unambiguous indicators.

When a real conflict comes, the pinstripe soldiers often return, many from reserve units. Fortunately, some with uncommon determination and patience remain in the military. Their critical leadership staves off defeat until the nation's strength can be militarized. In some cases there is enough time. In some cases there isn't, and defeat results.

The pool of good leadership in the active military varies for each service and nation. Navies and air forces retain more warriors. Flying an aircraft or running a warship is essentially the same in peacetime and wartime. The moments of combat are relatively few. Keeping the aircraft in the air or the ships at sea is a suitable challenge. A peacetime warrior can point to a deficiency in flying or seagoing operations and make a telling point. Sailors and pilots can test many of their wartime skills without a war.

Some nations give greater status to the military than others. If a country takes its military very seriously, then the military have an easier time attracting high-quality leaders. Also, political considerations carry much less weight than professional military judgment.

LACK OF AN EXTERNAL THREAT

One other factor will keep the professionals truly effective: a real external threat. Lack of such a threat is the next problem most nations face.

For centuries Germany was surrounded by warlike enemies. When unification occurred in the last century, Germany found itself with a powerful, effective and professional military. Kept constantly on the alert by surrounding enemies, the high-quality military tradition was maintained. Man for man, the most capable armed forces on both sides of the Iron Curtain today are German. Both the United States and Russia recognize this.

A nation of immigrants, a nation of people with no military tradition, the nation of Israel, has become the most effective military force in the Middle East. Person for person, Israel is probably the most capable military force in the world today. A nation faced with destruction is either destroyed or becomes stronger than those forces that would destroy it.

Nations with no such threat have less incentive to tighten up. A threat applied slowly enough, and which is recognized, forces the potential victim to become more militarily effective. America in the years before its entry into World War II is a good example.

The Russian Peacetime Army

In the Russian Army, appearance as opposed to reality is carried to an extreme. Most equipment is not only unused, it is not even in running condition. In regular units this equipment is brought out once or twice a year for large-scale exercises. The rest of the year, training is conducted with what little equipment is not in storage or with crude simulators.

Individual training is stressed. Seventy-five percent of the Russian Army consists of two-year conscripts. With 37 percent of the unit composed of new recruits and another 37 percent of one-year "veterans," there is a constant individual training program. Most days are a numbing routine of equipment maintenance, classroom instruction and running around with empty weapons shouting, "Bang, you're dead!"

Like many less affluent nations, Russia often calls out the army to perform nonmilitary duties. Bringing in the annual crops for several weeks is a usual activity. Helping out during natural disasters is standard, as the active divisions' thousands of trucks are a unique transportation resource in a nation where over 80 percent of the transportation system consists of waterways and trains.

Generally, however, the routine in the Russian armed forces is endlessly boring. Alcoholism is a major problem. The average recruit is paid barely enough each month to get drunk once or twice. This fosters a barter system whereby marketable military equipment is lost in exchange for vodka. This constant attrition of vehicle parts, tools, clothing and supplies does little for readiness.

Because of relatively severe discipline, the Russians can get their units into shape and on the road quickly, in spite of drunkenness, poor morale, missing supplies and spotty training. Fighting is another matter. Russian military journals complain that training has become all form and no substance. At best, units deploy and go through the motions of strict adherence to the regulations. At worst, even the simplest maneuvers are botched. With no combat experience since the "Great Patriotic War" (1941–1945), the Russian armed forces have too few veterans to warn of the perils of peacetime soldiering. Many years of apparent Russian strength have diminished the perception of an external threat. Experience gained from Afghanistan will shake things up somewhat, as did experience in the 1973 Arab-Israeli War with the Arabs using Russian equipment and methods.

Military officers are becoming a hereditary caste whose numerous privileges become a greater goal than real military skills. A historical cycle of long standing in Russia repeats itself.

The Industrialized West

The American peacetime army is unique and important, because so many other armies model some or all of their practices on the U.S. example. Its basic tenets were formed after World War II, when forces were maintained at a far higher level than any previous peacetime force. The men in the 1946 armed forces were overwhelmingly of recent vintage. Few could trace their military experience back beyond the 1930's. For all practical purposes, the entire American military tradition consisted of World War II, a war of untrained, inexperienced civilian soldiers waging a victorious war with the aid of bright ideals, steadfast courage, enormous material resources and high technology. Thus, the U.S. military are in love with technology. For a democratic nation that put a high value on human life and can afford technology, this attitude is natural.

Other Western armies often use the same technology, but their attitudes toward its use are somewhat different. These armies have a long military tradition. The current troops may never have fired a shot in anger, but they know the importance, in decreasing order, of leadership and well-trained, motivated and equipped troops. The importance of equipment is recognized, but only in fourth place after essentially human factors.

Therefore, American and other Western armed forces appear to do the same things during peacetime, but greater care is taken in leader selection and development among non-American armed forces.

The Third World

On one extreme are large and ancient nations like India and China. Both are quite poor. Both have long military traditions. Both also accord considerable prestige to those in the military. On paper their armed forces are less powerful than the West's because of their more modest hardware budgets. On the plus side, longer-term volunteers and more professional leaders get the most from their less capable equipment. Like most poor nations, China and India routinely assign their troops to civic duties. Agriculture, construction and manufacturing work make troops more self-supporting. With vehicle fuel costing over $500 a ton, and a mechanized division or squadron of aircraft capable of burning up a few hundred tons a day, there is not much opportunity for large unit exercises. Few units are mechanized, and the emphasis is placed on infantry training and operations. Because of poor economic conditions, the military present an attractive career. The recruiters can be more selective than in wealthier nations, and the general competence level of the troops is accordingly higher than in many wealthier nations. In the Korean War contingents from nations like Turkey and China demonstrated the superiority of their infantry. Vietnam likewise gave Korean, Vietnamese and other well-trained volunteer infantry a chance to show what can be done with slender resources.

On the other hand, poverty doesn't always cause the military to rise to the challenge. Air forces and navies require extensive practice to attain individual proficiency. For this reason the less wealthy nations usually have strong armies and less effective navies and air forces.

Nations without a military tradition often have armed forces in name only. The oil-exporting nations of the Middle East spend more per man on their armed forces than most Western nations, yet their troops are not considered effective. Jordan, which is not wealthy but has developed a military tradition, has the strongest Arab army, person for person, in the region.

Throughout Asia, Africa and South America there are armed forces that lack the leadership to create effective combat capability. They go through the motions and on paper look capable of delivering credible military force. But experience has shown they cannot deliver. They are peacetime armies that serve best as auxiliaries to the police forces. A country cannot wage war with police.

Part 5

SPECIAL
WEAPONS

19

THE ELECTRONIC BATTLEFIELD

COMMUNICATIONS IS one of the key tools that turn a potential mob into an organized fighting force.

During World War I, electronic communications first became common on the battlefield. The enemy listened in on radio messages and, if necessary, tried to break the code. On the battlefield the enemy planted microphones to overhear activity. The enemy also used direction finders to locate transmitters, and thereby fleets or headquarters.

By the end of World War II, electronic warfare had a big role. Major offensives were not prepared without drawing up a communications deception plan. It wasn't enough to declare radio silence; the enemy would know something was up. Dummy radio traffic was established, and radars and navigation devices were jammed or deceived.

All this happened nearly forty years ago. Since then, electronic warfare has matured into a decisive weapon. Tests have shown that ground units become aimless and unresponsive when hit with effective electronic-warfare weapons like jamming, target acquisition and EDM (electronic deception measures). Because no one has encountered such massive electronic warfare before, there is no traditional way to deal with it. This psychological response is somewhat similar to that toward nuclear and chemical weapons. But electronic warfare has no moral or psychological prohibitions. The United States used it extensively and successfully in its air war over North Vietnam. Similar experiences have occurred in the Middle East. There are numerous instances of electronic warfare's devastating effect during peacetime exercises.

The Russians take electronic warfare seriously. In theory they are less dependent on continuous radio communications than Western armies. Russian divisions are pointed in a direction and turned loose. They are expected to keep going without further direction until they reach their objectives or burn out in the attempt. On the tactical level the Russians still place great emphasis on signal flares and field telephones. On the other hand, lack of radio contact can stall Russian units once they have accomplished their immediate objective. In the Russian Army no one does anything without detailed instructions from the central command.

Compared with the Russians, Western armies are more lavishly equipped with radio equipment and are more dependent on it. The U.S. Army, with its love of gadgets, has become more vulnerable to electronic warfare than most other Western armies. Although Western nations have developed more powerful electronic-warfare equipment than the Russians, they are only now beginning to put much of it into the troops' hands. On the ground the West is winning the electronic-warfare battle in the laboratories, while the Russians are winning on the battlefield. Only at sea and in the air is the West winning in both areas.

Components of Electronic Warfare

Electronic warfare is not simple. It consists of a great number of activities.

ESM (electronic surveillance measures). Just keeping track of the enemy's electronic devices has become a major operation, especially since no one knows exactly how both sides' electronic equipment will interact until there is a sustained confrontation. The side that can do this first has a major advantage. All aircraft, ships, army units and ground vehicles have unique electronic signatures when detected by a particular type of electronic detection equipment. Some equipment may be more difficult to detect than others. With a little modification, a vehicle may be even harder to identify. As a reaction to the considerable ESM on Russian equipment, the U.S. Air Force is developing "stealth aircraft," which are extremely difficult to detect electronically.

Both Russian and Western forces also continually monitor each other's electronic transmissions, partially to pick up information from their content and partially to analyze the operation of the equipment. Once you know how the enemy equipment operates, you can jam or deceive it.

ESM is usually conducted with special monitoring units that have a wide range of receivers, recorders, signal processors, foreign-language interpreters and signal-processing equipment. These units either stay in one place or move about in aircraft or on ships. Transmission monitoring is very extensive in peacetime. The "Russian trawlers" are ESM craft. The U.S. Navy ships the *Liberty,* bombed by the Israelis in 1973, and the *Pueblo,* seized by the North Koreans, were ESM ships. Numerous ground stations in nations surrounding Russia perform this function as do many aircraft, like the SR-71, four-engine planes and the U-2.

Active sensors. Radar and sonar are the best examples of active sensors. Basically an active sensor sends out a signal that bounces off an object, telling where the object is, where it is going, how fast it is moving and what it is. The greatest drawback of active sensors is that they reveal their presence with the signals they emit.

Passive sensors. Unlike active sensors, passive ones just listen for enemy electronic signals. Shutting down electronic transmitters does not make the enemy undetectable. Passive sensors can pick up all sorts of emissions. Infrared detectors identify different sources of heat. Other sensors can pick up sounds, magnetic presence (any large mass of moving metal), and changes in shape and color. This is one of the areas where much work is being done. For example, some space satellites have extensive heat-detection equipment. They can pick up ships at sea, ground combat formations and aircraft. They can determine whether a cruise missile has a long or a short range by the engine's heat level. The stealth aircraft mentioned earlier is liable to be detected through emissions like heat, magnetic presence and atmospheric disruptions.

Radars and sonars without transmitters are the best-known examples of passive sensors. Electronic bugs are also passive sensors. These have proliferated enormously. On the battlefield they can be delivered by artillery shell, ship or aircraft, or simply placed in front of a ground unit's position. By broadcasting in short, high-intensity bursts of compressed data, they can burn through jamming and avoid detection. Their only problem is visual discovery by enemy troops, which is solved by designing them to look like local vegetation. In Vietnam they mimicked bamboo plants. Many were left in inventory after that war was over and were shipped to Europe. It is hoped that these have either been replaced or that Russian troops will not think it unusual that bamboo grows in West Germany.

ECD (electronic control devices). The "electronic battlefield" is also called the "automated battlefield." Automation means machines operating without constant human intervention. An example of automation is a sensor-controlled mine. If a vehicle with the proper signature passes, the mine automatically explodes. It might wait until a certain number of vehicles have passed before exploding, or it might do so only if vehicles of a certain size or weight pass. It might be not a mine but a sensor, which will collect information for a certain length of time and then transmit this information so that other weapons may do the destruction. Increasingly, electronic operations happen so quickly that ECD is essential. Using ever more powerful microprocessors, these devices not only move much faster and more accurately than a human operator, but also can be reprogrammed. Speed of change is a critical weapon. Any advantage gained will likely be temporary, so to retain the advantage you must constantly adapt to the enemy's actions.

ECM (electronic countermeasures). This wide range of techniques deceive or disrupt electronic devices. Jamming is one of the more obvious forms of ECM. This consists of broadcasting a loud noise on the same frequency the enemy is using for communications, sensors or whatever. Some jamming signals make the enemy think his equipment is defective. This type of jamming can only be countered by well-trained troops. Chaff jamming is done with strips of aluminum foil that form a cloud which active sensors cannot penetrate. Flares draw off missile sensors that home in on heat. Electronic noisemakers draw off radar homing missiles.

ECCM (electronic counter countermeasures). These are the various techniques for dealing with ECM. One of the simplest forms of ECCM is simply cranking up the transmitter and burning through the enemy jamming. A variation is burst transmission: The message is electronically compressed and transmitted in a very powerful and brief burst of energy. Automated frequency hopping also counters jamming and will soon become very common because of the growth of more powerful ECD's. For missile-guidance systems, the favorite form of ECCM is to have more than one type of guidance.

Signal processing. This has become quite a large field all by itself. Electronic data are worthless unless you can identify them. Originally signal processing consisted of photointerpretation and identifying a Morse-code transmitter by his fist, or style of hitting the transmission key. Intelligence used direction-finding equipment to locate transmitters. Headquarters, always choice targets, usually transmit quite a lot.

Today, with the extensive use of passive sensors, there is a critical need for fast and accurate signal processing. Signal processing has come of age through computers. Large numbers of vehicle signatures are maintained in the computer's memory. As signals are encountered they are quickly run through the computer until a match is made. This match is often not perfect, and the computer advises on the reason—signal too far away, damaged source, other possible interference. The computer also suggests the best course of action. The human operator can then intervene. This is the common procedure in submarines. In high-speed aircraft attempting to penetrate enemy defenses, the aircraft may automatically turn this way or that to avoid the most deadly enemy defenses.

In many cases signals collected from a variety of sensors are sent back to a central processing site, which analyzes them and dispatches units to destroy confirmed targets. An example is the U.S. Navy's worldwide underwater submarine-detection system. Constantly monitoring the depths, it sends all signals to processing stations. Here friendly and enemy ships, surface and submarine, are plotted. New signals are identified and operating characteristics of ships are determined. Friendly submarines are often sent out to acquire more detailed information. The U.S. Air Force processes information in its airborne AWACS (airborne warning and control system) aircraft and ground stations. Ground forces use smaller systems to process signals for local sensor nets.

Traffic analysis. This is a variation on signal processing. The patterns of enemy message traffic are studied and compared with known patterns for enemy activity. This technique can provide accurate knowledge of enemy capabilities and operations. You must be careful not to be deceived by phantom units or deceptive actions.

EDM (electronic deception measures). These are an assortment of techniques. Transmitters are set up to divert the enemy's attention or to simulate the presence of a unit. Simulated message traffic can indicate the unit is going to attack, when it is really going to retreat. Signal processing, microprocessors and computers in general establish and run the deceptive message traffic and determine which patterns are real and which are not. Finally, the oldest deception is sending messages in code. This field, cryptography, is almost entirely dependent on electronic devices. Computers devise the codes and attempt to break them. Sensors constantly monitor enemy codes.

How to Wage Offensive Electronic Warfare

In supporting offensive combat, electronic warfare supports the following functions:

Target acquisition. Sensors, ESM and signal processing identify the activity, strength and position of enemy units—and thus the most lucrative targets.

Disruption of C³ (command, control and communications). Outright destruction results or it prevents the enemy commanders from keeping track of and controlling their units.

Deception. EDM deceives the enemy about your real intentions before you lower the boom.

How to Wage Defensive Electronic Warfare

Electronic warfare does not disrupt enemy radio traffic continually. Much information can be gathered from it. Also, there is not enough jamming equipment to be everywhere at all times. This equipment also suffers high attrition rates, as antiradiation missiles have an easy time homing in on a continually broadcasting jammer.

Alternate communications. The most effective defensive electronic warfare is preventive. Not only must alternate means of communication be set up, but operations must be planned to reflect their use. If alternate communications, messengers, field phones, homing pigeons, regular telephone, signal flags, flares, etc. are slower than radio, operations must be prepared to slow down. A reduced number of communications devices requires either slowing down the rate of communication or adjusting operations to require less communications.

Communications discipline. Because ESM is the one electronic-warfare activity carried out continually, it is the one most in need of attention. If communications are undisciplined, the enemy will pinpoint key units and electronic-warfare equipment for destruction and more effective jamming.

ESM. Electronic surveillance measures are critical because if effectively done, they will let you know what the enemy is up to. This is the most valuable result of successful electronic warfare.

Equipment hardening. Nuclear weapons release a pulse of electromagnetic energy that will disable or destroy electronic equipment. Weapons detonated thousands of kilometers away would disable solid-state devices for minutes or hours. Closer detonations would destroy some equipment. Most of this can be avoided by shielding and/or redesigning solid-state equipment to resist this surge of electrical energy. Such measures increase the cost of equipment to 5 percent, but they are becoming common practice.

Who Can Do What to Whom?

Intelligence gathering is the most critical aspect of electronic warfare. The Western armed forces have always had an edge in this area. They currently lead in electronic technology and computers. The Russians have usually attempted to substitute espionage by agents for electronic eavesdropping. Because they are aware of their shortcomings, they have made greater efforts to use available tools. Radio-direction finding of enemy transmitters is pursued energetically by the Russians. Crude but powerful jammers are widely used. Most are barrage devices that jam a wide range of frequencies simultaneously. They require much more power, but the Russians have better access to portable electrical generators than they do to more sophisticated jammers. In the West a higher level of electronic technology created spot jammers that jam only selected frequencies. Microprocessor-controlled sensors detect and identify only those frequencies the enemy is using. Thus, more power may be applied to jamming only what has to be jammed. Barrage jammers tend to jam all communications, yours and the enemy's. Spot jammers are also smaller and can shut down frequently enough to avoid destruction by antiradiation missiles or counterbattery artillery fire.

In the final analysis it is not the equipment itself that will prove decisive, but how ably, intelligently and creatively it is used.

20

THE WAR IN SPACE

IN A FUTURE major war, the upper reaches of the earth's atmosphere will be a battlefield. This will happen for much the same reason that the air became a battlefield in 1914: the need for surveillance over enemy airspace. Additional uses make the control of space even more vital. Communications advantages alone are worth fighting for.

The space battlefield extends from 150 to 36,000 km above the earth's surface. The upper limits of conventional aircraft are about 36 km. As most spacecraft are unmanned, a war in space will be very much a robot war, a truly automated battlefield.

Military Uses of Space Satellites

Communications. Satellites are excellent relay stations for ground communications. They are much cheaper than relay facilities of equivalent capacity and transmission quality on the ground. Line-of-sight transmission is more secure than conventional radio.

The United States has about eight communications satellites. Russia has as many as three dozen. The disparity in numbers stems from geography, technology and doctrine. Russia is too far from the Equator, making it difficult to launch satellites into the most efficient orbits. Therefore, more satellites are needed so that one is always in position to relay data. The Russians' electronic technology is not as advanced, thus producing a higher failure rate. Their satellites are, one for one, less capable than Western models. Also, doctrine leads to additional missions, such as satellites that pick up transmissions from undercover agents and then relay them when the satellite next passes over Russia. The Russians also prefer more spare satellites in orbit at any given time.

Ocean surveillance. Radar and other sensors track surface shipping and, increasingly, submarine movements as well. Strategic satellites cover large areas by radar, and small-area-coverage satellites use electronic sensors. The radar is powered by a nuclear reactor. After it runs out of power, the reactor is boosted into a higher orbit to prevent its falling to earth. One such Russian satellite failed to boost higher and fell on Canada in 1978. The United States keeps three in orbit; Russia puts as many as three a year in orbit, each lasting about two months. Their powerful radar gives them better coverage than the U.S. satellites. However, no one as yet has a comprehensive ocean-surveillance capability.

Land surveillance. These satellites have infrared and passive electronic sensors. Data are transmitted to ground stations. High-resolution photographs from the cameras are sent back in canisters, usually to be picked up by aircraft over the ocean. The films, at least the U.S. ones, are so highly detailed that objects less than one foot in diameter can be distinguished. These are all short-endurance satellites, mainly because they consume film and require considerable fuel for maneuvering. Russian satellites last about two weeks, U.S. satellites four months.

Electronic intelligence. These satellites eavesdrop on transmissions over foreign territory. Called "ferrets," they often orbit low and carry a wide variety of sensors for electromagnetic transmissions. Russia launches four to six each year, the United States a few less.

Early warning. These surveillance satellites are specially designed to spot missile launches and give warning. They also collect information on missile performance during tests. They rely primarily on infrared and radio sensors. The latter pick up data from transmitters in the warheads. Russia launches about four a year. The United States launches one or so each year. Because the U.S. satellites have virtually unlimited endurance, new ones are launched only to put up more powerful equipment.

Navigation. Some satellites maintain stable, fixed orbits in order to give reliable position data to ships or ground forces on the earth.

Only the United States is currently installing such a system of twenty-four satellites. All should be in position by 1985.

Weather surveillance. These are civilian weather satellites, but the military have special requirements and their own satellites. Image and heat sensors plot weather movements. The United States has four in orbit with a life of three years. Russia has been launching two or three a year recently and apparently keeps four in orbit.

Scientific. These satellites conduct experiments with a wide range of military applications. A variable number are launched each year by the United States and Russia. Many are not announced.

Antisatellite. These destroy other satellites, and anything else within range (high-altitude aircraft, missiles). None has been permanently deployed yet. The only ones successfully tested maneuvered next to the defending satellites and exploded. More sophisticated versions will fire missiles at satellites. Eventually they will use lasers to inflict critical damage. This last type will probably not see service until the end of the decade. Russia has taken the lead, launching its first successful intercept satellite in early 1981.

FOBS (Fractional Orbiting Bomb Satellites). They carry reentry vehicles similar to the warheads of ICBM's. By launching their warheads in orbit, they gain an element of surprise ICBM's cannot achieve. These satellites are illegal but could be introduced quickly.

Battle stations. Very large manned or unmanned satellites could destroy large numbers of targets. Stations would contain numerous sensors, computers, communications gear, power supplies and laser or particle-beam weapons. The technical problems of detecting and destroying thousands of warheads within fifteen minutes are formidable. Dozens of stations would be required. None has yet been deployed, but they are being planned.

Getting Into Orbit

Launching satellites requires three things: sufficient thrust to escape the earth's gravity, accurate control to place the satellites in orbit, and effective communications with and control of the satellites.

The earliest satellites of the late 1950's had little beyond thrust and minimum thrust control. During the next ten years, thrust increased enormously, as did control. In the last ten years "smart" satellites have come of age. Expense is now a greater limitation than technology. The cost of putting satellites in orbit is enormous, up to $100 million per launch. The primary purpose of the U.S. space shuttle is to bring this cost down so more satellites can be put up.

Improvements in technology have made satellites much more capable pound for pound. It is now possible to launch eight or more on one vehicle. The primary limitation is the fuel required to maneuver satellites into the desired orbit. The launch-vehicle and satellite charts give a more detailed view of operating characteristics.

Orbits vary according to mission. A reconnaissance satellite requires a low orbit in order best to use its cameras and other sensors. A communications satellite ideally has a fixed orbit: the geosynchronous orbit, 36,000 km above the Equator. Lower orbits are less efficient, as less of the earth's surface is covered.

Limits to Satellite Endurance

If placed in a stable orbit, a satellite can stay up indefinitely. But a number of factors limit a satellite's useful life, if not its time in orbit.

Stability of the orbit. Satellites stay up longer in the higher orbits. Coming in close causes more drag and instability. A faulty initial launch results in an inefficient orbit. Although most satellites have some propulsion capability, it is intended only for fine adjustments.

Endurance of onboard maneuvering system. Most satellites have small rockets to adjust their position. They might turn around to catch more sunlight or use another instrument, or change or simply maintain orbit. When the limited supply of fuel is used up, adjustments are no longer possible and the satellites may enter the atmosphere and burn up.

Power supplies. At the very least, sufficient power is needed to transmit data to earth. The largest power load is usually for mission equipment, ranging from low power, passive sensors to high-power radar. The power source is usually a combination of solar panels and storage batteries or power cells. The more solar panels carried, the more power is generated. Batteries can store solar energy for the times the satellite is in the earth's shadow. Batteries are a less efficient sole source of energy. Russia has put small nuclear reactors on board, but even these have limited capacity, having been designed for high power output over a short period. If the satellite's designer accepts low power requirements and uses only solar panels, the satellites can last for hundreds of years.

Other expendable supplies. If film is sent back, only so much film and capsules can be carried.

Fatigue. This will eventually overtake any electromechanical device. Although satellites are built to last, usually with redundant systems, eventually some vital part will fail. Another primary reason for the space shuttle is house calls to perform repairs in orbit. Extensive self-test equipment on board allows some repair from the ground, often no more than shutting down a malfunctioning device so that it does not damage anything else.

Obsolescence. A satellite put up to last ten years may be replaced with a more cost-effective model before that time is up. Again, the space shuttle comes into play, either to modify satellites in orbit or to bring them down for rebuilding.

Vulnerability

Destroying enemy satellites in a war is a high-priority task. It would make enemy surveillance, communications and navigation much more difficult. There are various methods.

Destroy the ground stations. This is the simplest method, but it becomes more difficult as stations multiply. The saucer-shaped objects pointing skyward are becoming cheaper and cheaper. Many ground stations contain not only receivers but data processing, retransmission and satellite control instruments. Any attack on a satellite system would have to include as many stations as possible.

Destroy satellite launch bases. There are very few. The United States and Russia have two major bases each, which account for the vast bulk of their launches. Destruction of these bases would seriously impair, but not completely stop, launches of short-duration satellites, which are usually reconnaissance satellites and would be of crucial importance during a war or similar crises.

Jam or otherwise blind enemy satellites. Electronic jammers on the ground, in aircraft or in other satellites can do this. Blinding visual sensors is a recent possibility. A low-flying satellite can receive temporary or permanent damage from high-energy lasers on the ground. Another method, which combines blinding and jamming, is to detonate a nuclear weapon in the path of a low-flying satellite. The resulting pulse of radiation can have devastating effects on electronic devices. Reducing a satellite's effectiveness is the next best thing to destroying it.

Destroying the satellite. This is currently the most difficult method of putting satellites out of commission. A satellite is a very small object. Keeping track of one's own is difficult enough. Should a friendly satellite go out of control and shift position, it would be exceedingly difficult to pick it up visually or on radar. A maneuverable enemy satellite is even more difficult to locate. A stationary satellite is an easier target, but these are often the farthest out.

Once you locate the target satellite, you must have the means to destroy it. Russia has demonstrated a *kamikaze*-type system. This, however, presupposes an accurate knowledge of the target's location. There is a problem if you don't have very many attack satellites. The United States has developed a system for attacking low-flying satellites: A high-flying F-15 aircraft fires long-range missiles.

In the future satellite battle stations may combine sensors with laser or missile weapons. For the present, antisatellite weapons are primitive and not very abundant.

Satellite Defense

It is possible to increase a satellite's defendability. Components can be shielded from the electromagnetic pulse of a nuclear weapon or a laser attack. Missiles could home in on an attack satellite's radar. Satellites could be equipped with target-acquisition radar and missiles. Perhaps most effective would be to give satellites more maneuverability. Changing orbits frequently makes the attack satellite's job much more difficult. A satellite war would resemble a chess game in which satellites are moved and maneuvered.

During the past few years, Russia has been launching eighty satellites a year. The United States has been putting up less than a dozen. In wartime either side could easily convert some of its ICBM's into satellite launchers, as many of the current launchers are converted ICBM's. The U.S. Navy is considering equipping one missile in each Trident submarine with a navigation satellite to replace those lost in the opening stages of a war. A more critical lack would be a supply of additional satellites. These intricate devices are not mass produced. Whoever stockpiles the most will be in the best position. Currently Russia maintains the largest stockpile in order to support a larger launch program. In addition, Russia has over 1500 ICBM's not sitting in silos. American stocks are considerably lower.

The only advantage the West has is the additional capacity of its civilian satellite system. Over twenty civilian communications and weather satellites could easily fulfill military requirements.

The Prospects for War in Space

No nation has as yet an organized, tested and functioning space-warfare capability. Both the United States and Russia monitor each other's activities diligently and strive to develop equal capabilities. Because of the increasing importance of satellites, it is only a matter of time before the systems described above become reality. Before the end of this decade, war in space will be a real possibility.

Russia, if true to past performance, will deploy a large number of somewhat crude systems. The West will opt for a smaller number of more sophisticated systems. The West will also be more vulnerable to the destruction of satellite systems. As over half of all Western long-distance communications pass through satellites, a war in space would be very disruptive to Western economies.

Unlike nuclear weapons, there will be fewer moral restrictions on space warfare. In fact, space warfare increases the likelihood of a future nonnuclear war.

Satellite Launch Vehicles

DESIGNATION is the name of the vehicle. These are the most common vehicles. Most were originally designed as ICBM's. The space shuttle will eventually replace all other U.S. and many Western vehicles. This will happen only if the shuttle system is indeed capable of putting satellites up less expensively than current methods.

SATELLITE LIFT, LOW, HIGH is the amount of weight in tons each vehicle can put into a low (150 to 1000 km) or high (36,000 km) orbit. The latter is the geosynchronous orbit, in which the satellite stays over one spot on the earth.

LAUNCH WEIGHT (tons) is the total weight of the vehicle with payload. Most of the weight consists of fuel.

STAGES is the number of vehicle parts. Most of the fuel must be burned during the first few kilometers of lift-off. Rather than drag empty fuel tanks into orbit, vehicles are divided into separate stages, each consisting of a motor and fuel tanks. The final stage, containing the payload, comprises less than 10 percent of total vehicle weight.

USED BY is the nation making and using the vehicle.

SATELLITE LAUNCH VEHICLES

Most satellite launch vehicles are obsolete and/or modified ICBM's.

Designation	Satellite Lift Low (tons)	High (tons)	Launch Weight (tons)	Stages	Used By
DSV-3 Delta	2	1.2	132	3	US
Titan IIIC	13.1	1.4	633	3	US
Titan IIID	13.6	1.2	591	2	US
Titan 34-D	14.9	1.9	672	4	US
Scout	.2		22	4	West
Space Shuttle	29.4	2	1984	2	US
Soyuz	7.5	1.1	327		Russia
Zonda	22	1.6	1800		Russia
Salyut	22.7	1.7	1900		Russia
Proton	18.1	1.4	1600		Russia
Ariane	2.7	.5	160	3	EEC
N-1	1	0	91	3	Japan
N-2	2	.3	135	3	Japan
SLV-3	.04		17	4	India
CSL-X-3	10	1	600	3	China
CSL-2	2	.2	191	2	China

SATELLITES

The major difference among satellites is function. Size is restricted by the power of the launch vehicles.

Type	Typical Weight (tons)	Typical Orbit (km)	Endurance (days)
Navigation	.3	1000	2400
Communications	1	36000	Unlimited
Ferret	.3	200	100
Surveillance	12	200	80
Weather	.7	800	100000
Early Warning	1	36000	1000

Satellites

TYPE is the function of the satellite. See Chapter 20 for details.

TYPICAL WEIGHT (tons) is the weight of most satellites of each type. Most could be much larger if there were a cheap enough way to get them into orbit. Inflation will have its way with satellite weights as it does with everything else.

TYPICAL ORBIT (kilometers) is the height at which each type normally operates. Most orbits are not circular but elliptical. The work is usually done at the lower phase of the orbit.

ENDURANCE (days) is the typical useful life of the satellite. Endurance is usually a function of supplies of fuel and other materials. Often the spent satellite is sent toward earth, where it burns up while reentering the atmosphere. This is to prevent useless objects from clogging up valuable orbit space.

NOTE: A Ferret is a low-altitude recon satellite.

21

CHEMICAL, BIOLOGICAL AND NUCLEAR WEAPONS

WARFARE IS unpredictable enough without the uncertainty of chemical and nuclear weapons. This is probably a major reason for the tacit avoidance of these weapons. Only Russia and America possess them in significant numbers. Britain has nuclear weapons. France and China produce both chemical and nuclear warheads. Other nations may have both but aren't talking.

The Specter of Nuclear Escalation

The chief reason for avoiding the use of these so-called tactical weapons is the danger of escalation to the longer-range strategic weapons. There are enough poison gas and nuclear weapons lying about to destroy far more than just the fighting forces. Although both the Russians and the Americans hedge on tactical nuclear weapons, they are more strident in their desire to avoid using the really big nuclear weapons. In practice, it will be difficult to distinguish tactical from strategic. No rules or guidelines have been adopted, and it is unlikely they will be reached in the heat of combat. But both the Russians and the Americans maintain thousands of tactical nuclear weapons and have every intention of using them if they sense they are losing a conventional war. Indeed, the Russians emphasize so much winning early in a war that they may feel tactical nuclear weapons are an essential part of major conflict.

Our sixty-five-year combat experience with chemical weapons further muddles rational thinking on nuclear weapons. Their psychological effects confound military planners.

First used in 1915, these weapons quickly developed. By 1917, 15 percent of British casualties were caused by chemical warfare. To further confuse the issue, chemical weapons tended to be less fatal than conventional arms, artillery and machine guns. Only some 15 percent of gas casualties died compared with twice as many from other weapons.

For psychological reasons still not fully understood, chemical weapons were not used during World War II. There were a few isolated cases, but they were always against an opponent who could not respond in kind. The same pattern has persisted to the present.

Meanwhile, the effectiveness of chemical weapons was being increased. The Germans developed nerve gas just before World War II. It was truly a weapon of mass destruction, as there was no simple protection. Nerve gases were swiftly fatal and could enter the body through the skin as well as orally. Nerve gas could be present in far smaller quantities than the old mustard gas, and would kill on a massive scale. Only the Germans knew about it during the war. Their fear that the Allies also had nerve gas prevented them from using it. Germany feared a massive use of nerve gas on its civilians. Even if they had been sure of their monopoly, they would most likely have not used it. After all, they had developed it from U.S. and Russian technology. It would not take their enemies long to produce their own version and use it on a vulnerable German population.

Here is a historical precedent for restraint with a weapon similar to nuclear weapons. Does this historical lesson still apply? Available evidence indicates no. The Russians have equipped their troops with chemical weapons on a giant scale. They publish procedures for the extensive and immediate use of these weapons. They train their troops accordingly. In the face of such a commitment, it appears unlikely that the Russians will refrain. Can nuclear weapons be far behind?

The Russians first turned to the chemical weapons when they found themselves at a disadvantage in tactical nuclear weapons. In response to the Russians' buildup, Western armies have built up their own capability.

Unfortunately, the Western armies see not just more powerful conventional Russian forces, but potentially unbeatable chemical weapons. Western armies have only tactical nuclear weapons to even things up. The vicious circle could have dire consequences during a large-scale war.

The Nuclear Battlefield

Nuclear weapons are easy to use on the battlefield. Put simply, they are tremendous multipliers of firepower. A single 100-pound nuclear artillery shell more than equals the destructive effect of at least 8000 conventional shells (350 tons). In addition to the effects of the blast, they harm with radiation. Most armies attempt to minimize the radiation by using "clean" nuclear weapons.

With each side throwing several thousand at each other, soon there would be no one left to fight. Nor will there be anything left of the battle area. As the most likely arena for the massive use of Russian forces is Europe, you can imagine how the Europeans regard this type of victory.

The mathematics of this mutual slaughter are straightforward. The Western forces in Europe comprise some 30 divisions (360 or more combat battalions, 40,000 or more armored vehicles). Russian forces would typically amount to 60 divisions (900 or more battalions, 50,000 or more armored vehicles). Of the 7000 tactical nuclear weapons available to the Western forces, less than half would probably survive to be used. Two or three per battalion would be more than adequate to eliminate that unit's combat effectiveness. The Russians have half as many tactical nuclear weapons, which is more than sufficient to destroy all Western battalions. They have additional divisions in reserve, 30 within a month. The United States also has reserves of weapons outside Europe. But at that point, what would be left to fight over?

Four or five thousand tactical nuclear weapons used in densely populated Central Europe would leave over 50 million dead or dying. Which brings us to another grim aspect of tactical nuclear warfare: The targeting strategies for tactical nuclear weapons guarantee their use will quickly spread beyond the fighting front.

Only about a thousand of the Americans' tactical nuclear weapons in Europe are in artillery shells. The remainder are delivered by missile or aircraft. Almost all the Russian nuclear weapons are air- or missile-delivered. Therefore, the aircraft and their airfields are primary targets, not to mention the nuclear weapons' storage sites themselves. With short-range missiles raining down nuclear loads, it won't take much paranoia to prompt the targeting of additional strategic weapons on installations like railroad yards and supply centers. Russians and Americans stipulate the enemy's "nuclear means" as the primary targets of nuclear weapons.

A neutron bomb is a small nuclear weapon designed to emit the maximum amount of short-term radiation. A 1-kiloton neutron warhead has the explosive equivalent of 1000 tons of conventional explosive and would cause significant destruction for about 1500 meters in all directions. More important, it would release sufficient armor-penetrating radiation to kill or incapacitate all people within a radius of 800 meters. Troops to a distance of 1200 meters would still receive a dose that would be fatal after a few hours. Unlike from the radiation of a normal nuclear weapon, the thick metal of armored vehicles will not protect the crew. Against the radiation of neutron bombs, only a foot or more of concrete or dirt provides any significant protection.

The Chemical Battlefield

Chemical weapons also provide more firepower to conventional artillery and air power. Even more so than nuclear weapons, chemicals are area weapons. A chemical weapon attacks only troops. It can have immediate effects and then dissipate, or, by changing the formula slightly, its effects can linger for hours, days, weeks or years.

Troops can protect themselves. If the troops are properly equipped, well trained and have been warned, they can reduce chemical-weapon casualties to less than 2 percent. Otherwise, casualties can be as high as 70 percent to 90 percent with over 25 percent fatal. This is also a war of matériel. Gas masks require filters, and protective garments must be fresh. Prolonged injury from chemical weapons is a particularly unpleasant way to die.

Like nuclear weapons, chemicals are best used against targets far from your own troops. Enemy airfields are a favorite target. Also likely to be hit are supply dumps and other rear-area installations. A defending unit can be doused for a few days until weakened by fatigue and casualties.

Chemical weapons can be used defensively. This was done in

1917–1918. Flanks can be protected by spraying an area through which the enemy might advance. At the very least, this will slow attackers down.

Indeed, operations will slow down considerably on a chemical battlefield. The protective measures are simply cumbersome. Masks and protective garments are awkward, and they are stifling in warm weather. Vehicles offer some protection, although under the stress of normal use, vehicles have been known to leak. It would particularly devastate troop morale if casualties began to occur in supposedly safe vehicles.

Because the troops know there is a defense against chemical weapons, they will attempt to protect themselves. No matter how well trained and disciplined, troops will slow down in a chemical environment. During the initial use of a weapon, operations will be especially slow because of unfamiliarity.

Nerve gas can enter the body through the skin. Full protection means complete cover-up. Once infected, your only remedy is a syringe of antidote injected quickly into any large muscle. Upper leg muscles are preferred. If you inject yourself when not infected, the antidote will injure you.

Perhaps in recognition of these problems, most armies would try to stay out of contaminated areas as much as possible. When attacking, only nonpersistent chemical weapons will be used on the defending troops. Chemical monitoring teams will travel with all units to warn about entering an unanticipated contaminated area. These areas will be traversed as quickly as possible. If the enemy hits the assault unit's assembly area with chemicals, the area will be vacated whenever the tactical situation allows.

Actual holdings of chemical weapons are well-kept secrets. United States forces apparently maintain a stock of over 40,000 tons of chemicals. Not all are actually loaded into warheads, artillery shells or aircraft bombs. The stock immediately available to combat units probably amounts to no more than a few thousand shells, plus a smaller number of bombs and even fewer warheads for missiles. It would probably take a month or more to load the remainder of the U.S. chemicals into weapons. About half of all U.S. chemical munitions are stored in Europe. Given the Russians' emphasis on chemical weapons, their stocks are probably much larger, including tens of thousands of shells, warheads and bombs. It follows that they will attempt to use these weapons on a massive scale at the onset of a major war.

Complete surprise is no longer probable, as most Western armies have noted Russian intentions and have equipped themselves defensively. Another type of surprise is possible and quite likely: troop shock when encountering chemical weapons for the first time.

Prepared troops hit with an artillery barrage mixing chemical and nonchemical shells would experience 10 percent to 30 percent casualties, one quarter of them fatal. Anyone with a roof over his head would suffer half that rate. If the troops are fatigued and off guard, the casualty rate would rise to 40 percent to 50 percent. If the troops are untrained, 80 percent could be killed or permanently out of action, while 20 percent would be temporarily incapacitated. This rate would also apply to civilians, of whom there might be over 100 per square kilometer.

Surprise can also be achieved away from the fighting line. Aircraft and helicopters can spray lines of gas many kilometers long. If the wind is blowing the right way, a wall of gas rolls over the unsuspecting troops. Although the gas alarms will go off, a 10-km-per-hour wind can catch more than half the troops before they can put on their masks and protective clothing. Time-delay bombs and mines can be dropped near enemy positions. Set to go off at night, they will be particularly devastating.

Casualties: Physical and Mental

The chief characteristic of the chemical/nuclear battlefield is the increased number of wounded casualties. During the long wars of this century, there have been twenty or more nonbattle casualties for each man killed or wounded in combat. Chemical and nuclear weapons will increase the number of both combat and noncombat casualties.

Calculating the effect of nuclear and/or chemical weapons in a future war is part of the peacetime planners' art. World War I losses were 10 percent to 15 percent when troops advanced into a gassed enemy position. The defender's losses exceeded 60 percent, thus justifying the attacker's gas losses. Russian planners, who project a widespread use of chemical weapons, calculate daily loss rates of 20 percent. After five days of operations, they assume a unit, by now reduced to 33 percent of its original strength, will have to be withdrawn and replaced by a fresh one.

Radiation, in particular, has long-term effects. It's common knowledge that radiation, if not taken in large enough doses to be immediately fatal, will do you in within hours, days, weeks or months. Beyond that, the long-term effects can be sterility, birth defects, cancer and general unpleasantness. Radioactivity is odorless, tasteless and colorless.

The effects of chemical weapons are less insidious. Nerve gas has its effects and then either kills you or wears off, although some long-term damage is suspected. Some gas agents take effect immediately, but the effects persist. Blister gases leave wounds and scars; blindness and damaged lungs are common. Blood agents damage internal organs (kidneys, liver, etc.).

The constant danger of injury has given rise to a new class of casualty—combat fatigue. Consider the symptoms of radiation and some chemical sickness: listlessness, upset stomach, headaches, fatigue. The same symptoms can also result from stress. What could be more stressful than the knowledge that you might accidentally, and unknowingly, pass through a contaminated zone? Valium has seriously been considered as a means of calming everyone's nerves.

The fighting on a chemical nuclear battlefield may quickly evolve into an exhausted stalemate or a series of duels between the scattered survivors. Historical experience reveals that survivors of high-attrition combat either give up completely or fight only to survive. It will be difficult to carry on a war if the soldiers are either immobilized by shock or ready to fight only in self-defense.

Navy and Air Force

Naval forces have little to fear from chemical weapons. Seawater quickly absorbs chemical agents. If you can hit a ship, you want to destroy the hardware, not just the crew.

Once in the air, pilots are immune to chemical weapons. There may be some danger, however, to low-flying aircraft, particularly helicopters. Aircraft are a primary means of delivering chemicals.

Air forces are particularly afraid of the effect of chemical weapons on their ground crews and air bases. Airfield personnel must protect themselves against chemicals much as do combat troops. Ground crews must keep their exacting and fatiguing work schedule in order to keep the planes flying. Therefore, the primary danger of chemical weapons is fatigue. An aircraft that cannot take off on a mission because the ground crew cannot service it is almost as worthless as one shot down.

The effects of nuclear weapons on naval targets and airfields are much the same as on army targets. Nuclear weapons may be used even more freely at sea. As the saying goes, "Nukes don't leave holes in the water." Neither do dead fish prompt escalation like dead civilians.

Biological Warfare

Biological warfare is nothing new. For thousands of years, spreading disease throughout the enemy's army was considered a practical tactic. Until the last hundred years, disease always killed a far higher number of soldiers anyway than combat. The miracles of modern chemistry and medicine have made it possible to deliver diseases against an enemy much like chemical weapons. The attacking troops must first be immunized. Until the enemy can figure out the horror, and immunize his survivors, you have an advantage. The enemy can do the same to you.

Biological weapons can also be used strategically. Plant diseases can be inflicted upon the enemy's croplands. Herds and flocks can be decimated. Biological warfare sounds like divine retribution. No one really wants to unleash weapons that can so easily get out of hand. Both Russia and America have them, and neither seems eager to use them.

Offensive Strategies

Faced with the data shown on the four nuclear weapons charts, military planners have developed the following nuclear weapons tactics.

Go for key targets that can best be destroyed with tactical nuclear weapons. The primary ones are headquarters and units capable of delivering nuclear weapons. Headquarters are ideal targets because they coordinate the activities of subordinate combat units. They sit in one place a long time making a lot of electronic noise. After a number of headquarters go off the air in a blast of radio-active static, most units will tighten up. Initially only those disciplined by their commanders will survive. Even so, more than one tactical nuclear weapon will be successfully heaved at a suspected headquarters site.

Take advantage of the wide area effect of nuclear weapons. Artillery, missiles and rockets are significantly inaccurate. Artillery, firing by the map without visual observation of the target, can be off by over 200 meters at 20 or more km. If the target is in a bunker, or is otherwise resistant to blast (i.e., bridges), the mission may have to be passed on to rockets or missiles, which are even more inaccurate over the same ranges (300 meters or more off at 20 km). Heavier nuclear weapons, 10 to 100 or more kilotons, usually make the inaccuracy irrelevant. At extreme ranges (50 or more km) battlefield missiles can miss the target by as much as 500 meters. Again, a bigger bang covers up the error.

Beware the pulse. Current developments in electronics have made attacks close to your own troops completely unacceptable. Electronic chips use very small amounts of electric power. Gamma rays, produced in abundance by a nuclear explosion, carry a minute electrical charge. This minute charge is enough to cause the chip to go "tilt" and, as the computer people put it, "crash." Recovery from the interruption can take seconds or forever, depending on the device. There may be permanent loss of computer memory, especially damaging for radio, fire-control and electronic countermeasure equipment.

The range of the pulse is, at worst, hundreds of kilometers. The pulse effect depends on the type, design and shielding of the electronic component. Shielding, or hardening, is a very popular subject in the electronics business at the moment. There is a general lack of knowledge on the precise effects that pulse has on specific pieces of equipment. Work is being done, but it is far from complete.

Each nuclear weapon has to count. Only 8000 are available to NATO forces, about half that number to Russian forces. Worldwide, 30,000 are available, 75 percent of them in the West. At a cost of nearly a million dollars each, over two thousand times the cost of a conventional weapon, not many replacement weapons will be available during the first few months of fighting.

A package of nuclear weapons is used for each major operation. The enemy targets are analyzed for suitability to attack by nuclear weapons. Nuclear weapons are only as good as target acquisition. All the targets will be behind enemy lines. Reconnaissance determines what will be hit and with what. You must not only locate enemy targets, but keep them located until the attack.

The Russians place great emphasis on surprise, hoping to paralyze enemy resistance through massive, coordinated attacks on the largest possible number of targets. If they attack a U.S. corps, several targets are most likely.

Nuclear capable combat support units. Primarily missile, artillery and air force units and can amount to thirty or more units. The air force targets will often be outside the corps area, as will some of the missile units. Any support units with nuclear weapons within the attacking unit's area of responsibility are primary targets. If artillery units are included, this adds up to thirty or more targets. Locations will be difficult to pinpoint, as these units are highly mobile. Many are armored. They are generally spread over an area of ten or more square kilometers. Each unit is usually composed of four batteries, one a nonfiring headquarters-support battery. Each battery will occupy one square kilometer within the battalion area.

Headquarters. Thirty in all, including corps, division, brigade and support. Each covers a few square kilometers and will try energetically to avoid detection by means like dummy transmitter sites. So important are these targets that nuclear weapons will be targeted at possible locations. This is a common practice with conventional artillery.

Supply installations. Include the supply dumps and means of transportation. Perhaps two dozen. Except for preattack buildups, there should not be any large dumps. Most of the supply is either in transit or split up into small dumps near the units. More lucrative are the transportation networks: railroad yards, ports, airfields, bridges, tunnels and maintenance facilities. Many extend behind the corps sector.

Combat units. Half will not be in contact with the enemy. They furnish about twenty more targets.

The results. Assume a hundred nuclear weapons are used. This would not cover all the targets mentioned, but they will be more than sufficient. Russian tactical nuclear weapons have larger yields than those of the United States. Assume an average of 20 kilotons per weapon. Assume a division-level density of troops and equipment, about 50 percent the density used on Chart 21-3. This allows for accuracy of targeting and weapons delivery.

The hundred weapons would produce the following losses: 30,200 troops, 55 percent of corps strength, 20,250 caught in the open, the remainder under cover; 1050 nonarmored vehicles, 10 percent of corps strength; 600 armored vehicles, 18 percent of corps strength; and 141,800 civilian casualties plus 567,100 homeless civilians (nearly 200,000 dwellings destroyed or damaged).

This sort of attack leaves the front-line combat units intact while destroying all the support behind the fighting line. Once a conventional attack pierces this unsupported line, there is nothing to prevent the oncoming enemy from sweeping all before them.

The antidote for this grim scenario is good intelligence on the enemy's means, movements and probable intentions. This would produce sporadic use of nuclear weapons by both sides to forestall a massive attack.

There are 8000 or more tactical nuclear weapons present in Europe. Assuming 50 percent of them are destroyed before use, the remaining 4000 could cause the following losses: 1,210,000 troops (70 percent of the total); 43,000 nonarmored vehicles (17 percent of the total); and 25,000 armored vehicles (29 percent of the total). Proportionately more armored than nonarmored vehicles would be destroyed because the armored vehicles tend to be concentrated, while the trucks are on the road most of the time. Additional losses could include 5,600,000 civilians; 23,000,000 civilians made homeless, and 8,000,000 dwellings destroyed, including one third of West Germany's economy.

These conservative estimates include only immediate casualties. Lack of medical, support and maintenance facilities would eventually double all the personnel losses. Similar lack of support will increase vehicle losses by four or five times.

Most casualties will be among support troops. The combat units will still be battling away with conventional weapons, including chemicals.

Tactical Nuclear Weapons' Effects on Ground Forces

TYPE OF TARGET. Down the left side of the chart are listed the various situations in which troops will find themselves when a nuclear weapon goes off.

TROOPS IN THE OPEN. It is assumed that most troops will be in this situation when a nuclear weapon goes off. A tactical nuclear weapon depends on surprise, because most of its energy goes into creating blast and heat (or flash). The heat will burn exposed skin. The flash will be diminished by clouds, haze or fog. Naturally clothing, or any obstacle between the bomb and potential victim, will offer protection from flash.

In any situation away from the fighting, most troops will be in the open. If the weather is bad, more troops will be under shelter and wearing more clothing. This will reduce casualties by more than half. Depending on the situation, flash burns can comprise the majority of casualties. Troops in the open are also liable to injury from the side effects of the blast, which produces a 130-km-per-hour wind at the maximum distance shown and 380-km-per-hour winds at half that distance.

TROOPS PROTECTED IN OPEN EARTHWORKS represents troops in fox-holes, trenches, vehicles and other light structures that shield them from much of the flash and blast. There would still be substantial casualties from fire and debris.

NONARMORED VEHICLES AND AIRCRAFT ON GROUND would have enough components damaged to be unusable. The primary cause of damage will be blast. Winds of over 500 km per hour (240 mph) will do terrible things to trucks and parked aircraft.

ARMORED VEHICLES are generally too heavy to be damaged by high winds. Antennas, searchlights and other protrusions can be damaged. Heat will also damage nonmetallic components, fire control and sighting gear, etc. At this range radiation will kill crew members inside the vehicles. Any crewmen outside the vehicle at this range will also become casualties. As crews spend 85 percent of their time outside their vehicles, their loss will also render the vehicles useless.

HEAVY STRUCTURES OF CONCRETE, ETC. This represents substantial commercial buildings as well as military bunkers. Personnel inside will be protected from most radiation effects.

WEAPONS SIZE is represented across the top of the chart in equivalent kilotons (thousand tons) of TNT. The most common sizes for tactical nuclear weapons are from 1 to 1000 kilotons, with the preferred range under 100. The larger ones (1000 to 50,000 kilotons) are used in strategic missiles. Even these larger sizes are falling out of favor and are being replaced with weapons in the 100- to 1000-kiloton range. Note that a 20-kiloton weapon was dropped on Japan in World War II.

THE EFFECTS. The top figure is the distance from the explosion at which 50 percent of the troops or vehicles will become casualties. One third of the troop casualties will be fatal immediately. Another third may be fatal eventually without adequate treatment. Casualties may increase a further 50 percent if radioactive fallout is not avoided or decontamination does not take place.

The range of effect is measured from the explosion. An airburst is assumed. The height of the airburst varies with the size of the weapon. The only reason for a ground burst is to increase radioactive fallout or to ensure destruction of hard targets.

The bottom figure is the area covered by the effects of various weapons under the conditions shown (square kilometers). This is convenient when comparing effects that depend on the density of troops or equipment in the area.

TACTICAL NUCLEAR WEAPONS' EFFECTS ON GROUND FORCES

The area of fires and general destruction of a tactical nuclear weapon is much larger than the area in which combat units will be hurt.

Type of Target	Distance from Point of Explosion Where 50% Casualties Will Occur						
	1 Kiloton	10 Kiloton	20 Kiloton	100 Kiloton	1000 Kiloton	10000 Kiloton	50000 Kiloton
Troops in the Open	1000	2154	2714	4642	10000	21544	36840
Sq Km>	3.14	15	23	68	314	1459	4266
Protected in Open Earthworks	700	1508	1900	3249	7000	15081	25788
Sq Km>	1.54	7	11	33	154	715	2090
Nonarmored Vehicles & Aircraft	600	1293	1629	2785	6000	12927	22104
Sq Km>	1.13	5	8	24	113	525	1536
Armored Vehicles	450	969	1221	2089	4500	9695	16578
Sq Km>	0.64	3	5	14	64	295	864
Heavy Structures of Concrete, etc	200	431	543	928	2000	4309	7368
Sq Km>	0.13	1	1	3	13	58	171

TACTICAL NUCLEAR WEAPONS' EFFECTS ON SHIPS

Ships are more likely to suffer debilitating damage than sink.

Type of Effect on Ship	Range of Effect (in meters)			
	20 Kiloton	200 Kiloton	2000 Kiloton	20000 Kiloton
Sink or Permanent Disable	800	1724	3714	8000
Sq Km>	2	9	43	201
Temporarily Disable Mobility	1500	3232	6962	15000
Sq Km>	7	33	152	707
Temporarily Disable Sensors and Weapons	2000	4309	9283	20000
Sq Km>	13	58	271	1257

Tactical Nuclear Weapons' Effects on Ships

SINK OR PERMANENT DISABLE indicates sufficient damage either to sink the ship outright or to make repairs at sea impossible. Immediate radiation casualties at this range will also be high, putting as much as 50 percent of the crew out of action. This is also the range for damage to modern submarines from an underwater explosion of a nuclear weapon. Older subs would be destroyed at somewhat longer ranges, perhaps 25 percent farther.

TEMPORARILY DISABLE MOBILITY indicates sufficient damage to the ship's power plant to impair or completely shut it down. On aircraft carriers any aircraft on the deck would be destroyed, and landing or takeoff operations would probably be impossible for at least a few hours. Any aircraft in the air would have to find an alternate landing field, quickly, as U.S. carriers fly most of their aircraft in 105-minute cycles. Repairs on the power systems could take from hours to days. Meanwhile, the ship would be vulnerable to additional enemy attacks. This is also the extreme range of temporary damage to modern submarines. Older subs would be hurt at somewhat longer ranges, perhaps 25 percent farther. Beyond this range subs would most likely be unaffected. Surface ships within range would be damaged, by an underwater nuclear explosion, to the same extent as submarines.

TEMPORARILY DISABLE SENSORS AND WEAPONS. Physical damage to antennas and viewing devices as well as light deck structures—missile launchers especially—would be significant. Within this range aircraft in the war would also suffer severe, often fatal damage. Ships, usually Russian, without sufficient onboard repair capability would have permanent damage. The electromagnetic pulse of the explosion would extend for many hundreds of kilometers. This would be particularly damaging to aircraft outside the blast range. Submerged submarines are unaffected at this range from underwater explosions.

Across the top of the chart is shown the weapon size (in kilotons). Naval nuclear weapons tend to be larger than land weapons. The 2000- to 20,000-kiloton weapons would be found in strategic missiles, which could easily be used at sea.

THE EFFECTS. The top figure gives the range of that effect.

The bottom figure is the area covered (square kilometers). An average modern task force of 8 to 12 or more ships would occupy an area of 1000 square kilometers. A merchant marine convoy of 30 to 50 ships, plus 8 or more escorts, would cover the same area. The more important ships occupy the center of such an area.

National Differences

The Western nations possess larger navies and vastly superior surveillance capabilities than the Russians. In response, the Russians can use larger nuclear weapons, especially strategic missiles. Their numerous cruise missile-armed subs, using tactical nuclear weapons of up to 200 kilotons, would have to either be very accurate or fire a large number of nuclear weapons. Given Russian doctrine, they would probably opt for more firepower applied in the general area of the target.

The Western navies prefer more accurate attacks with nonnuclear weapons. Of the 22,000 tactical nuclear weapons in the U.S. arsenal, 2500 are deployed with combat ships.

Number of Losses Assuming Indicated Densities

This chart shows the losses in troops, vehicles or structures, depending on the nuclear weapon's size and posture of the target.

DENSITY PER SQUARE KILOMETER is the assumed density for calculating casualties. The targets are assumed to be battalion size. The battalion is the basic combat unit in all armies. Containing an average of 400 to 1000 men, these units will be distributed throughout the battle area. Each battalion will cover an area of 5 to 12 square kilometers. A nuclear weapon covering an area larger than 10 square kilometers will be much less effective. Actual losses inflicted with these larger weapons will be one half to one third the number indicated on the chart. For example: Instead of 405 personnel casualties from a 20-kiloton bomb, there would be 135 to 202.

NUMBER OF LOSSES ASSUMING INDICATED DENSITIES

Civilian targets will take a greater beating than the less numerous and better-protected soldiers.

	Density per Square km	1 Kiloton Meters Distant	10 Kiloton Meters Distant	20 Kiloton Meters Distant	100 Kiloton Meters Distant	1000 Kiloton Meters Distant	10000 Kiloton Meters Distant	50000 Kiloton Meters Distant
Troops in the Open	35	55	255	405	1185	5500	25529	74646
Protected in Open Earthworks	35	27	125	199	581	2695	12509	36577
Nonarmored Vehicles & Aircraft	5	3	13	21	61	283	1313	3839
Armored Vehicles	5	2	7	12	34	159	739	2159
Heavy Structures of Concrete, etc	20	0	1	2	7	31	146	427
Civilian Losses	250	192	894	1418	4147	19250	89351	261263
Dwelling Rendered Uninhabitable	85	262	1215	1929	5640	26180	121517	355317
Civilian Made Homeless		770	3573	5671	16583	76969	357259	1044632
Distance at Which Housing Is Rendered Uninhabitable (in meters from bomb)		1400	3016	3800	6498	14000	30162	51576

TROOPS. Half the troops are assumed to be outside, the remainder under cover.

NONARMORED VEHICLES density is of a division area (heavy military activity).

ARMORED VEHICLES density is of a combat battalion (infantry or tank).

HEAVY STRUCTURES density is of a heavily urbanized area.

CIVILIAN LOSSES density is for West Germany. Nations like Belgium and the Netherlands have a higher density. The rest of Europe (east and west) has about half the density of West Germany. Civilian casualties will vary considerably throughout a country. The heavily inhabited areas will have densities of over 5000. Even prime agricultural areas will have densities of over 400. Nonagricultural rural areas will have densities of less than 100 (down to 10 people per square kilometer).

If particularly dirty nuclear weapons are used, eventual casualties will be more than doubled due to the delayed effects. Because of the magnitude of the civilian casualties, the medical facilities will be overwhelmed and otherwise nonfatal injuries will kill. Fifty percent of all casualties will be fatal.

Many of the civilians may have fled from a combat zone. However, they have to go somewhere. The nations having nuclear capability state that their nuclear weapons doctrine is to hit primarily targets behind the fighting line. Civilians would do well to stay away from military targets as much as possible.

DWELLINGS RENDERED UNINHABITABLE. Given the density of population, this is the number of dwellings rendered uninhabitable by each type of explosion.

CIVILIANS MADE HOMELESS is the average number of civilians whose residence is uninhabitable. Up to a point, these civilians can be accommodated in other homes. The average dwelling in Europe has four to five rooms with less than one inhabitant per room. Still, each lost dwelling is a significant loss for the inhabitants.

DISTANCE AT WHICH HOUSING IS RENDERED UNINHABITABLE (IN METERS FROM BOMB). This is heavy damage. It includes broken windows, minor fires, roof-tile damage and the like. Without repairs this housing is only marginally habitable. Well, it's better than staying outside.

Density of Troops, Vehicles and Weapons on the Battlefield

This chart shows the average density of troops and vehicles in units of the United States Army (or armies of other Western nations) and the Russian Army.

UNIT DESIGNATION represents the most common units of both armies. The U.S. corps and the Russian Army are roughly equivalent. The U.S. corps contains 2 divisions, an armored cavalry regiment and support units. The Russian Army contains 4 divisions and support units. The figures for both armies are average infantry and tank divisions. Also included are the normal allowance of nondivisional units attached directly to the division. The battalions are also composites of infantry, reconnaissance and tank units. Each division contains 12 to 16 battalions.

LOCAL CIVILIANS gives the average density of civilian personnel, dwellings and vehicles to be found in the battle area.

AREA OCCUPIED is the area in which the unit is spread out, in square kilometers. Generally a square or slightly rectangular area. The corps-army area is 50 by 60 km; the division area is 25 by 24 km or 20 by 17 km; the battalion area is 3 by 4 km. These are averages; they are often as much as one third lower in an attack.

TROOPS TOTAL are the total number of troops assigned to that unit. The divisions usually are in contact with the enemy and thus occupy much of the forward portion of the corps-army area. The empty space at the front line is covered by corps-army reconnaissance troops. The corps' rear area has the least dense concentration of troops.

ARMORED VEHICLES are the total number of armored vehicles in the unit. This includes armored vehicles of all types (tanks, personnel carriers, artillery).

TRUCKS TOTAL is the total number of nonarmored vehicles in the unit. This includes aircraft.

TROOPS PER KM are the average number of troops per square kilometer in the unit's area.

ARMORED VEHICLES PER KM are the average number of armored vehicles per square kilometer of the unit's area.

TRUCKS PER KM is the average number of nonarmored vehicles per square kilometer of the unit's area.

DENSITY OF TROOPS, VEHICLES AND WEAPONS ON THE BATTLEFIELD

Troops can spread out to avoid damage much more easily than civilians.

Unit Designation	Area Occupied (sq km)	Troops Total	Armored Vehcls Total	Trucks Total	Troops Per km	Armored Vehcls per km	Trucks per km
US Corps	3000	55000	3300	10000	18	1	3
Russian Army	3000	61000	4400	9000	20	1	3
US Division	600	20000	1300	3500	33	2	6
Russian Division	350	14000	1100	1500	40	3	4
US Battalion	12	900	60	10	75	5	1
Russian Battln	10	600	50	2	60	5	0
Local Civilians	All				250 Civilian	85 Dwellings	90 Cars and Trucks

CHEMICAL WEAPONS

There are two types of chemical agents—nerve gas and all the others. As deadly as the nerve agents are, most casualties will be caused by nonlethal doses.

"Gas" Name	Code Name	Physical Effect	Used to	Inhaled Agents			Skin Contact Agents			
				Persistence (hours)	Time to Take Effect (minutes)	Minimum Dosage Level	Time to Take Effect (minutes)	Minimum Dosage Level	Can It Be Smelled?	Tons to Cover (sq km)
Tear	CS	Irritation, tears, nausea	Harass	.5	1	1157	NA	NA	Yes	NA
Vomiting	DM	Headache, cough, nausea	Harass	.5	1	4080	NA	NA	No	NA
Blister	CX	Severe skin blisters	Harass	36–1300	NA	NA	100	5	No	NA
Mustard	HD	Severe skin blisters	Harass	36–1300	NA	NA	300	6	Yes	10
Choking	CG	Cough, suffocation	Kill	.1	600	89	NA	NA	Yes	NA
Blood	AC	Convulsions, suffocation	Kill	.1	8	139	NA	NA	Yes	NA
Nerve	GD	Convulsions, suffocation	Kill	.2–50	8	2	8	143	No	1
Nerve	VX	Convulsions, suffocation	Kill	1–2700	6	1	6	28	No	.3

Chemical Weapons

"GAS" NAME, the common name for the chemical agent, is taken from its effects on personnel.

Tear gas, commonly used by police and military forces, has been produced in many variants. Other names are CS, CN, etc. Because of its generally "nonlethal" nature, tear gas has become accepted de facto as not being a chemical agent, but it is, and some of its more powerful variants induce severe coughs, involuntary defecation and vomiting. These effects can render victims quite helpless. Tear gas is often used as a powder, to serve as a persistent harassing agent on the battlefield.

Vomiting gas is a super tear gas. It is also known as adamsite. It is a favorite Russian chemical weapon. Like tear gas, it is ideal for clearing out enemy troops in built-up areas, caves or fortifications.

Blister gas is a Russian development, an improvement on the World War I mustard gas. It is also known as phosgene oxime. It acts much more quickly than mustard gas and completely destroys skin tissue. Very ugly.

Mustard gas is an updated version of the harassment agent used extensively during World War I. It takes a while to act, but once it does it leaves ugly blisters. Many elderly veterans still carry scars (not to mention blindness and lung injuries) from this gas.

Choking gas is one of the first modern chemical agents. Also known as phosgene, it caused 80 percent of chemical-agent fatalities during World War I. Still used, although being overtaken by more efficient products.

Blood gas also had its origins in World War I. It was most valued for its ability to act quickly. This made it the ideal surprise agent. The original was called cyanogen chloride (CK). Modern versions are prussic acid and hydrogen cyanide. This gas is much favored by the Russians.

Nerve gas was first developed just before World War II. It has gone through many reformulations and is known by a variety of names: Tabun (GA), Sarin (GB), Soman (GD), CMPF (GP), VR-55, VX, etc. It comes in persistent and nonpersistent forms, and can be used in lethal and harassing concentrations. This is perhaps the most widely stockpiled agent and would probably be the most widely used. Very deadly.

CODE NAMES are the two-digit U.S. Army code names.

PHYSICAL EFFECT on victims. Most chemical agents are fairly simple elements that primarily irritate the skin. Any that get into the lungs have a more pronounced effect. The only chemical agents that go beyond these simplistic effects are the blood and nerve gases. Blood gases interfere with the absorption of oxygen by the blood in the lungs. Nerve gases interfere with the transmission of messages in the body's nervous system.

All these agents are potentially fatal. The fatal ones will not kill in smaller doses. This phenomenon exacerbates any hypochondriac tendencies. Sickness in general tends to be higher among troops in the field than the population in general. Less than 5 percent of the casualties in an army are the direct result of combat. Chemical agents will blur the distinction as the real, or imagined, minor side effects of gas put more troops out of action.

USED TO indicates whether the chemical agent is intended primarily for harassment or killing. Harassment agents are popular because they can be used more aggressively without endangering your own troops. A harassment chemical agent that is nonpersistent (see next column) can be used without much fear of causing any injury to one's own troops. This makes it an ideal weapon for close to friendly troops. Usually harassment chemical agents are fired upon troops about to be attacked. This tactic, if carried out for a period of days, can substantially weaken the defenders. Harassment gases are also barriers to enemy movements. Lethal chemical agents are most often used on enemy targets far to the rear.

The effects of death and wounds on an army's effectiveness can be only roughly quantified. Consider, for example, the effect on a soldier's fighting spirit after he has been gassed once or more than once. A harassment chemical agent is not used just to be humane; it is an attempt to discourage the troops from fighting by forcing them to wear cumbersome protective gear and by inflicting painful injuries. Painful nonfatal injuries make more of an impression than fatal ones. The victims, instead of being buried, live to tell of their experiences.

PERSISTENCE (hours) is the length of time the chemical agents remain, after release, in an effective concentration. The least persistent form is a gas or vapor, which, like common smoke, quickly dissipates in the atmosphere. Depending on the concentration, the amount of wind and the humidity, the potency of the agents may be gone in minutes.

There are other factors that affect chemical persistence. Persistent agents will last longer in vegetation. Porous soil will retain them longer. Nonporous soils allow precipitation to wash the agent away more quickly. Sunlight by itself reduces area coverage or nerve gases by more than 60 percent. The sun causes the gas to degrade much more quickly. Temperature affects chemical agents in two ways. Cold weather decreases the speed at which it spreads out. Thus, the gas stays in a smaller area at higher concentrations. A high temperature gradient holds chemical agents to the ground, making inhalation less likely. Moisture washes the agents away. A bad side effect is the contamination of any nearby water supplies until the chemical agent is diluted enough to become ineffective.

With gas, it's one damn thing after another. Wind both aids and hinders the gases. It dilutes them to an impotent level more quickly. A 20-km-per-hour wind reduces area coverage over a 4-km-per-hour wind by more than 60 percent. Wind also creates a downwind hazard. Depending on the time of day and strength of the wind, nerve gases can cause nonfatal casualties up to 120 km away. A high wind, at any time of day, can carry the gas up to 75 km. On a sunny day, with only light winds (under 10 km per hour), the agent will travel no farther than 1 km. At dawn, dusk or on a heavily overcast day, the range will be to 10 km. The worst is a calm night. Even with a light wind (4 to 5 km per hour) the gas will travel to 45 km. Maximum travel would be up to 120 km. At night when troops are off guard, asleep or driving, the damage caused by impaired vision, dizziness and other nonfatal effects could be considerable. Gases tend to flow along the contour of the ground, collecting in low areas. One bright young officer, while on a training exercise, set off some tear-gas grenades on high ground overlooking an enemy headquarters. As the cloud charged down the hill, the lieutenant and his troops advanced behind it to mop up.

Altogether, the above factors can decrease the area coverage of chemical agents by more than 90 percent. For example, in most cases a 155mm artillery shell would spread a lethal dose (to 50 percent of unprotected troops) of nerve gas over an area 13 meters from the point of burst. On a day with high winds (38 km per hour), the fatal dose extends only 4 meters from the point of burst. With a 28-km-per-hour wind, it's only 6 meters. With still air, a sunny day and light wind (4 km per hour), the radius of effectiveness is 7.5 meters. With subzero temperatures, the radius is 8 meters. Sunlight, heat and wind combine for a 4.6-meter radius with subzero weather and a 19-km-per-hour breeze.

The most effective time for chemical agents is at night. Not only will the fatal dosage cover the largest area, but nonfatal concentrations will travel farthest. Daylight windy and wet conditions are the least effective. For persistent agents, winter gives the longest duration time. However, this is the period when troops wear the most clothing and move around the least. The next best time is warm, dry weather. Mustard gas will last up to eight weeks during the winter, seven days during the summer and only two days during a rainy period. Nerve gases are similarly affected: GD will last six weeks in the winter, five days in dry summer weather and no more than thirty-six hours in wet weather. VX, a liquid, persistent nerve gas, will last sixteen weeks in winter, three weeks in summer and only twelve hours in the rain.

INHALED TIME TO TAKE EFFECT (min.) is the shortest average time for the agents to take effect through inhalation, in minutes. This assumes a sufficient concentration. The quickest gases are those that work on the nervous or respiratory systems. Nerve gas is by far the fastest. If inhaled, its effect can be within seconds. Even exposure to nerve agents through the skin often takes effect in minutes. Blood gases act quickly to block the absorption of oxygen by the body. This is the equivalent of suffocation. Tear gases act upon the sensitive eyes as well as the skin. Blister and mustard gases can be inhaled, but require a concentration eight times the one required for skin contact.

In less than lethal doses, which will be quite common, agents will take much longer to have an effect. Normally the gases will continue to diminish and their effects will not get worse.

MINIMUM DOSAGE LEVEL is the *relative* amount of the gas in milligrams that must be present in a cubic meter, during a one-minute period, to kill 50 percent of unprotected personnel. Multiply by 36 to obtain the lethal dosage.

SKIN CONTACT TIME TO TAKE EFFECT (min.) is the shortest average time for the agent to take effect through skin contact, in minutes. For mustard and blister gases, this is the time required to cause blindness. The eyes are the most sensitive external part of the body; mustard and blister gases attack the eyes first. The substance can enter the eyes if a soldier gets some of it on his hands and then rubs his eyes. Over ten times more chemical agents are required to blister the skin. To inflict fatal casualties, fifty times as much is required.

MINIMUM DOSAGE LEVEL is the minimum amount required to have the desired effect on 50 percent of unprotected personnel. A much higher dosage is required for skin transmission than inhalation. This is worthwhile, as a mask alone will no longer provide protection. Nerve gases, in particular, need not be fatal to put a soldier out of action. Nonfatal doses of nerve gas, either inhaled or absorbed through the skin, have a very debilitating and demoralizing effect.

CAN IT BE SMELLED? This is an important consideration. If the agent cannot be smelled, its presence will be announced either by scarce "chemical warning instruments" or by troops starting to become casualties. The gases that can be smelled can also usually be seen. Many are odorless and invisible. This makes detectors all the more important. These devices can often detect agents in sublethal doses.

TONS TO COVER SQ KM indicates the tonnage of the chemical agents required to cover a square kilometer. Method of delivery is aircraft spraying. This is not only the most effective method of delivering chemical agents, but likely to be the most widely used in the opening stages of a future war.

Spraying will be widely used by the Russians because of their emphasis on surprise. The United States will have to resort to spraying, because there is no other way for them to rapidly deploy chemical agents. NATO artillery forces do not normally carry any chemical ammunition. Further, many ideal targets for chemical agents are not within range of artillery.

A MiG 27 can spray 4 tons of GD over a 6-km frontage in less than a minute's flying time. One kilometer downwind, this 6-km wall of gas will cause 50 percent fatalities. Five kilometers away (in open terrain, with a light wind, etc.), unwarned but gas-trained and equipped troops will most likely suffer 20 percent fatalities and 70 percent nonfatal casualties. Four tons of "drizzle" (VX) will not travel far, a kilometer or so until hitting the ground. A line of VX 2 km wide and 130 km long will kill 50 percent of any unprepared troops entering it.

Why bother to use anything but nerve gas? The main problem is decontamination. Nerve agents are potent, persistent and unable to tell friend from foe. Protective clothing and masks cause a significant loss of efficiency. For example, voice communication and vision are reduced by 25 percent to 50 percent. When the temperature rises above 60 degrees, troops cannot be active for more than a few hours without suffering from heat prostration. Prolonged wearing of full protective gear causes additional problems, as it is very difficult to sleep, eat or drink.

The only way out of this mess is to decontaminate. Even leaving the area will do no good, as you take the gas along. Decontamination of nerve agents means washing everything down with a lot of water, or a lot less of a water-decontamination-chemical solution. Even when you use the solution, 320 pounds of liquid are required to decontaminate one vehicle. Other methods, faster and using less liquid, are available but are not 100 percent effective. What will happen to troop morale and effectiveness if casualties are caused by chemical agents on previously "decontaminated" vehicles?

22

THE STRATEGIC NUCLEAR WEAPONS PEOPLE

No MATTER HOW you slice it, nuclear missiles are weapons of *mass* destruction. They are a much-feared, never-before-used weapon. No one knows how leaders will react to their use. Even with restraint, the destruction will be enormous. Even worse, many leaders of nuclear nations are talking now of how it may be possible to win a nuclear war. But a nuclear war would have no winners.

The Conventional Option

A variation on the above strategy would be to arm the missiles with nonnuclear warheads—chemical, even biological. The biological weapons would most likely be avoided, because they take a while to get really going. Also, once a new disease has taken hold, it could spread to the nation of origin. Worst of all, the victim could launch its own plague via missile or aircraft. This option is not to be completely dismissed; it is possible.

Chemicals would not be as devastating as biological or nuclear weapons. A ton or more of persistent nerve gas spread over a few hundred square kilometers would cause tens of thousands of deaths. More devastating would be the disruption. Panic, and the resources needed to deal with the casualties and the cleanup, would shut down the target area for weeks or months.

The crucial difference between these alternatives is the lack of massive damage. Nuclear retaliation would be difficult to justify. The attacker would have to be careful to launch only a few missiles (if not one at a time) so as not to provoke a nuclear response.

There are indications that Russia has already tested such a system. It could be used with great effect if the victim was not immediately prepared to retaliate in kind. It might take a month or more to ready chemical warheads. Meanwhile, Russian gas missiles would systematically work over NATO military and civilian targets.

With missiles coming over one at a time or even in small groups, current surveillance systems could give the targets about fifteen minutes' warning. Casualties could therefore be reduced, but not the massive disruption caused by the need to wash away the nerve-gas droplets.

High-explosive conventional warheads could be used against large, vulnerable targets like oil and gas fields, refineries, nuclear reactors, chemical plants, and ammunition storage and fabrication facilities.

Russia certainly has the capability to use missiles in a conventional mode. Their missile silos use cold launch, which does not damage the silo A new missile can be reloaded in a day or two. Also, Russia has 1500 spare missiles. Finally, the Russians have stated repeatedly that they consider chemicals a conventional weapon.

The critical problem with nuclear weapons is that they are all or nothing. A less lethal warhead in less massive applications could still be decisive without ending civilization. This is advantageous for the Russians, as they are well aware that they might be able to negotiate victory over nations on their borders without making a direct threat to the United States. Turning Western Europe into a collection of Finlands would be to their advantage. Life would go on as before, with Russia getting a larger slice of the global economic pie and the Western industrialized nations still relatively affluent.

Naval Options

Nuclear missiles might be used against naval targets without triggering attacks on national economies. Missiles can reach a target in half an hour, and ships don't move fast. By calculating a probable future position of a task force and then launching three or four MIRV (multiple independent reentry vehicles) missiles, you could almost guarantee target destruction.

Submarines are currently difficult to destroy and can be moved around. Yet, in some respects, subs are the most vulnerable of all, as each one carries sixteen or more missiles. The U.S. Navy is more confident of destroying Russian subs than the Russians are of destroying U.S. subs. A technological breakthrough in submarine detection could enable one side or the other to destroy all enemy missile subs.

Proliferation

The first use of nuclear weapons will most likely be between two smaller nations. One nation might well have a nuclear weapon and the other not. The weapon could be used either after a threat or in the face of defeat in a conventional war. Threats of intervention by the major nuclear nations may not be sufficient to prevent a few weapons going off. The damage may be restricted to a local area and provide all concerned with new incentives not to use nuclear arms. Terrorist use is also possible. The Western nations devote considerable police resources to seeing that this does not happen.

The Middle East has high potential for first use of nuclear weapons. Strong evidence exists that Israel has such weapons. Other Middle Eastern nations are attempting to obtain them.

India and China also possess nuclear weapons. China needs them to give Russia pause before providing a "nuclear solution" to some future China crisis. Should Taiwan obtain such weapons, and if China again insists on reunification, the situation could be ugly.

Over a dozen smaller nations could obtain nuclear weapons before the end of the century. Currently they could be delivered by aircraft or short-range missiles. As longer-range missiles become more widely available, the danger of their use beyond the smaller nations' immediate neighbors will increase.

Ah, But Will They Work?

Yes and no. As the charts indicate, serviceability and reliability vary from missile to missile. Like complex machinery everywhere, missiles can be depended on to work only some of the time. This much can be learned from previous experience with the missiles. But what of those aspects of missile operations for which there is no experience?

Bias. It is already known that certain navigational problems, called bias, degrade the accuracy of missile-guidance systems. These problems were discovered during hundreds of peacetime tests with missiles fired from east to west or from west to east. In wartime most missiles will be fired north, over the Arctic. Vagaries in gravitational effects, magnified by the unpredictable forces of wind and weather, affect the missile warhead reentering the atmosphere. These bias problems degrade accuracy to the point where their CEP's (circular error probable) may increase by hundreds of meters.

A 1-megaton weapon will destroy a circular area 14 km in diameter. If general destruction is the goal, missing the aiming point by a few hundred meters will not mean anything. If destruction of enemy missile silos is the objective, this inaccuracy renders the hit worthless. It is theoretically possible to build more intelligent guidance systems that use stellar navigation or navigation satellites to eliminate most bias. During the reentry phase, the warheads could be equipped with sensors to allow terminal homing. None of these systems has yet been proved or deployed. For the present, many scientists consider average errors of thousands of meters quite likely.

Fratricide. If more than one nuclear weapon is aimed at the same target to strike within, say, half an hour, the first weapon's side effects will destroy the warheads that follow. When a nuclear weapon explodes, it releases large amounts of radiation. The pillar of radioactive debris, the mushroom cloud, rises 20,000 meters into the sky. The approaching warhead travels at such high speed that even these small particles of earth and rock are sufficient to break it up. The radiation can also disable a warhead's electronic arming system.

Multiple missile hits are often required to assure complete destruction of large urban areas. This problem also plagues silo bashing, as you must hit each silo with at least two missiles within less than fifteen minutes to assure destruction and prevent launch of the missile.

This means delaying the launch of many missiles so that the multiple hits arrive half an hour apart. The delayed launch, second and subsequent wave missiles are put at risk to enemy silo attacks.

Fratricide has implications for defense. By launching defensive missiles into the upper atmosphere, it might be possible to create a radioactive shield. It might even be possible to create a sand-particle cloud to break up warhead heat shields on reentry.

READINESS

Bias and fratricide are only two problems that may render the world's 4000-plus ballistic missiles less than fully effective. Perhaps the most difficult potential problem is maintaining readiness. The missiles are marvels of engineering and technology, but consider the conditions under which they operate. For most of their lives they sit in underground or underwater silos, inactive and barely alive. Once launched, the missiles undergo enormous stress as they escape earth's gravity and plunge back through the atmosphere. These stresses can produce any number of catastrophic system failures.

GUIDANCE-SYSTEM UNRELIABILITY

The guidance mechanism is the most remarkable system on the missile. Over a range of 10,000 km, the missiles are aimed at a 100- to 2000-meter circle. This works, most of the time, during tests. But what of the thousands of guidance systems that just wait? Of course, these installations are constantly monitored. Repairs and replacements are made when the monitors indicate an existing or potential failure. Experience to date has shown that defects can develop in the guidance system which the monitoring equipment will not detect. Experience with the similar but simpler guidance systems of commercial and military aircraft indicates that constant feedback from users is necessary to provide failure-proof systems. Currently a few percent of the missiles are fired each year to test reliability. These tests do not involve the warhead. The United States has conducted only *one* test of a missile with a live warhead. System reliability and accuracy haunt every missile user.

MIRV (multiple independent reentry vehicle) further compli-cated missile guidance. The MIRV missiles carried not one war-head to reenter the atmosphere but a "bus," which released three or more warheads. Each warhead has its own guidance system to the target. This development made the missiles more expensive and more prone to failure. On the plus side, MIRV's will probably deliver more warheads per missile, even with a higher guidance-system failure rate. Each warhead will carry a nuclear warhead of less power, but for most targets this is of small consequence, especially if warhead accuracy can be increased.

POLITICAL DISTORTIONS

Doctrinal requirements—what the missiles are required to do and how they are to do it—have become an increasing burden. Initially users were grateful to see the missiles lift off and go in the general direction of the target. Success begat excess, and before long the strategy of first strike appeared. This notion sprang from the theoretical accuracy of multiple warheads launched from MIRV missiles. With more than one warhead coming from each missile, it was theoretically possible to launch enough warheads to destroy all the enemy missiles in their silos. Enemy missiles left at sea in their submarines would presumably submit to some equally devastating technological breakthrough.

These tantalizing technological possibilities have put the missile people in an embarrassing position. To admit that their weapons are not capable of such feats is not politically prudent. Governments will not spend enormous sums on weapons that cannot match similar enemy system performance. A vicious circle develops as each side suspects the other of superior technical performance. Lacking any means to validate this performance, the claims become even more outrageous and expensive. More pressure is put on the commanders of the missile forces. In Russia, where the spirit and practice of the Potemkin village (a false front, as in motion picture sets) still lives, the national mania for secrecy only makes the problem worse. The possibilities are endless, as is the expense. Even more dangerous is a national leader believing the illusions and attempting to use them.

The Post-Holocaust World

All of this is rather irrelevant from the standpoint of our society surviving any massive use of these weapons. Assuming there is no restraint and all available weapons are used, a nuclear war will kill tens (if not hundreds) of millions of people within days of the war's start. Many, if not most, of the survivors will perish during the following few years from disease, starvation and exposure, not to mention the lingering effects of radioactivity. The remaining survivors, perhaps 20 percent to 30 percent of the pre-holocaust population, will live in a society that is technologically in the nineteenth century and socially part of a much more primitive era.

The above assumes a nuclear exchange that hits enough population centers to cripple terminally the major industrialized economies. Most, if not all, of this destruction will occur in the Northern Hemisphere, in the industrialized nations. The less developed nations will then be cut off from supplies of technology and food. The global realignment of power will be interesting to contemplate.

Even in this worst-case situation, there will be survivors. We have not yet developed the capacity to destroy all of the world's population. For now, you have an opportunity to become one of the unfortunate survivors.

THE RADIATION PROBLEM

In a post-holocaust world, radiation will not be a major concern. This problem has to be considered in terms of how much additional radiation individuals will receive as a result of the nuclear exchange.

Radiation is a natural event. We are all exposed, on the average, to 150 mrem (milliroentgen equivalent to man) per year. This exposure causes the following annual health problems per million population: 41 fatal cancers, 25 nonfatal cancers, 46 genetic defects (not all of which are obvious). For every additional mrem per person per year, the above rates will increase .67 percent (75 cancers and genetic defects per 100 million population).

About a third of natural radiation is received from the sun, an ongoing thermonuclear explosion. About 15 mrem come from proximity to building materials, stone being the worst culprit. Living inside a stone building adds 50 mrem per year. The things we eat and drink contribute another 25 mrem. The remaining 60 mrem come from such manmade sources as: chest X-rays (9 mrem), air travel (1 mrem per 1500 miles), watching TV (1 mrem if you watch 6.67 hours a day), fallout from previous tests of nuclear weapons (4 mrem). Spending full time next to a nuclear power plant adds 5 mrem, less than 1 mrem 2 km away, and zero, 8 or more km distant.

Nuclear weapons are noted for instant blast and heat damage and longer-lasting radioactivity damage. The unit of radiation for nuclear weapons is the rad, which is equal, for our purposes, to 1000 mrem.

In most cases radiation kills over time. If enough radiation is received in a short period, it can kill immediately or within days, weeks or months, depending on the dose. Six hundred or more rads can kill within hours, they can disable immediately; 500 to 600 rads are always fatal, often within days; 200 to 500 rads will kill 50 percent or more of those affected; 100 to 200 rads will kill 5 percent to 50 percent, and at this point, the long-term effects (cancer) become more of a problem. Under 100 rads there will be a few prompt fatalities; the primary casualties will be long-term. A 50-rad dose will induce nearly 2 percent early deaths in the affected population from cancer and genetic defects.

These high levels of radiation exist for a short period of time, seconds in some cases. In the area closest to a ground-level explosion, which is preferred for destroying military bases, there will be hot spots of intense, longer-lasting radiation. One year after the explosion of a 1-megaton bomb, the 100-rad zone will be 46 square kilometers. Nearly 4 percent of the population living in this area will die early; many others will suffer radiation sickness. This should be the forbidden zone. The point of highest radiation is the explosion crater, 360 meters wide and 120 meters deep. The 50-rad zone, 67 square kilometers, will not have much radiation sickness, but as many as 1 in 50 inhabitants will die early. The 10-rad zone, 300 square kilometers, could be lived in, if every thousand inhabitants were willing to accept three or four early deaths. Within a few years, even the crater will be under the 10-rad limit. By that time most of the potential inhabitants will have died of starvation or disease.

By the way, you can forget about giant spiders and two-headed mutants. Insects are far more resistant to radiation than mammals. Most mutations are either fatal or unnoticeable. The sun's radiation has been responsible for more mutation than any nuclear war could ever produce.

HERE COMES THE SUN

One of the theoretical side effects of a massive nuclear exchange is the partial destruction of the ozone layer of the upper atmosphere, which screens the earth's surface from much of the sun's harmful radiation.

Partial destruction of the ozone layer could subject all inhabitants of the earth to perhaps hundreds of additional mrem per year. Worse, the additional ultraviolet radiation could blind land animals. In theory, a major nuclear war could destroy most of the ozone layer in the Northern Hemisphere, and somewhat less in the Southern Hemisphere.

But wait, worse things can happen. All these nuclear weapons will throw enormous amounts of dirt into the upper atmosphere. This would cause a buildup of heat on the earth's surface, cause the polar ice cap to melt and eventually trigger another ice age. Maybe it won't happen that way. But who wants to take the chance?

PUTTING IT ALL IN CONTEXT

The overall health effects of all this post-holocaust radiation should be put into context. Medical care, standards of living and nutrition will decline to levels today found in only the most primitive nations. While current death rates, per 1000 population, vary between 6 and 10 for developed nations, they climb to 20 or more for less developed nations. The initial blast and heat destruction of nuclear weapons will produce ten times more deaths than the long-term effects of radiation on the death rate. The destruction of our machines, structures and trained workers will kill even more people in the short run, not things that glow in the dark.

Strategic Weapons

WEAPON. This is the designation. For the Russian weapons, the NATO designations are used. Often a missile is modified over the years, resulting in variants with often very different capabilities. All weapons are arranged in order of the equivalent megatonnage they can deliver, in other words, in order of destructive capability.

MISSILES DEPLOYED. This is the number of missiles ready for use in underground concrete silos or on board submarines.

WARHEADS. The number of warheads per missile.

TOTAL WARHEADS. Missiles times the number of warheads per missile.

ON TARGET. Nothing is perfect. A certain number of missiles will be out of service for repairs or maintenance. Once launched, a certain number of missiles or warheads will not perform as planned. This number is obtained by multiplying the number of warheads by the serviceability and reliability levels. For example: The Minuteman II would have 450 warheads but only 90 percent would be serviceable, and only 80 percent of the remainder would perform reliably (450 × .9 × .8 = 324).

EQUIVALENT MEGATONS (EMT). Compares the destructive effects of nuclear weapons on all targets except hardened ones, namely underground missile silos. The formula is number of weapons times yield of each in megatons, to the two-third power. Thus, a 9-megaton weapon is equal to 4.34 EMT, a 1-megaton weapon is equal to 1 EMT, a 170-kiloton weapon is equal to .31 EMT, and a 40-kiloton weapon is equal to .12 EMT.

The effects of a 1-megaton weapon are as follows. If detonated in an airburst 2000 meters high, nearly every building within 7 km of the explosion will be destroyed or damaged beyond repair. Nearly everyone within this area will be killed or severely injured, without hope of medical aid. How large an area is this 14-km-diameter zone? Pick a ground zero, the point on the ground directly under the blast. Drive about 13 minutes in any direction at 20 mph. Or walk for an hour and a half. That's 7 km, or a total area of 154 square kilometers. Population densities in central cities vary by size of city, layout and time of day. Densities go from 3000 to 4000 (most new cities) to over 100,000 (New York City) per square kilometer. A single EMT would put any metropolitan area back in the Stone Age.

"K" FACTOR. This is the warhead's silo attack value. To destroy an enemy missile silo, the attacking warhead must explode as close as possible. The "K" factor is derived using the formula K = weapons yield to the two-third power divided by CEP to the second power. As you can see from the yields and the CEP's of the various weapons, accuracy is far more important than yield. Look at the difference between Minuteman II and III. The II has three times the yield as the III, 1000 kilotons versus 335 kilotons, but the III has about one third the CEP, 220 meters for the III, 630 meters for the II. This makes all the difference, giving the Minuteman III a "K" that is nearly four times larger than that of the Minuteman II. Look also at the MX and cruise-missile warheads. See the section below on SILO "K" DEFENSE.

TOTAL "K". Warheads reaching target times each warhead's "K" factor.

SILO "K" DEFENSE. This is the missile "K" factor needed to disable a missile silo with 97 percent probability. Thus, a 100-psi (pounds per square inch) silo requires 20 "K"; 300 requires 45; 1000 requires 108; 2000 requires 150; 3000 requires 200. The higher the "K" value, the more likely the silo will be rendered unusable. With the "K" factor, weapon explosive force is not nearly as important as accuracy. The nuclear weapon explodes near ground level to transmit the maximum shock effect to the underground silo. From the air the silo appears as a 30-meter circular object. The silo itself can be hardened to withstand blast pressures of up to 3000+ psi. But this is expensive. Hardening to 3000 psi raises the silo cost to over $5 million. Even going from a 300-psi silo to a 1000-psi silo will cost $1.2 million per silo. Not applicable (NA) to submarine-based missiles.

TOTAL "K" DEFENSE. Missiles deployed times silo "K" defense. For one side to defeat the other in a nuclear war without suffering unacceptable damage, the attacker would have to be able to destroy the defender's nuclear forces before the defender could use them. At the moment this is impossible because no one has developed a way to quickly destroy the SSBN's (missile-carrying submarines). Also, no matter how high an individual missile's "K" factor is, it does not absolutely guarantee destruction, because of reliability and other problems.

CEP (circular error probable). This is the measure of a missile's accuracy. The CEP is expressed in meters from the target point. This circular area represents the area in which 50 percent of the missiles with that CEP will fall. Farther out, the circle eventually covers an area in which more than 99 percent of the missiles will fall. The CEP represents a convenient midpoint for measurement. Given the fact that the destruction zone for an urban target of a 1-megaton weapon is a circle 14 km in diameter, CEP's of under 1000 or even 2000 meters are not that critical. If you want to destroy hardened pinpoint targets, like enemy missile silos, accuracy becomes more critical. Even when these electromechanical wonders perform to their theoretical maximum, they cannot get below a CEP of 150 meters (400 meters for submarine launch). The 150- to 400-meter barrier is to be broken with terminal guidance, usually a form of radar in the warhead that identifies and homes in on the target as the warhead plunges to earth. In spite of all theoretical technical improvements, past experience shows that equipment of this complexity rarely performs to its specified level. Many defense analysts degrade CEP estimates by at least 200 to 400 meters. To see what effect this has, look at the Russian SS-9 (CEP 750) and SS-18 Mod 3 (CEP 400): there's a large difference in "K" factor (45 versus 207).

RANGE. The maximum range of the missile. There is also a minimum range of up to a few hundred kilometers for any missile.

WARHEAD YIELD. The destructive power of nuclear weapons is expressed in terms of kilotons (thousand tons) of conventional high-explosive TNT.

RAW MEGATONS. As opposed to equivalent megatons, Raw megatons are just that, the number of arriving warheads times the yield of each.

PERCENT SERVICEABLE. This is the percentage of missiles that will be available for launch at any given time. Each missile is a very complex piece of machinery. Today most use solid fuel instead of the earlier liquid fuel, eliminating the problems of maintaining all that plumbing and volatile liquid. Numerous mechanical and electronic components can still, and will, eventually go bad. Keeping a high percentage of these weapons ready starts with a reliable missile system. Newer systems are less reliable until all their bugs are discovered and put right. Because these weapons are never actually tested, it is quite probable that bugs will go undetected. Then, readiness requires competent maintenance personnel and a dedicated inspection policy. The serviceability figures are estimates based on the little information that has been made available to date.

PERCENT RELIABLE. Once launched, a certain percentage of the launched systems will not work. This often happens during testing. In actual combat, the unreliability will probably be much higher. For aircraft and cruise missiles this

STRATEGIC WEAPONS

Accuracy, which has never been tested thoroughly, is the most important factor. Reliability in large-scale operations is another murky area.

Weapon	Missiles Dep.	Warheads	Total Warheads	On Target	Equiv. Megatons	"K" Factor	Total "K"	Silo "K" Def.	Total "K" Def. x1000	CEP meter	Range (km)	Warhead Yield (kt)	Raw MT	% Serviceable	% Reliable	Throw Wght tons	Year Deployed
US Total	1708		7272	4067	1275		38094		110				999				
Russia Total	2387		7145	3804	3267		89502		189				3390				
United States																	
Land Based																	
Minuteman III	300	3	900	689	332	34	23560	108	32400	220	12800	335	231	90	85	.99	1970
Minuteman II	450	1	450	324	324	9	2803	108	48600	630	12800	1000	324	90	80	.81	1965
Minuteman III	250	3	750	574	176	11	6093	108	27000	315	12800	170	98	90	85	.99	1970
Titan II	36	1	36	20	88	7	137	45	1620	1482	11665	9000	182	75	75	4.1	1963
CSS-4 (China)	18	1	18	8	24	4	37	0	0	1500	6000	5000	41	70	65	4	1970
CSS-3 (China)	12	1	12	7	7	1	6	0	0	2000	6000	1000	7	80	75	1.5	1970
MX	0	10	0	0	0	204	0	108	0	90	12000	335	0	90	90	1.5	1986
Minuteman I	0	1	0	0	0	4	0	0	0	927	12000	1000	0	85	75	.7	1962
Total	1066		2166	1622	951		32636		110				883				
Submarine Based																	
Poseidon (C3)	480	10	4800	2304	269	2	4316	NA		463	4600	40	92	60	80	1.2	1971
Polaris A3	80	1	80	36	26	3	103	NA		926	4600	600	22	60	75	.5	1964
MSBS M-20 (FR)	96	1	96	43	43	4	173	NA		926	3100	1000	43	60	75	1	1977
Polaris A3 (UK)	64	1	64	29	20	3	82	NA		926	4600	600	17	60	75	.5	1964
Trident (C4)	16	10	160	77	17	12	909	NA		250	4600	100	8	60	80	1.6	1979
Polaris A2	0	1	0	0	0	3	0	NA			2800	200	0	60	65	.5	1962
Cruise Missile	0	1	0	0	0	1305	0	NA		30	2400	200	0	60	40	.25	1985
Trident II (D5)	0	14	0	0	0	24	0	NA		200	7400	150	0	60	80	2.6	1987
Total	736		5200	2489	375		5583						182				

Russia

Land Based

SS-19 Mod	1	280	6	1680	945	634	14	13614	150	42000	400	8000	550	520	75 75 3.75	1974
SS-18 Mod	4	201	10	2010	1131	712	35	39130	150	30150	250	8800	500	565	75 75 8.35	1979
SS-18 Mod	2	107	8	856	482	449	20	9632	150	16050	400	8800	900	433	75 75 8.35	1976
SS-19 Mod	2	100	1	100	72	335	255	18423	-150	15000	250	8800	10000	723	85 85 3.5	1978
SS-17 Mod	1	160	4	640	360	297	18	6377	150	24000	400	8800	750	270	75 75 3	1975
SS-11 Mod	3	470	1	470	282	273	2	477	110	51700	1400	9700	950	268	80 80 1.1	1966
SS-18 Mod	3	0	1	0	0	0	207	0	150	0	350	12000	20000	0	85 85 8.25	1977
SS-17 Mod	2	20	1	20	14	45	71	964	150	3000	400	9000	6000	82	80 85 3	1977
SS-13		60	1	60	41	29	1	28	110	6600	1900	8000	600	24	85 85 .75	1969
SS-7		0	1	0	0	0	2	0		0	1900	10000	4000	0	75 75 2	1962
SS-8		0	1	0	0	0	2	0		0	1900	11000	3000	0	75 75 1.75	1967
SS-9		0	1	0	0	0	45	0	110	0	750	11000	20000	0	80 80 5.5	1971
SS-16		0	1	0	0	0	10	0	110	0	500	8800	650	0	85 85 1	1978
SS-18 Mod	1	0	1	0	0	0	179	0		0	400	9600	24000	0	80 85 8.25	1974
Total		1398		5836	3326	2774	179	88645	189					2885		

Submarine Based

SS-N- 18		160	3	480	234	234	2	410	NA	NA	1400	7200	1000	234	65 75 2.6	1977
SS-N- 6		468	1	468	106	140	3	283	NA	NA	1300	2500	1500	160	35 65 .75	1968
SS-N- 8		289	1	289	122	101	1	154	NA	NA	1500	7700	750	92	65 65 2.4	1973
SS-N- 5		57	1	57	12	16	0.57	7	NA	NA	2800	1400	1500	18	35 60 .75	1963
SS-N- 17		12	1	12	3	2	1	3	NA	NA	1400	3200	500	1	35 70 1.6	1977
SS-N- 4		3	1	3	1	1	0.44	0	NA	NA	2800	500	1000	1	35 55 .5	1961
SS-N-6 Mod 3		0		0	0	0	3	0	NA	NA	1300		1500	0	35 .75	1974
SS-N- 20		0	6	0	0	0	2	0	NA	NA	1000	9000	500	0	65 70 2.2	1983
Total		989		1309	478	492		858						505		

percentage also includes systems shot down on the way to their targets. Being shot down en route is not yet a problem for ballistic missiles.

THROW WEIGHT. The total weight of the missile that can be delivered to the target. In the beginning this was all allotted to the nuclear weapon. As MIRV's were introduced, some of this weight was taken by the multiple warhead-carrying bus. A warhead can also carry penetration aids, decoys and electronic devices to help the warhead past future enemy defenses. Future guidance systems may also rely on terminal guidance, which require more equipment in the warhead. It's getting crowded in there.

YEAR DEPLOYED. The year in which the missile was first available for use. Like any piece of machinery, a missile can be made to last forever by replacing worn parts. Although the missile is used only once, it is alive while sitting around waiting. The guidance system is always on whenever the missile is available. The electrohydraulic systems that work the mechanical controls, flaps, fins and air brakes must be exercised periodically. The exercise will eventually fatigue them. The fuel, even if solid, is a volatile chemical that deteriorates over time. The longer the missile sits in its hole in the ground, the lower its reliability and serviceability rates.

Strategic nuclear weapons are weapons of mass destruction. Their purpose is to destroy a nation's ability to wage war by destroying military forces, bases and means of producing weapons and raising troops. In other words, they are designed to destroy people and their economies.

The primary weapon for this destruction is the ICBM, intercontinental ballistic missile. ICBM's are launched from underground concrete silos or from submerged submarines. Most are fired over the North Pole between Russia and North America. As they leave the earth's atmosphere, they orbit until the warhead reenters the atmosphere at speeds up to 14,000 meters per second. At this high speed the heat shield of the warhead glows brightly enough to be seen for miles. Total travel time: up to half an hour from launch.

Aircraft could also carry nuclear bombs and missiles. Shorter-range ballistic missiles can also be used for nation smashing. Coming into use are nonballistic weapons, cruise missiles. Weapons such as these have never been used. Aside from the two nuclear bombs dropped on Japan in 1945, no one has dared use them.

Both Russia and the United States have radar and satellite systems that would give at least fifteen minutes' warning of an enemy attack. This would be sufficient time to launch one's own missiles if the decision were made quickly. At least five minutes are required to launch once the word is given. It is debatable whether a decision of such magnitude would be made that quickly. This is one system that is difficult to test under realistic conditions. Instant response has become known as MAD (mutually assured destruction). Balance of terror is a more euphemistic term.

A selected response strategy has recently become fashionable. It is based on the untried premise that the missiles could be used a few at a time in a nuclear chess game. A discussion of nuclear weapons effects further on will allow you to decide the practicality of such a strategy.

There is no effective defense against ballistic missiles. Theoretically, laser or charged-particle beams could damage the warhead heat shield (to induce over-heating on reentry) or damage the warhead electronics (to prevent detonation or accurate guidance). The chief problem is fire control. Sensors must track a small object over thousands of kilometers. You must then hit that target and sense if you have destroyed it before going on to the next target. No one is very close to having a working model of this system. Missiles launched to intercept the ICBM's have the same problems. This system has worked in tests but is more expensive than the missiles it is designed to shoot down. Not highly reliable.

Tactics and Strategies

Both Russia and the United States have about ten times more nuclear warheads than they need to guarantee mutual destruction. In such a situation, nuclear "tactics" becomes irrelevant.

For example, each warhead guidance system can be set for a specific target. In the more advanced U.S. systems, the missiles can be retargeted rather quickly, within minutes in some cases. But this becomes a factor only if you want the option of destroying only military installations or going after everything. As the damage to nonmilitary population and structures from a purely "military" strike may not be much less than a mass strike, this is a moot point.

Missile Construction Techniques

The basic principles of ballistic-missile construction were developed and tested over fifty years ago. The first practical application was during World War II, the German V-2 rocket. Technology has become more refined, but not radically different.

A rocket must attain a speed of 6000 to 7000 meters per second to escape the earth's gravity. This is achieved by stacking a series of rockets on top of each other. The first rocket, or "stage," comprises more than 75 percent of the total vehicle weight. Once in the upper atmosphere, this stage is dropped and the second stage (15 percent to 20 percent of vehicle weight) puts the warhead into an orbit that will take it over 10,000 km to the target. The third stage is the unpowered warhead, usually less than 3 percent of vehicle weight. The warhead is equipped with a shield to prevent burnup on plunging through the earth's atmosphere at speeds in excess of 13,000 meters per second.

The U.S.-developed, powerful solid-fuel motors have changed missile design by eliminating much plumbing. It was possible to have smaller missiles and more stages. The warheads were also smaller, although 15 percent to 20 percent heavier in relation to total missile weight than warheads in liquid-fuel rockets. With more efficient warheads and guidance systems, the United States was able to build missiles weighing less than one fifth as much, but having equal range, warhead yield and superior accuracy. No less critical was the lower cost of such a system.

Higher manufacturing and maintenance costs lead the Russians to spend three times more money per missile. Russia was apparently forced to use liquid-fuel missiles by their less capable technology. They have developed solid-fuel rockets for their submarines, where weight and space are restricted. But their earlier sub-launched missiles (SS-N-4 and SS-N-5) were liquid fueled. Of their land missiles, only the SS-16 uses solid fuel. The earlier SS-13's were apparently not very successful; few were deployed. Neither is the more recent SS-16 being deployed, although a chopped version without the third stage is in use as the SS-20.

Liquid-fuel rockets are used primarily for lifting satellites. The only liquid-fuel rocket in U.S. ICBM service (the Titan II) is retained primarily as a bargaining chip in arms-limitation talks with the Russians. This same missile type launches satellites.

NATIONAL DIFFERENCES

As is obvious from the chart, Russian missiles are larger and less accurate than their U.S. counterparts. Russian missiles were originally heavier because the technology did not allow them to manufacture nuclear weapons as powerful, pound for pound, as their U.S. counterparts. This heavier throw weight gave them room to develop MIRV warheads, whose additional mechanical and guidance systems were also heavier than equivalent U.S. devices. The larger-yield nuclear weapons on Russian missiles were a practical solution to the lower accuracy of their guidance systems.

The Russians have been developing a cold-launch technique in which the missile is propelled out of its silo by nonexplosive gases before the rocket motor is ignited. The silo can be reloaded in a day or two for another shot. A hot launch, in which the missile motor ignites in the silo, makes the silo unusable for a period of weeks. Currently, Russia's SS-17 is cold-launched.

Russia depends largely on its land-based ICBM's for waging strategic warfare. The chart shows the history of Russian ICBM development. A number of systems were developed during the 1960's, with the SS-11 finally taking the lead. The Russians often deploy many systems for the same job and let experience point out the best contender.

Another aspect of Russian ICBM development is the deployment of a heavy and a light ICBM in each generation. The current generation (SS-17, -18, -19) is characterized by higher throw weights and more sophisticated guidance systems and warheads. In addition, the Russians continue to produce shorter-range missiles. This makes sense, considering the number of potential enemies they have as neighbors.

The United States' technological superiority has allowed it to put much more of its nuclear force at sea in submarines. Given the Russians' track record, it is quite likely that their submarine-based missiles are not nearly as effective as their land-based models. Time and money spent on development will slowly remedy this situation.

Aircraft and Tactical Missiles and Artillery

WEAPON is the system designation. Weapons are listed in order of their equivalent megatonnage. More destructive systems are ranked higher. Most of the non-superpower systems show up on this chart: UK = United Kingdom; FR =

France; NATO = NATO countries that possess vehicles to carry nuclear weapons but not the weapons themselves (West Germany, in particular); China = People's Republic of China; CV based indicates U.S. aircraft flying from aircraft carriers; ALCM indicates a bomber equipped with air-launched cruise missiles—this will eventually be the same missile described under the missiles-artillery section. Currently bombers carry SRAM, described in the aircraft weapons chart. Russian bombers all can be equipped with a variety of ALCM, also described in the aircraft weapons chart.

VEHICLES DEPLOYED are the number of aircraft or missiles in use. The "0" next to Cruise Missile indicates it has not yet been deployed.

WARHEADS PER VEHICLE. For aircraft, indicate the number of nuclear weapons carried on one sortie. Western aircraft have sufficient nuclear weapons for two or three sorties. Russian aircraft, perhaps half that number. For larger missiles, those with a range of over 1400 km, it indicates the number of warheads carried by each missile. For all other weapons it covers the total number of nuclear missiles or artillery shells available.

TOTAL WARHEADS, WARHEADS REACHING TARGET, EQUIVALENT MEGATONNAGE, RAW MEGATONNAGE AND WARHEAD YIELD are described under the strategic missiles chart notes.

RANGE is the maximum distance a missile or an artillery shell will travel (in kilometers). For aircraft it is the outbound leg of a round-trip mission.

SERVICEABILITY is the percentage of that system that will be available for use at any instant. This will decline once the fighting starts. In peacetime large quantities of spare parts are kept aside as the war reserve. Additionally, wartime maintenance standards are lower than those of peacetime. Once the fighting starts, many of the peacetime rules are broken in order to keep the aircraft flying.

SURVIVABILITY is the percentage of serviceable systems that will make it to the target and deliver their nuclear weapons. The survivability figure also includes losses taken before missiles can even be launched, as well as reload missiles and artillery shells that are destroyed before they can be used.

FIRST DEPLOYED is the year the system was first available for use.

National Differences

The U.S. nuclear warheads are split somewhat equally among land-based ICBM's, submarine-based missiles and aircraft. The current U.S. ICBM's were first developed twenty years ago and have been steadily upgraded. The same approach has been followed with bombers. The venerable B-52 has been rebuilt and reequipped many times. The airframes are showing their age. The required

AIRCRAFT AND TACTICAL MISSILES AND ARTILLERY

More apt to be used, better tested and more numerous, these weapons are likely to be the ones that deliver the holocaust.

Weapon	Vehicles Deployed	Warheads per Vehicle	Total Warheads	Warheads Reaching Target	Equiv. Mega-tons	Raw Mega-tons	Raw Warhead Yield (kt)	Range (km)	Service-ability	Surviva-bility	First De-ployed
NATO Totals	4330		9143	5477	2068	1954					
Russia Totals	4385		9050	4655	1523	1148					
NATO											
Aircraft											
B-52G/H	240	12	2880	1512	1303	1210	800	12000	75	70	1959
B-52D	75	4	300	98	285	488	5000	10000	65	50	1959
FB-111A	60	6	360	230	79	46	200	4000	80	80	1969
F-4	520	1	520	195	67	39	200	750	75	50	1962
F-111E/F	156	2	312	187	64	37	200	2400	80	75	1967
A-7E (CV based)	144	2	288	97	33	19	200	900	75	45	1966
Buccaneer (UK)	60	2	120	39	25	20	500	950	65	50	1962
F-104 (NATO)	300	1	300	74	25	15	200	800	70	35	1958
A-6E (CV based)	60	2	120	59	20	12	200	1000	75	65	1963
Vulcan B-2 (UK)	48	2	96	31	20	19	500	2800	65	50	1960
Jaguar (UK,FR)	80	1	80	42	14	8	200	720	80	65	1974
Mirage IVA (FR)	33	1	33	11	11	11	1000	1600	65	50	1964
Tu-16 (China)	50	1	51	11	11	11	1000	2100	60	35	1968
Super Etendard(FR)	36	2	72	22	7	4	200	560	75	40	1980
Mirage IIIE (FR)	30	1	30	9	3	2	200	600	75	40	1964
Aircraft Total	1892		5562	2615	1966	1939					

Missiles/Artillery											
CSS-2 (China)	90	1	90	60	60	60	1000	2500	70	95	1971
Pershing (NATO)	180	3	540	462	34	9	20	720	90	95	1964
203mm How (NATO)	300	4	1200	972	28	5	5	16	90	90	1962
Lance (NATO)	108	6	648	467	22	5	10	110	80	90	1976
155mm How (NATO)	1800	.5	900	729	12	1	2	16	90	90	1964
Honest John (NATO)	40	5	200	144	7	1	10	40	80	90	1953
Pluton (FR)	42	5	126	85	6	2	20	120	75	90	1974
SSBS S-3 (FR)	18	3	18	14	4	2	150	3000	90	85	1980
Cruise Missile	0	1	0	0	0	0	200	2400	45	40	1985
Missiles/Artillery	2578		3722	2932	172	85					
Russia											
Aircraft											
SU-24	600	2	1200	546	187	109	200	1600	65	70	1974
TU-22M (Backfire)	200	4	800	420	144	84	200	4000	70	75	1974
TU-16	560	2	1120	328	112	66	5000	2100	65	45	1955
TU-95	100	1	100	32	92	158	5000	6200	70	45	1955
MiG-27	600	1	600	254	87	51	200	720	65	65	1973
TU-22 (Blinder)	165	2	330	119	41	24	200	750	60	60	1962
MYA-4	40	1	40	13	37	63	5000	4800	70	45	1956
Aircraft Total	2265		4190	1710	699	554					
Missiles/Artillery											
SS-20	315	3	945	567	160	85	150	4300	75	80	1977
SS-12	120	2	240	143	143	143	1000	900	70	85	1966
SS-4	300	1	300	117	117	117	1000	2000	65	60	1959
FROG-7	480	3	1440	842	114	42	50	70	65	90	1967
SS-23	90	2	180	105	105	105	1000	350	70	90	1980
SS-21	240	2	480	302	65	30	100	120	65	90	1971
SS-22	200	2	400	234	50	23	100	1000	70	90	1979
SS-1C (SCUD B)	180	2	360	227	49	23	100	300	70	90	1962
SS-5	35	1	35	18	18	18	100	1400	70	75	1961
180mm Gun	160	3	480	389	2	8	20	30	90	90	1969
Missiles/Artillery	2120		4860	2945	824	595					

constant training has made it difficult, and expensive, to keep them flying.

An additional major system, the cruise missiles, will be introduced in the next five years. The U.S. cruise missile is something of a strategic bomber without a crew. It is so small, and can fly so low that existing air-defense systems are much less able to deal with it. Russia also fears that these missiles will eventually be equipped with penetration aids in the form of miniature electronic counter-measures.

Cruise missiles can also be used against tactical targets with conventional or chemical warheads. Ultimately this may be the major practical use for these missiles. Cruise missiles have a terrain-recognition navigation system and are capable of extremely accurate terminal homing. Thus, they are the best weapon for going after point targets deep in enemy territory (headquarters, communications sites, transportation bottlenecks, etc.).

Western aircraft can be divided into strategic and tactical according to range. Anything over 1500 km is generally considered strategic. From the Russian and European point of view, such arbitrary distinctions are meaningless. Most major Russian and European cities are within range of tactical Western aircraft. United States targets are largely immune to enemy aviation. Most Russian long-range aircraft are intended for naval targets. The other nations' aircraft listed are basically tactical aircraft equipped with nuclear weapons targeted against threatening neighbors (Russia, for the most part).

Russia deploys a large number of shorter-range missiles against European cities. Most U.S. short-range missiles are intended for battlefield targets. French, British and Chinese missiles are targeted largely against Russian cities and military bases; they give some leverage against the large number of Russian missiles and aircraft pointed toward Europe and China.

All aircraft can deliver conventional bombs and missiles. In a war most would be used this way most of the time.

Missiles can deliver only nonnuclear weapons if nonnuclear warheads are available. If, as usual, they aren't, then it's either nukes or nothing. Missiles equipped with nonnuclear warheads are becoming increasingly common as they become accurate enough to destroy a target without nuclear weapons.

Chinese missiles are not mobile. Neither are the larger Russian models (SS-4, -5, -20). The French SSBS also operates from a silo. All the other missiles are mobile, either self-propelled or towed.

Part 6

WARFARE BY THE NUMBERS

LOGISTICS

LOGISTICS IS THE art and science of supplying the troops with weapons, ammunition, fuel, spare parts, replacement equipment, food. As this is not a very glamorous aspect of warfare, it is often ignored by soldiers. Such lack of dedication often leads to disasters.

Grim Numbers

Logistics disasters are quite common in military history. They occur not only because commanders ignore logistical matters, but also because the enemy disrupts supply arrangements. Most of the military effort expended in the Vietnam War was an attempt to disrupt North Vietnamese supply. The results of the largest bombing campaign in history were mixed, although the bombing did cause the North Vietnamese enormous casualties and considerable trouble. A similar campaign was waged against the Chinese in Korea (1952–1953). The results were similar to the bombing of Italy in 1943–1944; the transportation facilities were so large and the military supply requirements so small, that air power was not able to completely choke off the flow of supply.

Consider the situation. A double-line railroad can, under wartime conditions, move at least 50 or more trains (400 tons each) in either direction each day. That's 20,000 tons a day. A double-lane, hard-surface road can handle at least as much traffic, although at much greater expense.

A nonmechanized army requires only 15 to 30 pounds of supply per man per day. Every 1000 tons of supply keeps 100,000 men in combat. If one road or rail line enters the area occupied by 100,000 men, over 95 percent of the transportation capacity must be destroyed before the flow of supply would be appreciably hurt. Worse still, combat capability won't be reduced until over one third of the requirements are denied. Once that level is reached, for every percentage point of supply denied, 1 percent of the unit's combat power is lost. Even completely cut off from supply, the average unit still retains one third of its combat power. Unit mobility would be damaged more seriously. The more motorized the unit, the more it would have to choose between immobility or losing combat power through abandonment of vehicles.

Once combat power is lost, the affected unit takes a beating in combat and, if the condition persists for weeks or months, medical problems further decrease personnel strength and morale. But that unit is still capable of combat even after complete loss of supply. How is this possible? Troops with a lot of supply will find a way to use it. This has long been recognized by commanders. When automatic weapons were first developed a hundred years ago, many commanders opposed their introduction on the grounds that they would encourage the troops to waste ammunition and create supply problems. This attitude is not without foundation. The old saying "make every shot count" usually applies only when the troops have little ammunition and no other choice. Necessity is the mother of efficient use of supplies. When supplies dry up for any reason, expedient methods are found to get by with less.

When the enemy begins to attack the supply lines, the dumps and means of transportation, the cost of logistics goes up. Both supplies and transport are destroyed and replaced at no small expense. When resources are available, the attacker runs up the greater cost. A single aircraft sortie costs up to half a million dollars. A truck costs less than $50,000. While a train can cost millions of dollars, the usual targets are the railroad bridges and tunnels and repair yards, which are either repaired or detoured. Moving matériel by truck around a railroad cut or by ferrying rolling stock across a river by barges are but a few of the ways in which a determined logistics force can continue delivering the goods. Worst of all, the attacker is hard pressed to measure accurately whether significant damage is being done to the enemy supply position.

The Modern Major General's Dilemma

Few of us have much logistical experience beyond getting the groceries from the supermarket to the kitchen. But consider some of the basics for contemporary soldiers. Each person needs about 6 pounds of food daily plus 20 pounds of water, which in the field often has to be delivered. For a division of 15,000 men, this comes to 177 tons to be moved each day. Such basic necessities are the least of the supply officer's worries. For all operations, over 60 percent of the weight of supply will be fuel. The next largest category, ammunition, takes up some 20 percent of the weight to be transported.

This is in sharp contrast to World War II supply needs. The German Army, somewhat of a cross between the largely motorized Western armies and the largely unmotorized Russian Army, required an average of 28 pounds per man per day of which 40 percent was ammunition, 38 percent fuel (one quarter being fodder for the horses) and the remainder everything else. United States units required 55 pounds per man per day, 50 percent ammo and 36 percent fuel. Current U.S. divisions require an average of 200 pounds per day per man, 20 percent ammo and 60 percent fuel. The air forces' supply needs have increased much like the armies' have, while the navies' needs have not. This is due to nuclear propulsion, smaller crews and ships, and missiles (fewer rounds, lower rate of fire equals lower total weight).

Although modern armies are burning more fuel and firing more ammo, this does not always mean an equal increase in combat performance. All that extra fuel is needed to move the vehicles around. Many of the armored ones are not truly combat vehicles but transports. Ammunition does relate to combat power, but its effectiveness has been decreased by the proliferation of armored vehicles. Thus, the 40 percent increase in ammunition weight since World War II has not done much, unless you are fighting a World War II type of army. The primary difference between World War II and today is that the logistics people have to move five times as much matériel.

Rules of Thumb

When dealing with logistics, it is best to start with the larger numbers and then work into the details. Every soldier outside his own country needs 100 pounds of supply per day. This includes combat and noncombat troops. Each sailor needs four to six times that amount, and each airman up to 1000 pounds a day. Troops serving on land can be cut down to under 50 pounds a day if they are not doing anything. Sailors at sea still require 300 pounds a day just to keep the ships operational.

When supply is moved by sea or rail, the fuel required for transport is not a significant factor. To move a ton of matériel 100 km, a train uses 14 ounces of fuel, a large ship uses about half that. When matériel is moved by truck or air, it's a different story. By truck, 1 percent of the weight moved will be consumed as fuel for each 100 km moved. By air, the cost will be from 2 percent to 5 percent, depending on the type of aircraft. Large civil-type jets will use the least amount. Helicopters are notorious fuel hogs and can consume to 10 percent of their cargo weight for each 100 km traveled. Moving supply by animal, including humans, will have the same fuel cost as aircraft in terms of food. A recent innovation is the portable fuel pipeline, which is more than twice as efficient as trucks but more vulnerable.

Obviously a lot of supply will be in transit at any time. There will be transportation problems, and the supply demand will vary with the tempo of operations. Therefore, readied reserves for 30 to 90 days are always maintained. For land operations, over 10 tons of supply are stockpiled for each man in the area. For an army of 250,000 men, this will be a fairly large quantity of supply to store and track—two and a half million tons, more or less. The dimensions of the problem are becoming clearer.

Running the Store

A general or an admiral has been educated in a number of disciplines. But logistics will be known intimately more than any of the others. Even in peacetime, large quantities of supply must be moved. Valuable experience is thus gained for the more massive demands of wartime.

In peacetime the logistics officer is concerned primarily with maintaining the war reserve stocks and the regular flow of spare parts, food and fuel. It is important to realize the critical importance of these two supply items.

War reserve stocks are stockpiles of supply which will get units through the first thirty or more days of combat. They are absolutely essential to a unit's wartime effectiveness. Immediate resupply from the civilian economy is unlikely, and these reserves are all the fighting units will have initially. The war reserve stocks contain everything needed: ammunition, fuel, food, spares and supplies of every description. A three- to five-day supply is carried with each division at all times. The remainder is stockpiled at the rear. "Getting into the enemy's rear" often means that the enemy is soon going to be hungry, immobile and fighting with empty weapons.

The placement of war reserve stocks is critical. Ammunition is often stored in bunkers or under some sort of cover. Because so much of it consists of missiles, this degree of protection is necessary. If it is stored in bunkers, a degree of protection against enemy air attack is also provided. Fuel, because of its chemistry, is difficult to store and tends to degrade with time. The fuel stocks must therefore be rotated regularly. The same problem exists, to a lesser degree, with ammunition. The war reserve stocks should be stored far enough in the rear to avoid being overrun by an enemy advance. At the same time, these stocks shouldn't be so far to the rear that a lot of fuel has to be burned getting them to the troops. While railroads would eliminate this fuel cost, the rail network would be a prime enemy target, so only trucks could be depended on.

Spare parts and replacements for lost equipment have always been a problem for armies. With sophisticated weapons, equipment and munitions, spares become a critical problem in peacetime also. Most military equipment is similar to civilian goods, like automobiles, electronic equipment and sundry gadgets, but more complex and more prone to breakdown. Designing military equipment for fast and simple replacement of parts works as long as there are sufficient stocks of spares. Much progress has been made in this area lately, and many recent designs actually have fewer replaceable parts than previous ones. A good example is the F-15 fighter, which has far fewer electronic components than the F-4 aircraft it replaced. The F-15 components are also easier to identify and replace when they fail. This all works very well as long as you have the spares. A spares kit to keep an F-15 squadron of 24 aircraft operating for 30 days in wartime costs $40 million, or $150 million if components are just replaced and not repaired by unit technicians. The weight of the parts is not great, just their value. Their enormous cost often limits the quantity provided. When budget decisions have to be made, there is always the temptation to ease up on spares in order to get more equipment into the field.

Under the best conditions, the spares situation presents a number of massive problems. A huge number of individual parts must be kept in stock. Literally millions of individual items must be tracked. The spares themselves must often be maintained. In the case of electronic spares, the part is often a complex piece of equipment itself. The quantity of each spare to be maintained is determined only precisely by experience. Many parts will fail at a different rate during wartime than during peacetime. Consumption rates are always inaccurately predicted for new weapons. When the spares problem becomes critical, the users will have to fire it less intensively and strip slightly damaged equipment for parts. As much as fuel and ammunition, these parts are essential to keep a mechanized army functioning.

Some forms of supply can often be obtained locally. Food, water and fuel are commonly foraged. Unlike the good old days when the troops were instructed to grab whatever wasn't nailed down, living off the local economy must be conducted in an organized manner for the best results. Much of this planning is done in peacetime. In Europe much of an American army's requirements can be supplied by European industry, but it may be disrupted or destroyed during a war. In some parts of the world, a strong economy is close enough to supply much war matériel. Japan really got started in its postwar reconstruction during the Korean War; a wide range of materials were made in Japan rather than being shipped 11,000 km from the United States.

The widespread introduction of computers has helped the logistics planner stay on top of the situation. Yet the very complexity of logistics has made it more vulnerable to disruption. There has been no real experience in maintaining such technically sophisticated armies in a major war. Not only do logistics planners face unprecedented problems, they also have to fight the ever-present general attitude that the supply situation will somehow take care of itself. It won't.

Us vs. Them

In modern warfare two different styles of logistics have developed. The Western style, best exemplified by the United States, can best be compared to a small, highly automated factory capable of mass-producing firepower, but only if constantly fed enormous quantities of raw materials and energy. Take away the supplies, and there is a lot of expensive, idle and useless machinery.

The other style could be called Russian or traditional, since this is how logistics functioned in the past. The Russian armed forces have fewer munitions for each weapon. They have a larger factory using less raw material. The Russian style emphasizes munitions, fuel and everything else, in that order. Food and other nonmunition/fuel supplies may never reach the troops. Units are encouraged, by hunger if nothing else, to live off the land. Russia, operating with a less efficient economy, has learned to get by with less.

Russia's inability to move supply as efficiently as Western armies has forced it to try to begin a major operation with the combat units carrying all the supplies to complete the mission. This gives Russian operations a stop-and-go quality. Units advance for a few days and then stop for resupply. At that point they make another jump. Western armies attempt to maintain continuous operations, fueled by a continuous flow of supply.

Whichever supply method is used, a division maintains supplies at three different levels. The lowest level is the weapon. This load of ammunition is calculated in units of fire, what a weapon normally carries. A tank carries about a ton of munitions and up to a ton of fuel. Artillery units usually carry several tons of munitions per gun. Infantry weapon units of fire are lighter, because the weapons are smaller and the infantrymen often have to carry the ammunition themselves.

The battalion of the weapon carries additional units of fire and fuel fills in trucks, usually only one additional unit of fire and fuel fill. The division supply trucks carry the bulk of the additional ammunition and fuel. Often the division will carry all the additional fuel. Supply held at higher levels is often not mobile, but is stored in dumps until needed. Divisions carry less than a week's worth of supply; as trucks unload, they go to these dumps for more. During all this travel, the supply trucks are most vulnerable to destruction.

A Russian division attempts to carry five units of fire and five fuel fills weighing 1000 tons of fuel per fill and 1000 tons per unit of fire. If possible, more artillery ammunition is carried—up to a dozen units of fire. The limiting factor is transport. Most Russian divisions, when mobilized, seize civilian trucks. If experience is any guide, this will be a rough-and-ready exercise. Each division will get all the trucks and drivers it can smoke out on short notice.

A Western division is less independent of its supply lines of trucks. The units of fire weigh a bit more, 1500 tons, but a fuel fill is still 1000 tons. In addition, Western divisions expect regular shipments of fuel; spare parts; and numerous other supplies: paper for the mimeographs, wire for the signal troops, clothing, etc. The Russians are expected to go without.

The important point is that both logistical systems rely largely on unarmored trucks to move most of the goods. With today's much larger supply requirements, more trucks are running around. Even if the average supply truck hauls 5 tons, 400 trucks are needed to resupply all the ammunition racks and fuel tanks of a division. All that very explosive cargo makes the trucks vulnerable. Modern tactics also make much of the "fluid battlefield," with friendly and enemy units wandering around and about each other. This is rough on all those truck drivers.

Incidentally, the age-old practice of soldiers looting has not disappeared, even though it is illegal. Arming a man still seems to change his concepts of property rights, even though modern warfare keeps a soldier more occupied than in the past. Raping and plundering are regarded as serious disciplinary problems.

Doing Without

Contingency planning is an article of faith with logistics people. There always comes a time when things don't work out as planned. At this point they must either do without or pull an alternative plan out of the hat. Often they end up doing both.

An intelligent and capable commander will always attempt to deny the enemy supply while preventing his opponent from doing the same. Most combat aircraft sorties are flown against enemy supply lines and dumps. If one side gains permanent air superiority, his opponent can usually forget about victory. With an enemy air force overhead, supplies are constantly being hunted down and destroyed. In past wars it was possible to hide much supply from enemy observation. Destruction from the air was a problem, but not a fatal one. Today, airborne sensors, especially infrared or heat detectors, can uncover just about anything, and precision weapons can then destroy it.

When faced with supply deprivation, there are two choices. First, operations can be either halted or reduced to the level that can be supported by available supply. This only works if the decision maker is the attacker or otherwise in control of the tempo of operations. The second and more likely choice is to accept higher losses in both matériel and manpower. This is an accepted technique when a less well-supplied army meets one with far greater firepower. To give a historical example, consider the situation in Korea (1950–1953). The Chinese-North Korean Army was outfitted primarily with small arms, mortars, good leadership and courage. The UN forces opposing them were fully motorized, heavily armed divisions similar to those deployed today in the West. The Chinese found that they required twice the manpower suffering twice the rate of casualties in order to match the UN units. In other words, for every UN casualty there were four Chinese casualties. The Chinese traded lives for supplies, for it is not the weapons themselves that kill but the ammunition they fire at the enemy.

All armies have to make this trade if they lack supply. In terms of ammunition, the most efficient form of combat is infantry going in by itself. Not quite hand-to-hand combat, but often face to face. This type of operation causes the largest number of casualties. Artillery can inflict more casualties at less loss of life. But inflicting one casualty takes up to five tons of artillery and bombs. If large supplies of munitions are available, they will also be used to block travel, which will very indirectly cause casualties or lower your own. It appears that beyond the first casualty caused by five tons of munitions, the point of diminishing returns rapidly approaches. The recipient of all this attention soon learns to change his life-style. The shovel is ultimately more effective than the mountain of munitions.

For modern armies the chief result of losing supplies is immobility. Enormous quantities of fuel are required to support the large number of armored and support vehicles. When supplies no longer arrive, any movement means abandoning many vehicles. This destroys combat power just as effectively as enemy firepower. The fully motorized unit is thus much more vulnerable to a lack of supply than the nonmotorized unit. The modern, fully motorized, high-technology combat division is a very fragile unit. It has fists of iron, and feet of clay.

Air Force and Navy Logistics

Air force operations consist primarily of generating sorties. Hundreds of man-hours and 10 to 20 tons of supplies are needed to launch one sortie. An average 100 to 200 aircraft are available for each division. Each aircraft can fly up to three sorties a day for short periods. Three hundred sorties means 4000 tons of supplies, over twice the daily rate of a division. Air bases tend to be more static than combat divisions. Air units' primary problem is just getting the supply to the aircraft.

Most naval supply in the U.S. fleet takes place at sea, particularly refueling. Carriers generally carry sufficient fuel and munitions to fly as many as 1000 sorties. Intensive air operations would require frequent resupply of the carriers. As most large ships stay at sea for months at a time, a continuous flow of tankers and supply ships must reach the carriers and other ships in order to maintain operations.

Like the U.S. Army, the Navy prefers to keep much of its supply mobile. The Army never has enough trucks to do this, neither does the Navy ever have enough ships. In wartime civilian tankers and cargo ships can be pressed into service, but specialized Navy technicians and equipment have to conduct at-sea replenishment.

Divisional Daily Supply Requirements

This chart shows divisional supply requirements for each day of operations. Operations are divided into four categories: offense, defense, pursuit and reserve. Reality is not as neat as this chart. Offensive operations will definitely use the most supply. Defensive operations must respond to the intensity of offensives, but will generally use less supply. A pursuit is similar to simply moving around a lot. Reserve is sitting around using as little supply as possible, usually in contact with the enemy.

The supply norms for Russian and U.S. divisions were estimated from each nation's policies. Divisions can easily consume three or four times as much ammunition in a day. Often this will occur. For planning purposes norms are established so that usage does not outrun supply.

DIVISIONAL DAILY SUPPLY REQUIREMENTS

When in combat, divisions require more tonnage of fuel and ammunition than any supply system can likely deliver on a sustained basis.

Supply Type	Type of Division							
	Russian Tank Division (tons)	%	Russian Rifle Division (tons)	%	US Armor Division (tons)	%	US Infantry Division (tons)	%
Offense								
Ammo	432	40	462	38	1125	59	1290	63
Fuel	554	52	620	51	618	32	660	32
Food	26	2	31	3	40	2	51	2
Spares	61	6	110	9	137	7	55	3
Total	1073	100	1223	100	1920	100	2056	100
Lbs/Man	215		199		248		258	
Defense								
Ammo	550	59	620	58	1200	74	1350	74
Fuel	303	33	330	31	336	21	366	20
Food	29	3	33	3	41	3	49	3
Spares	50	5	83	8	40	2	50	3
Total	932	100	1066	100	1617	100	1815	100
Lbs/Man	187		174		209		227	
Pursuit								
Ammo	69	7	73	7	198	15	207	19
Fuel	858	88	937	86	1044	78	816	73
Food	17	2	21	2	42	3	50	4
Spares	33	3	55	5	46	3	44	4
Total	977	100	1086	100	1330	100	1117	100
Lbs/Man	196		177		172		140	
Reserve								
Ammo	132	34	145	32	390	53	438	59
Fuel	211	54	237	53	264	36	240	32
Food	23	6	26	6	41	6	48	6
Spares	25	6	39	9	44	6	20	3
Total	391	100	447	100	739	100	746	100
Lbs/Man	78		73		95		93	

SUPPLY TYPE is the class of supply.

AMMO is munitions, primarily artillery ammunition.

FUEL is all types of fuel, for vehicles, aircraft and power generators. Usage is highest in pursuit operations because every vehicle is moving. In combat operations, combat vehicles do most of the moving as they maneuver about the battlefield. In reserve operations most movement is by noncombat support vehicles.

FOOD is just that.

SPARES is all the spare parts to keep equipment going and the troops in good health. Also includes medical supplies, normal replacements for equipment, and so on.

TONS is the number of tons of supply for each class and type of operation.

% shows the percentage of each class of supply for each type of operation.

LBS/MAN is the pounds per man of supply for each type of operation.

Armies other than those of the United States and Russia have similar supply norms. Western armies use, if anything, somewhat higher norms than the United States, especially for ammunition. Armies using Russian equipment generally use somewhat lower norms. All other armies have lower norms, more on the Western model. Norms will be modified by the quantities of men, weapons and equipment in divisions, as well as the tempo of operations.

It is still possible to fight a low-budget war. It depends on whom you are fighting and where. An ill-armed and equipped opponent requires less ammunition to fight. More fuel may be required for running around a lot, as in an antiguerrilla war. Modern armies fighting guerrillas also use a lot of munitions trying to keep the little buggers out of mischief.

Ground-Transport Characteristics

This chart shows the characteristics of the combat and noncombat vehicles most frequently found in divisions.

VEHICLE USER is the nation using that vehicle, also the manufacturer.

NAME is the designation of the vehicle.

VEHICLE WEIGHT is the empty weight in tons.

VEHICLE LOAD is the average load carried on roads. Off roads, load capacity is reduced about 50 percent. Also some trucks will pull trailers.

VEHICLE RANGE is how far the vehicle can go, on road, with one tank of fuel.

LITERS OF FUEL CARRIED is just that. One gallon equals 3.79 liters.

TONS FUEL PER 1000 KM is the tonnage of fuel the vehicle will need to travel 1000 km on roads. Up to twice as much fuel is needed to travel off roads, depending on the roughness of the terrain.

IN DIVISION is the number of each vehicle found in an average division.

TONS OF FUEL 100 KM is the tonnage of fuel required to move all the vehicles of each type in the division 100 km. The armored vehicles account for the bulk of the fuel.

TONS OF LIFT TOTAL is the tonnage the division's transports can move. U.S. and Western divisions possess more transport than Russian-style divisions. Russia is attempting to close this gap, but economic considerations and a shortage of manufacturing capacity prevent dramatic progress.

GROUND-TRANSPORT CHARACTERISTICS

The farther you have to truck supplies, the less you can deliver because of the heavy fuel needs of the trucks and other vehicles.

Vehicle User	Name	Vehicle Weight (tons)	Vehicle Load (tons)	Vehicle Range (km)	Liters Fuel Carried	Tons Fuel per 1000 km	# in Division	Tons of Fuel 100 km	Tons of Lift Total
	Trucks								
US	M-54	11.6	9	563	380	0.61	400	25	3600
US	M-34	5.5	4.7	563	189	0.31	1500	46	7050
US	M-37	2.6	.9	362	91	0.23	700	16	630
US	M-151A2	1.1	.45	482	56	0.11	400	4	180
	Transport Tot Tons	15150			474		3000	91	11460
	Armored Fighting Vehicles								
US	M-113	11.5	0	480	310	0.59	1100	65	
US	M60A3	52	0	480	950	1.80	351	63	
	AFV Total	30902			613		1451	128	
	Division Tot Tons	46052			1087		4451	218	
	Trucks								
Russia	Ural-375	8.4	6	750	360	0.44	300	13	1800
Russia	Zil-131	6.7	5	850	340	0.36	500	18	2500
Russia	Zil-157K	5.8	4.5	510	215	0.38	400	15	1800
Russia	Gaz-69	1.5	.4	430	60	0.13	300	4	120
	Transport Tot Tons	8666			347		1500	50	6220
	Armored Fighting Vehicles								
Russia	BMP	12.5		455	400	0.80	1000	80	
Russia	T-62	37		500	950	1.73	400	69	
	AFV Total Tot Tons	27300			709		1400	149	
	Division Tot Tons	35966			1056		2900	199	6220
	Other Armored Fighting Vehicles								
France	AMX-30b	36	0	500	970	1.76	150	26	
Germany	Leopard I	40	0	600	955	1.45	300	43	
US	M1	52	0	440	1900	3.93	351	138	

Classifications of Vehicles

There are transports for carrying supplies and equipment, and combat vehicles that carry troops and fight. Combat vehicles also include specialized vehicles such as APC's, which carry infantry and a wide range of specialized combat support equipment like communications gear, mobile headquarters and artillery.

Transports also include jeeps for carrying commanders and light weapons and for running errands.

Not all vehicles are covered in the chart. Russia has a wide range of tracked transports with the same general load capacities and fuel requirements as wheeled vehicles. Because of the more complex track-laying mechanism, maintenance loads are much higher. These vehicles are almost as difficult to keep operational as their cousin tanks and APC's.

Because the Russians use more tanks in their divisions, their fuel requirements are not much less than those of Western divisions. Fewer transports in Russian divisions leads to more supply and maintenance problems.

The fuel situation is likely to get worse as armies introduce heavier armored vehicles. The United States is bringing out the heavier M-1 tank, which uses a gas turbine engine. The new engine, plus the greater weight, more than double fuel requirements. The heavier M-2 APC will further increase fuel needs. The heavier T-72 and T-80 Russian tanks being introduced will not increase fuel requirements more than 10 percent.

Supply Requirements for Aircraft

AIRCRAFT is the aircraft designation.

PRINCIPAL USER is the nation of manufacture and the principal user. West is many Western nations (including Israel and South Africa). The Chinese F-6 is a copy of the Russian MiG 19.

FUEL is the average tonnage of fuel carried on one sortie.

WARLOAD is the average load of disposable weapons carried on one sortie. It is assumed that some of the air-to-air missiles carried will not be used and be brought back.

AVERAGE SORTIES PER DAY is just that.

TONS PER 100 SORTIES is the total tonnage of supply required for 100 sorties. This includes an allowance for maintenance supplies and the supplies needed to maintain the air and ground crew.

SUPPLY PER 100 AIRCRAFT PER DAY is the total tonnage of supply for 100 aircraft of each type if they are flying the maximum number of sorties.

SUPPLY REQUIREMENTS FOR AIRCRAFT

Primarily fuel; aircraft require an average of over five tons of supply for each sortie.

Aircraft	Principal User	Fuel	Warload	Average Sorties per Day	Tons per 100 Sorties	Supply per 100 Aircraft per Day	Percent of Supply Warload
Russian Landbased Aircraft							
MiG 21	East Bloc	2.1	1.5	3	383	1148	39
MiG 23	Russia	4.7	1.5	2	675	1349	22
MiG 25	Russia	15.1	2	.5	2051	1025	10
MiG 27	Russia	4.7	4.5	2	975	1949	46
Su-17	Russia	3.3	4.5	3	810	2429	56
Su-24	Russia	24	4.5	1	2980	2980	15
Tu-22M	Russia	13.4	8	.66	2450	1617	33
Western Landbased Aircraft							
F-4	West	5.7	7.2	2	1478	2955	49
F-16	NATO	3.2	6.9	2.5	1110	2775	62
F-15	US	6.1	7.2	2	1468	2935	49
A-10	US	6.1	7.2	4	1380	5520	52
Mirage 3	West	2.7	1.5	2	608	1215	25
Jaguar	NATO	3.4	4.5	2	915	1830	49
Starfighter	NATO	2.8	3.4	2	920	1840	37
Harrier	NATO	3.5	3.6	3	835	2505	43
Tornado	NATO	5.2	7.2	2	1540	3080	47
Alpha	NATO	1.5	2.2	3	437	1310	50
F-6	China	1.8	.5	1	344	344	15
F-111	US	15.4	10	1	3040	3040	33
US Carrier Aircraft							
F-14	US	7.5	6.5	3	1547	4640	42
A-7	US	4.5	6.8	2	1245	2490	55
A-6	US	7.2	8.1	2	1705	3410	48
F-18	US	5.1	7.7	3	1350	4050	57

PERCENT OF SUPPLY WARLOAD is the percentage of the total supply requirements that is disposable weapons. This shows the greater efficiency of Western aircraft. Even the most recent Russian models (MiG 23, SU-24) do not approach the efficiency of twenty-year-old Western aircraft (F-4, A-7, etc.).

Helicopters are generally smaller than combat aircraft but have the same voracious supply requirements. An attack helicopter will fly three sorties a day and consume six tons a day. A hundred attack helicopters use up 600 tons per day. Transport helicopters are only slightly less profligate, as attack helicopters carry only about a ton of weapons. The larger transport helicopters use up to 600 tons per day per 100 aircraft.

24

ATTRITION

WARFARE IS dangerous. People can easily get killed or injured out there. But warfare's greatest losses are not to the sudden and violent carnage of a battle, but to the day-to-day losses brought on by wear and tear.

Attrition is people and their machines wearing out. Even in peacetime more than 2 percent of aircraft a year are lost to accidents. Rarely more than 90 percent of armored vehicles will be in running condition at any one time. Those vehicles that are running will likely break down after going less than 500 km. More important, people wear out, too. Without people to tend to them, the machines wear out even faster.

What Really Destroys Armies

Annually, disease and noncombat injuries often cause far more loss than the dangers of combat. Most major wars go on for years. Battles are relatively infrequent. As long as the troops are in the field, they are prone to disease and injury. The annual loss rates in the wars of the twentieth century, expressed in terms of average daily losses per 100,000 men, bear this out. Battle losses, killed and wounded but not prisoners, varied from a low of 6 per day in World War II theaters such as North Africa to over 200 Germans a day on the Russian front. Russian casualties were sometimes double the

German rate, but this was a style of warfare even they do not wish to emulate or repeat. Anyway, most armies are now so heavily equipped with armored vehicles that the equipment, and the combat power, will be disabled long before the casualty rate reaches 1941–1945 Russian-front levels. What are more likely are the casualty rates reached on the Western front during the same period. Here the daily rate rarely went beyond 40 men per day per 100,000 strength. Chemical or nuclear warfare could push these rates higher, but the armies would probably disintegrate first from the loss of support troops and the vital equipment and supply systems they operate.

Nonbattle casualties, primarily from disease and particularly in tropical or winter conditions, regularly reached rates of 200 to 500 men per day per 100,000 strength. Malaria alone caused nearly 200 casualties a day. Another constant menace was venereal disease, which could render ineffective as many as 40 men per day. Injuries, part of nonbattle casualties, often exceeded battle losses. The troops tended to get careless in the combat zone. Vehicle and weapons accidents were so common that the rate often reached 20 men per day.

The Forms of Combat Injuries

Injuries as a direct result of combat—the effects of enemy weapons as opposed to indirect effects such as trench foot, malaria or pneumonia contracted from sleeping in a wet trench—fall into three categories. Fatal: the victim is dead. Wounded: the victim is injured but not fatally and has a good chance of returning to combat. Mental: the victim suffers a mental breakdown from the stresses of combat. In addition, troops taken prisoner are also losses.

FATAL/WOUNDED RATIOS

The rate and lethality of combat casualties vary with such factors as the amount of opposing artillery versus bullet-firing weapons. Artillery will cause more but less lethal casualties than bullets. Closed terrain allows more bullet wounds by reducing the effects of artillery. Fast-moving operations prevent treatment of severely wounded troops, who then die. Better-prepared and better-led troops avoid casualties. Armored vehicles, fortifications or protective clothing will reduce casualties. Finally, the availability and efficiency of medical care make a difference.

If prisoners are ignored, and assuming a heavy use of fragmentation weapons, shells and bombs, historical experience suggests there will be one fatality for every three wounded troops. About 80 percent of these injuries will be caused by fragments. About 12 percent of these wounds will occur in the head (43 percent immediately fatal), 16 percent in the chest (25 percent fatal), 11 percent in the abdomen (17 percent fatal), 22 percent in the arms (5 percent fatal). In the past some 20 percent of all wounds were multiple and over half of those combinations were fatal.

Modern lightweight plastic armor in the form of kelvar cloth or rigid plate will reduce fatalities and injuries by 10 percent to 20 percent. This material does cause heat buildup in warm weather, leading to heat exhaustion injuries. If it is used selectively for troops in exposed situations, its beneficial aspects are retained without injurious side effects. Although the jacket covering the chest and abdomen costs about $100, the savings in troops, not to mention the positive morale boost, are well worth it. In addition to flak jackets, improved helmets, boots and curtains for vehicle interiors are available. The U.S. Army has taken the lead in this area, with other Western, but not Russian, armies following.

MENTAL CASUALTIES

Combat is an extremely stressful activity that causes a number of nervous breakdowns. The rate of breakdown is highest in poorly trained and led armies. During World War II, the U.S. Army had three combat fatigue cases for every two combat wounded troops. For every 100 men killed, 125 were discharged because of mental breakdown. The average combat fatigue victim was out of action half as long as a man who was physically wounded.

The German Army had only 13 combat fatigue cases for every 100 wounded. This was primarily a result of better training and leadership. Most other armies fell somewhere in between these extremes. A contemporary war would produce higher levels of combat fatigue loss because of the higher intensity of fighting and the lower level of training and experience. The Germans were well prepared during World War II because they had carefully studied their World War I experience and planned accordingly.

PRISONERS

The number of prisoners lost varies considerably depending on how badly you are losing. Even a victorious force lists a few percent of its total losses as MIA (missing in action). About 50 percent of the MIA will be dead. Some are men who were killed or badly wounded and did not recover. Historically all men who surrender are not captured alive by the enemy. Up to 50 percent of those surrendering do not survive the process. They are either killed on the spot or die in captivity. Troops in combat quickly learn this, which explains why surrenders are not more common. When they do occur, they tend to be in large numbers or by negotiation.

Wasting Away

It is perfectly possible for an army to waste away to nothing without ever having come in contact with an enemy force. Historically natural causes have killed or disabled far more soldiers than combat.

Many wars have been won by the side best able to maintain a fighting force. Perceptive military commanders have long recognized the substantial assistance of General Winter, Colonel Mud and the carnage wrought by pestilence, poor climate, thirst and starvation.

An armed force may be an impressive sight. Yet people have to live. They must eat, sleep and escape the elements. Disease and injury are ever present. Medical care prevents minor afflictions from becoming major ones. More important is public sanitation. Many diseases thrive in careless accumulations of human waste. Public sanitation eliminates the cause of most disease. (From 1900 to 1940 in the United States, the average life expectancy of males increased twelve years [31 percent] as a result of improved sanitation. Since 1940, the introduction of the many wonder drugs and procedures has lifted life expectancy another nine years [15 percent].)

The American army's history of disease deaths is illustrative. In 1846, 10 percent of the troops died from disease; in the 1860's, 7.2 percent; in 1918, 1.3 percent; and in the 1940's, .6 percent.

Although deaths due to disease have declined markedly, the incidence of disease has not. As the chart on noncombat losses demonstrates, armies are never far from a disaster of uncontrolled disease.

Useful Combat Life

Experience has shown that it is often more valuable militarily to injure rather than kill troops. A dead soldier is no longer militarily useful, but he is also no longer a burden on the medical services or on the army in general. An injured soldier is still a burden, as well as not being useful.

Look at it from the viewpoint of a soldier's normal useful combat life. Once in combat, a soldier is effective for about 200 days of action. After that point if he hasn't become a physical casualty, he will be a psychological one. If taken out of combat before he breaks down, he will still be able to serve in a noncombat or civilian job for the duration of the conflict. Let us assume that this is another 400 days. Assuming a soldier is killed halfway through his 200 combat days, the armed forces loses 500 days of service.

During World War II, although 65 percent of all cases of time lost were from noncombat injuries and sickness, these resulted in very few deaths. Still, each case put a soldier out of action for ten days. The average combat injury kept a soldier out of action for 100 days. Twenty percent of combat injuries resulted in death. For every day a soldier is out of action due to wounds or disease, one or more additional soldiers are assigned to taking care of him. For this reason nonfatal casualties comprise two thirds of the days lost due to injury or illness. Therefore, taking 100 injured soldiers we have the following pattern of lost days:

Combat deaths—one (500 days lost, 21 percent of total days lost. Only 13 percent of days lost if the time of the medical personnel is included). *Combat wounded*—four (400 days, 17 percent. Twenty-two percent if medical personnel time is included).

Noncombat deaths (mostly from accidents)—one (500 days, 21 percent of total days lost. Thirteen percent of days lost with medical personnel time). *Noncombat illness and injuries*—ninety-four (940 days, 41 percent; some 52 percent if medical personnel time is included).

Chemical and nuclear casualties tend to be more severe than usual noncombat "illness," but not as devastating as combat wounds from shot and shell. Perhaps only fifty days are lost per casualty. With medical personnel time added, this comes to over one hundred days lost per incident. Since each soldier may be wounded by chemical agents several times, the total time lost approaches that of a combat death.

The Rate of Return

Again, depending on the quality of training and leadership in an army, the rate of troops returning to duty after combat injury will vary. During World War II, the German Army achieved an 80 percent return rate, while the U.S. Army returned 64 percent. The higher German rate was partially the result of returning slightly disabled troops to less physically demanding duty, and partially attributable to better administration.

Generally 60 percent of the combat wounded who eventually recover return to duty within three months, 85 percent return within six months and over 95 percent within a year. Fifty percent of the noncombat casualties return within a month, 85 percent within three months and nearly all by six months.

Naval and Air Casualties

Generally navies and air forces suffer far fewer casualties, in absolute and relative terms, than armies. In a full-scale war, when naval and air bases are attacked, the naval and air force casualty rate can be expected to be one quarter to one half an army's rate.

Most naval casualties are suffered at sea. About a third of the deaths are from noncombat injuries. Combat deaths generally equal combat injuries, because ships suffer catastrophic damage—being blown up or sinking quickly and killing nearly all the crew. Most modern ships are heavily armed and unarmored, which makes them prime candidates for massive losses.

Naval actions tend to be sporadic. Submarine operations are more intense and tend to have relatively higher casualty rates than land operations.

During World War II, the loss rate—dead, wounded and captured—for strategic air operations averaged nearly four men per aircraft lost, and one aircraft was lost for every 100 sorties (66 sorties for bombers, 145 for fighters). Altogether, 40,000 aircraft were lost during the three-year strategic bombing campaign in Europe. Currently 9000 aircraft face each other in Europe. Each can fly two sorties a day and would probably last less than two months in combat. Most have a crew of one. Modern aircraft require over twice as many men per aircraft for support: 140 versus 60 during 1939–1945. Crews are also smaller, so that the average number of men lost per aircraft will also decline. Modern aircraft can generate four times the number of sorties as a World War II aircraft. Yet the aircraft can only be shot down once, and the pilot has a good chance of surviving and either landing in friendly territory or avoiding capture.

Therefore, if all 9000 aircraft are lost, with 7000 crewmen killed or captured, there will be 7000 losses for over a million men engaged in air operations. Air warfare is unlikely to be this kind to ground crews. Raids on air bases will cause casualties. The ground crews will be ready for such attacks and will probably still suffer relatively fewer casualties than the ground forces.

Basic Daily Loss Rate

This chart shows the basic daily personnel loss rates of modern armies as well as the factors that will increase or decrease these rates.

ATTACKER is the basic daily combat loss rate (3 percent of personnel) for the attacking force. This is a daily loss rate derived from historical experience since 1940. It includes losses from all combat-related sources—dead, wounded, mental breakdown, prisoners, desertion, etc.

DEFENDER is the basic daily combat loss rate (1.5 percent) for the defending force.

BASIC DAILY LOSS RATE

Although the average divisional losses are some 2% a day, a number of other factors can increase or decrease that by a factor of ten.

Modified by the Following Factors	Max Effect	Min Effect	Attacker 3 Cumulative Max Effect	Min Effect	Defender 1.5 Cumulative Max Effect	Min Effect
Size of Force	2.00	.5	6.00	1.50	3.00	0.75
Posture	1.00	.3	6.00	0.45	3.00	0.23
Force Ratio	1.60	.7	9.60	0.32	4.80	0.16
Time of Day	1.00	.5	9.60	0.16	4.80	0.08
Main Effort	1.50	.7	14.40	0.16	5.80	0.08

Each factor can increase or decrease casualties, depending on the situation. The examples following these notes explain this phenomenon.

MAX EFFECT, maximizing effect, is the most that a particular factor can increase casualties.

MIN EFFECT, minimizing effect, is the most that a particular factor can decrease casualties.

CUMULATIVE MAX EFFECT. The maximum effect times the basic daily loss rates for attackers or defenders.

CUMULATIVE MIN EFFECT. The minimum effect times the basic daily loss rate.

SIZE OF FORCE takes into account that larger forces devote a smaller proportion of their total manpower to combat troops. A brigade, or smaller, force has double the basic daily loss rate. A force of over 100,000 troops can cut the basic daily loss rate 40 percent (.6 in the table). A division, 10,000 to 20,000 men, has no change.

POSTURE indicates type of operation the unit is engaged in. Normal attack or defense is a 1. Various degrees of retreat cut the basic daily loss rate by up to 70 percent (.3). Any form of retreat in the face of the enemy has to prevent the other side from swarming over the withdrawing troops; otherwise, it is a trade of space for casualties. A more deliberate withdrawal or delaying action lessens defender casualties.

FORCE RATIO. All other things being equal, 10,000 troops of one army are equal in combat power to 10,000 troops of another army. The force ratio, the ratio of one side's troops to another, increases the basic daily loss rate for the side with the smaller force. At a 3+-to-1 ratio the force's basic daily loss rate is decreased 30 percent (.7). At a 1-to-7+ ratio the force's basic daily loss rate is increased by 60 percent (1.6). At 1 to 1 there is no effect. Examples: a 3+-to-1 ratio would be a force of 30,000 (or more) versus a force of 10,000. A 1-to-7+ ratio would be a force of 1000 against a force of 7000 or more. But all things are not equal. It is rare for troops of two different armies to have equal combat value. Before force ratios can be calculated, a unit's combat power must be calculated. This is described in chart 24-3.

TIME OF DAY has not always been an important consideration. In more civilized times battles almost always took place during the day. Fighting at night is safer, however, and cuts your basic daily loss rate in half.

MAIN EFFORT indicates the intensity of combat. This is usually dictated by the attacker. If the attacker increases the level of activity, the defender is forced to respond. This increased level of activity is usually more fire or keeping up the action around the clock. For this the attacker increases its basic daily loss rate 50 percent. The defender has its basic daily loss rate increased 20 percent. Withdrawal actions increase the basic daily loss rate 30 percent to 50 percent.

Worst Case

Looking at the chart, you can see that wide differences in casualty rates are possible between attacker and defender. Let us examine in detail the forces that would comprise the extreme case of attacker, 14.4 percent a day, defender .1 percent a day. The attacking force could be a brigade of, say, 4000 men. It would make a main effort, in daylight, against a defending force of 100,000. The defending force would retreat. Total casualties per day: attacker, 576, defender, 100.

While there are tactical situations where such a large force would withdraw before such an aggressive small force, it is more likely that the tables would be turned. In this case the attacker's daily loss rate would be 1.58 percent, while the defender would be losing 1.73 percent a day. Daily casualties would be 69 for the defender and 1580 for the attacker. The attacker is making a main effort. This assumes that the defender takes the most prudent action, running. If the defender stands and fights, the loss rate goes up to 5.76 percent a day.

The above examples are rather extreme. A force of 4000 outnumbered twenty-five to one would not last long. In terms of actual casualties, though, the above figures are historically accurate. It is also historically accurate that the defending force could be run right off the battlefield (the unit could run only so far before the campaign would be over), surrounded by a portion of the superior force (perhaps 10,000 men) and bypassed, or convinced by the unequal struggle to surrender.

Other factors could enter into the equation. If geography provided the defender with a narrow front (a peninsula or a mountain pass), it could stand and be destroyed piecemeal. Such sieges are not unknown. But this demonstrates that there is more to warfare than simple attrition.

A Multitude of Exceptions

Some armies are more prone to attrition than others. Some wars are likewise more prone to higher casualties. Such an unfortunate matchup occurred during World War I. Some armies consistently produced attrition rates three and four times the above rates. No one during World War I had less than double those rates. It is feared that the next big war will again see the basic rates doubled, even without chemical or nuclear weapons.

Armies that put a premium on skill and/or technology in place of masses of troops tend to plan with the basic attrition rates. In other cases, the basic rate is doubled or tripled. The Japanese, Chinese, North Vietnamese, Koreans (North and South) and Russians all used doubled rates until the last time they fought major wars.

The armies of the industrialized nations are more sparing in their use of manpower. During World War II, even though Germany lost millions of troops, the loss was according to the above rates. The Russians lost at twice the rate. Of course, the Russians won the war. But so did Britain and the United States. And the Japanese, who had a somewhat medieval attitude toward casualties, lost. There is no gainsaying that the road to victory, or defeat, is paved with dead bodies.

UNMODIFIED HISTORICAL CASUALTY RATE

World War II casualty experience, when less lethal weapons were used, demonstrates the expected pattern of losses. The units were infantry battalions, .6% per day noncombat losses included.

Type of Action	Attacker	Defender	Ratio Atckr:Def
Meeting	7.5	4.9	1.53
Attack of Position Day 1	11.5	6.1	1.89
Attack of Position Day 2+	6.1	3.5	1.74
Attack Fortifica- tions Day 1	18.7	9.8	1.91
Attack Fortifica- tions Day 2+	9.8	5.2	1.88
Pursuit	4.3	3.2	1.34
Inactive	2.6	2.6	1.00

Unmodified Historical Casualty Rate

This chart gives the United States' casualty experience in World War II. It demonstrates the rapidity with which combat units melt away due to normal casualties. The figures are for infantry, tank or reconnaissance battalions; combat units of 500 to 1000 men. These units comprised 50 percent of a division's strength but incurred 80 percent to 90 percent of the division's casualties. This means that these rates are 20 percent higher than those for a regiment (3200 men) and 50 percent higher than those for a division (15,000 men).

TYPE OF ACTION names the types of combat activities that would produce different rates of loss.

MEETING is a meeting engagement. Both sides are marching, on foot or in vehicles, when they encounter each other. The side that takes the initiative and becomes the attacker suffers proportionately less than in other types of engagements. These actions are rather confused affairs with little artillery and a lot of infantry action.

ATTACK OF POSITION DAY 1 is the first day of a normal attack, that is, an attack in which the defender is prepared. This is generally the day of heaviest fighting, as the attacker would like to reach a decision as soon as possible. The defender would like to avoid this.

ATTACK OF POSITION DAY 2+ represents subsequent days of fighting if the first day's push fails to decide the issue. This is attrition fighting, which favors the defense.

ATTACK FORTIFICATIONS DAY 1 is similar to the position attack, but the defender is better prepared and the attacker is making a more substantial effort.

ATTACK FORTIFICATIONS DAY 2+. Grinding into enemy fortifications is even more expensive, and risky, than going after normal defensive positions.

PURSUIT is combat between a rapidly advancing attacker and a defender attempting to delay this pursuit. There are ample opportunities for defenders to prepare ambushes. The attacker has such a high degree of initiative that the defender is not able to take full advantage of the attacker's reckless movement.

INACTIVE is opposing forces in contact but not actively fighting. There is still a lot of artillery and small-arms activity. Patrolling also takes a heavy toll as both sides attempt to keep tabs on each other.

ATTACKER AND DEFENDER indicate the daily loss rates, as a percentage of current personnel strength, for the attacking and defending force, respectively.

RATIO is the ratio of attacker to defender losses. The attacker usually loses more men. This ratio shows that the attacker is better off in some types of actions than in others.

What Does It Mean for Today?

Since World War II, firepower has increased. Soldiers noted those weapons that were most effective during World War II. Since then armies have equipped themselves with greater and more technically improved quantities of these weapons.

Automatic rifles. At the end of World War II, armies were providing an ever-increasing proportion of their infantry with fully automatic rifles. These were, in effect, machine guns. Whereas at the beginning of the war, 13 percent of German infantry operated automatic weapons, by 1944 this had risen to 23 percent, and at war's end to 46 percent. Five years after the war, all Russian infantry operated either machine guns or automatic rifles (the well-known AK-47). Vietnam forced this practice on the U.S. Army. Overall, in the space of forty years the infantryman has come to face five times as many automatic weapons.

Artillery. The bigger is better concept applies to artillery. Seventy years ago the caliber of the standard artillery piece was 75mm, forty years ago it was 105mm, now it is 152mm. Heavier shells are deadlier. Also, more, not less, of the larger-caliber shells will be fired.

Instead of just filling a projectile with explosives, 30 percent to 60 percent of projectile weight, more deadly cargoes have been developed. The most notable ICM (improved conventional munitions) are: napalm (jellied gasoline), fuel air explosives (a mist of flammable liquid is spread and then exploded), guided projectiles and submunitions. The submunitions, or bomblets, are small projectiles that function as antipersonnel or antivehicle weapons. Often they are equipped with timers or sensors so that they can also act as mines. The timed bomblets cause significant damage to enemy morale, plus occasional physical losses. Guidance equipment on projectiles increases their accuracy and is particularly effective with all the AFV on modern battlefields.

334 HOW TO MAKE WAR

Armored vehicles. There are a lot more in armies. At the end of World War II, the average Western division had about ten AFV (armored fighting vehicles) per thousand troops. Today there are close to one hundred AFV per thousand troops. These vehicles are primarily weapons carriers.

Air power. This had never been a significant source of injury to combat troops, although it was the scourge of support units behind the fighting line until the introduction of the helicopter. The U.S. Army has more helicopters available for combat support than the air force has aircraft for the same purpose. Not to be outdone, the air force has built special ground support aircraft (A-10) and installed fire-control equipment that delivers firepower very accurately in all weather conditions.

Electronic weapons. Many of these new weapons depend on electronic controls or communications. This has made electronic warfare a significant weapon. From a purely destructive point of view, artillery radars add considerably to the carnage. They also can locate an artillery weapon by examining the flight path of projectiles.

Chemical and nuclear weapons. There are considerable inhibitions against nuclear weapons, fewer against chemical weapons.

Considering these developments, what is to prevent casualty rates from doubling, tripling or worse? There are a few additional defensive measures. Some infantry now possess flak jackets and most troops are assigned to armored vehicles. But recent experience has shown that this has not lessened the casualty rate appreciably.

Offensive actions are also discouraged by the fact that casualty rates among AFV are four to ten times higher than for personnel. That is, if a unit loses 5 percent of its personnel, it will lose 20 percent to 50 percent of its AFV. The result is that for a few days, somewhat higher casualty rates can be expected among personnel, 10 percent to 20 percent higher. But because so much of a unit's combat value is tied up in AFV, the much higher loss rate of these vehicles will quickly reduce the unit to impotence without causing a majority of the troops to become casualties.

In combat without AFV, the remaining increased firepower will produce considerable casualties, if the troops can be motivated to attack into ever more deadly firepower. World War I demonstrated that up to a point troops could be made to get themselves killed off during ineffective advances against superior firepower. But most of the armies that went through the meat grinder eventually mutinied, disintegrated or adopted less debilitating tactics.

Factors Modifying Unit Combat Power

This chart shows how various factors can lower the combat power of an army unit. The adjusted combat values show that units of equal size and equipment are not all equal.

A unit's basic combat power resides in the destructive power of its weapons, quantified by various formulas. For one calculation of combat power, see the charts in this section that show casualty rates by branch. For more detail on the relative combat power of weapons, see Chapters 28 and 29 on armies and weapons.

FACTORS MODIFYING UNIT COMBAT POWER

Surprise and leadership will usually be more critical than any other factors.

Factor	Min Effect	Cumulative Effect Worst Case	Average Case
Natural Elements			
Terrain	.5	50	95
Weather	.8	40	90
Command Elements			
Air Superiority	.9	36.00	84
Leadership	.6	21.60	79
Posture	.6	12.96	71
Surprise	.6	7.78	60
Supply	.5	3.89	52
Training	.3	1.17	35
Command, Control & Commo.	.3	0.35	21
Morale	.2	0.07	10

There is nothing fundamentally complex about calculating basic combat power. Most nations have the same standards of organization and levels of equipment. The most powerful individual army weapons today are tanks and artillery. Assigning each tank a value of up to 100 and each artillery piece a value of 30 to 70, a rough basic combat power for a unit can be rapidly calculated. To put this in perspective, an infantry squad, rifles, machine guns, etc., would have a value of 1. If they had an armored personnel carrier, this vehicle itself would have a value of 5 to 40 depending on the weapons it carried. An antitank missile system would have a value of 8 to 25.

Calculating precise capabilities of weapons is an inexact science. A precise calculation would be irrelevant anyway, as the basic combat value of a unit is a small fraction of its *eventual* combat value. Many other factors modify the basic values, and they are the keys to success in combat.

NATURAL ELEMENTS are the elements over which man has no control. The effects of natural factors on combat performance will persist only as long as the natural factor is present.

TERRAIN is the effect of geography. Some terrains are more difficult to fight in than others. Terrain becomes difficult by creating three conditions favorable for the defender. Mobility is cut by rough ground (mountains, hills, river banks) as well as by soft ground (swamp, sand dunes) and numerous obstructions (forests, built-up areas). Weapons ranges are cut by obstructions, particularly by forests and numerous hills and buildings. Observation is cut by obstructions that allow the defender to conceal himself more easily and to prepare the classic ambush attacking forces fear so much.

Terrain can cause these losses of attacker effectiveness: mountains, 25 percent to 50 percent; swamps, 20 percent to 40 percent; hilly terrain, 0 percent to 20 percent; flat terrain (soft ground and sand make it worse), 0 percent to 20 percent; built-up areas, 10 percent to 20 percent. Each type may also contain forests, which aid the defender.

WEATHER affects performance in three ways (in order of importance). Rain, snow and excessive humidity cut mobility and the efficiency of weapons and troops. Fog, clouds and mist obstruct observation. What they can't see, they can't shoot. This element, unlike the other two, favors the attacker, because it enables him to get close to the defender without being observed or shot at. Extremes of temperature cause both troops and equipment to fail more frequently.

COMMAND ELEMENTS are the human factors. These are more important, because human factors persist. The first two factors are more controllable than the rest, the "battlefield factors."

LEADERSHIP is defined here as the quality of the unit commanders. Good leadership, given enough time, will train troops properly. On short notice, good leadership gets the most out of poor troops. Good unit leaders often overcome the debilitating effects of an inefficient high command. Good leadership and good training usually go together.

TRAINING is the extent to which troops are taught effectively to use their weapons and equipment. Not until this combination of men and machines is put into combat is the training conclusively revealed.

POSTURE is the nature of the unit's activity. Other things being equal, it is easier to defend than to attack. The stronger the defensive posture, the more it will decrease the attacker's combat power. The weakest defensive posture is a retreat, either just to escape or to delay the attacker's advance. Then, in ascending order, are hasty, prepared and fortified defense.

AIR SUPERIORITY is the impact of air control and the ability to go after the enemy with aircraft. For the side without air superiority, mobility is impaired, as units move and operate more cautiously to avoid air strikes. For the side with air superiority, operations proceed more smoothly as a result of superior reconnaissance.

SURPRISE is another often underrated factor. Surprise comes in various degrees. It increases the effectiveness of the attacker's weapons and mobility while decreasing the defender's.

SUPPLY—the lack of it—stops the most successful army from doing anything. Normally this is not a critical factor, as many commanders pay attention to logistics. At least they avoid reasonable risks for fear of supply problems.

COMMAND, CONTROL, COMMUNICATIONS (C^3) is the commander's ability effectively to control his army. If control is lost, the ability to respond to enemy action and coordinate one's own forces is also lost. *Blitzkrieg* warfare is aimed at destroying the enemy's C^3, rather than undertaking the more formidable task of destroying the army itself.

MORALE is often much underrated. Large relative differences in morale have a devastating effect on combat performance. Morale is usually modified by leadership, training, situation, politics, weather and numerous other factors, probably including the phases of the moon.

MIN EFFECT is the maximum minimizing effect or deflation of the unit's basic combat power. The basic combat value of a unit is multiplied by these deflators to determine the actual combat value. For example, if a unit had a basic value of 400 and it was operating in the worst possible terrain conditions (a deflator of .5) then the unit's value would be 200. As the other factors are applied, this value can decline still further.

CUMULATIVE EFFECT is the percentage of the unit's original strength remaining after the factor has affected it. With each factor, the worst case is given, to show the extremes to which a unit's theoretical strength can be reduced. The average effect is also given.

Calculating Who Is Winning

The above values allow you to calculate the effective combat power of each side in a conflict. These values are used on the basic daily loss rate chart to determine casualties. We know how many people are likely to be injured. But who will win?

Defeat goes to the side that quits first. This means that some wars are much bloodier than others. The elusive "resolve" often determines the victor.

For a rough and ready rule for who will win, consider the "force ratio," that of the stronger force to the weaker one.

Assume two forces each contain 10,000 men, 200 tanks, 300 APC's, 50 artillery pieces, etc. Assume each one's basic value is 30,000. Let us further assume that the attacking force has lost nothing to natural elements. Assume the attacking force loses only about 20 percent to command elements. This gives them a value of 24,000.

The defending force is less well trained and led (with deflators of .7 and .8, the value of the force is now 16,800). Their morale is not as high (.8 = 13,440). They do not have air superiority (.9 = 12,096). They are surprised (.6 = 7257). After a few days of combat, they develop supply problems (.7 = 5080). The enemy gets into their rear area and C^3 problems develop (.5 = 2540).

After two or three days of combat, the attacking force has a combat value of 24,000, the defender's is 2540. Dividing 24,000 by 2540 gives 9.45. This is the force ratio. The attacking force has 9.45 times as much effective combat value as the defender. We have not calculated combat losses or further damage to the defender's morale. Also at this point surprise is no longer a factor. Time and space elements come into play.

A battle is a dynamic activity. A lot of movement takes place. On an open battlefield, where the attacker is not forced to take on the defender in frontal assaults, the more mobile force will be able to get to the defender's support units, causing supply, C^3 and morale problems.

History has shown that when the force ratio approaches 10, the weaker force begins to see the handwriting on the wall. There is little chance of recovery from such an unfavorable ratio. There are bloody-minded commanders who will attempt to fight to the last man. As the force ratio goes over 10, even this often becomes impossible. The defending force ceases to function. If the whole doesn't surrender, the parts usually do. That's how battles normally proceed. This description could very well have been that of Russia in 1941, or of the Sinai campaigns of 1967 or 1973. Or it could be of somewhere else tomorrow.

On the Ocean and in the Air

Most of the above factors also apply for naval and air forces. Weather is the same. Terrain has some interesting similarities; water is not all the same. Read the chapters on naval warfare for more detail on the differing composition of water. Additionally, operating near land favors defenders, particularly those with a lot of small, fast ships. In some areas the land just beneath the water, the shoals and reefs, becomes a critical factor. During violent storms the sea itself assumes awesome shapes. Yes, there is terrain at sea.

In the air there are basically three flavors of terrain: high, medium and low altitude. The high altitudes, over 10,000 meters, have thin air, which harms engine performance and the maneuverability of aircraft not built for operating at those heights. The low altitudes, less than 1000 meters, have a much thicker atmosphere. On the deck, debris can be sucked into the engine. A bird can hit the canopy and kill or injure the pilot. These are only some of the risks encountered flying close to the ground. An often ignored risk is sheer fatigue. Much more concentration and effort are required speeding along at 200 meters a second at an altitude of 100 meters.

All the other factors apply. Supply is immediate for aircraft. If you run out of fuel, you hit the ground. Not much ammunition is carried, perhaps four missiles and twenty seconds' worth of cannon shells. Ship supplies are similarly limited. Some ships require refueling once a week or more. Carriers have sufficient aircraft fuel and munitions for less than ten sorties per plane.

Attrition is similar. In the air losses are calculated in losses per sortie. A loss of more than 2 percent is dangerous to unit integrity and morale. Aircraft may fly more than one sortie per day. See the chart on aircraft attrition.

Naval losses tend to be more catastrophic, as there are fewer "vehicles." Unlike previous ships, modern vessels are not armored. Peacetime accidents indicate that wartime losses are likely to be higher than in the past. Training, particularly in damage control, becomes a critical factor. Training, leadership, morale, C^3 are all areas in which Western navies have a comfortable, but not invincible, lead over potential opponents, namely, the Russians.

For these reasons most Western naval commanders are not too perturbed, unofficially at least, about the prospect of Russian aircraft carriers. Most Western aircraft carrier admirals see the Russian carrier as a golden opportunity for them to play "keep up with the Russkies." Legislatures vote Navy when they are properly terrified. How do you think the Russian admirals got their carriers?

NONCOMBAT CASUALTIES

It is still possible for disease and accidents to inflict more damage than an armed enemy.

	Multiplier (maximum)	%	Percent Permanent Losses Daily	Permanent Daily Losses per 100000	Monthly Losses per 100000
Average (constant)		2	.3	300	9150
Climate	1.5	3	.45	450	13725
Living Conditions	1.5	4.5	.675	675	20588
Medical Care Level	2	9	1.35	1350	41175

Noncombat Casualties

AVERAGE (CONSTANT) is the percentage of a unit's strength that will be out of action due to noncombat injuries at any one time. This is an average of operations in all climates and conditions. Disease, including venereal, and accidents are the cause. Eighty-five percent of those affected will return to duty after an average 10-day absence, compared to 100 days for combat casualties. About 50 percent will be out of action for a month or more. Medical services will devote over 35 percent of their efforts to treating noncombat injuries. As long as medical services function, there will be daily, cumulative losses of only 15 percent of those afflicted. Because the average time in hospital is 10 days, only 200 men out of 100,000 need get sick each day to represent a 2 percent daily loss. Some 15 percent (30 men) will not return. This represents a permanent daily loss rate of 30 men per 100,000, or .03 percent.

CLIMATE can modify the loss rate considerably. Temperate climates can reduce the loss rate. Deserts, jungles, severe cold and other unhealthy environments will cause more disease and injury. Tropical rain forests are the worst. Any area is a bad climate if disease conditions are harmful to the troops operating in it. For example, troops from tropical areas would suffer in temperate zones. Any troops going from their own "disease pool" to an unaccustomed one will suffer. Most modern industrialized nations, because of a steady flow of travelers, share the same diseases. If their troops go to a less traveled area, such as parts of the Middle East, Asia or Africa, they will have less resistance to the local afflictions.

LIVING CONDITIONS represents the level of sanitation and general living conditions. Living in tents is more injurious to health than living under more substantial cover. Sleeping on the ground is not healthy. Regular, nutritious meals and clean, dry clothing are critical.

LEVEL OF MEDICAL CARE is the most crucial factor. Without medical services, minor afflictions become major ones. Even in temperate zones, lack of medical services, particularly public sanitation, will rapidly increase the level of losses.

MULTIPLIER is the multiplier effect of climate, living conditions and level of medical care on the average noncombat casualty rate. At the extreme end, all the troops will be afflicted with disease severe enough to require hospitalization. Under such conditions, few, if any, will receive medical attention. This occurred during World War II in the Pacific theater, particularly among Japanese troops, who had a lax attitude toward battlefield medical care. The Germans, during their first winter in Russia, were similarly unprepared and suffered accordingly.

PERMANENT LOSSES (DAILY) is the percentage of disease and injury that results in permanent loss. These losses include death and permanent incapacity for military service. When these losses are high, many troops will simply become ineffective. Without good leadership and training, most units will disintegrate or cease to function.

PERMANENT DAILY LOSSES PER 100,000 is the number of noncombat dead, or permanently incapacitated, per day for a force of 100,000 troops.

PERMANENT MONTHLY LOSSES is the number of dead, or permanently incapacitated, for a force of 100,000 over a period of a month. These compilations show that without effective measures to control noncombat casualties, an army will disappear without ever seeing the enemy.

Navies and Air Forces

More than that of armies, naval and air force combat power is represented by the readiness of their machines. For air forces an annual aircraft loss rate of 2 percent is "normal." Most non-Western, including the Russian, air forces have substantially higher rates. Rates over 10 percent are common if less capable forces attempt to use their aircraft at the same rate as Western forces, 500 or more flight hours a year. Noncombat personnel losses are much less, because air force personnel normally operate from fixed installations. In wartime new air bases are thrown together quickly. When this occurs in an unhealthy part of the world, primitive living conditions take their toll.

Navies have a higher "attrition" rate. Normally, in peacetime 20 percent or more of a fleet's ships are laid up for overhaul or repair. Losses at sea are less than 1 percent. Personnel losses are about the same. Disease is a lesser danger because of the controlled shipboard environment. Accidents are somewhat more frequent as a result of the combination of sea conditions and high proportion of machinery jobs.

National Differences

Some armies are more efficient, or callous, than others at dealing with noncombat casualties. During World War II, 89 percent of German hospital admissions were noncombat casualties, while in the U.S. Army the figure was 96 percent. Precise data for the Russian and Japanese armed forces are lacking, but available information indicates that noncombat admissions were closer to 60 percent. The Russians and the Japanese suffered more frequent and overwhelming casualties. Superior training, inadequate medical facilities, overwhelming losses and harsher discipline can lower hospital admissions. Combat casualties are given priority, although at least 15 percent of potential noncombat admissions will result in a dead or an incapacitated soldier.

AIR COMBAT ATTRITION

More than ever before, aircraft will be destroyed far faster than replacements can be built.

Period	Loss Rate per 1000 Sorties	Air Force	Aircraft Type	Loss Caused by Enemy
1939-45	9	Allied	All	All
1942	2	Allied	Bombers	Flak
1943-44	4	Allied	Bombers	Flak
1945	6.5	Allied	FtrBmbrs	Flak
1950-51	4.4	USA	All	Aircraft
1966	3.5	USA	All	All
1967	3	USA	All	All
1968	1.5	USA	All	All
1971	12.5 (approx)	India	All	Aircraft
1971	17 (approx)	Pakistan	All	Aircraft
1973	8	Israel	All	80% Flak
1973	10-15	Israel	A-4	Flak

Air Combat Attrition

PERIOD is the time frame in which air combat took place. The years 1939–1945 include all air operations in Europe; 1942 was the beginning of the large-scale bombing offensive in Europe; 1943–1944 was the height of the bomber offensive; 1945 was the height of tactical fighter-bomber operations; 1950–1951 was the Korean War; 1966–1968 were the three critical years of the air war in Vietnam; 1971 was the India-Pakistan war; and 1973 was the Arab-Israeli War.

LOSS RATE PER 1000 SORTIES is the number of aircraft lost per 1000 sorties (one aircraft flying one mission).

AIR FORCE is the nationality of the air force that incurred the losses. Allied means Britain and the United States.

AIRCRAFT TYPE is the type of aircraft that took the losses. *All* means all types. *Bombers* are primarily four-engine bombers (B-17–B-24). *FtrBmbrs* are fighter bombers (P-47, Typhoon, etc.). *A-4* is an American light bomber used by Israel.

LOSS CAUSED BY ENEMY is the type of enemy weapon that caused the loss. *All* is aircraft and antiaircraft artillery *(Flak)*. *80% Flak* means 20% of losses caused by enemy aircraft.

Changes Through History

The majority of losses in World War II were caused by aircraft. As the German Air Force became weaker, its flak effectiveness increased. This did not make up for the lost air-defense aircraft.

After World War II, the attrition rate continued to decline. In Korea there was much less flak than in World War II. In addition, political considerations restricted the intensity of the fighting. Additionally the Communist pilots were greatly overmatched by the veteran U.S. pilots. This was not a high-intensity air war.

In Vietnam the attrition rate declined still further. There, however, the strongest flak defenses were constructed against U.S. aircraft. The United States responded by devoting considerable resources to the systematic destruction of these flak defenses, lowering its losses. This was not a representative situation, however, as this was a war of unequal opponents. The North Vietnamese were in no position to force the issue. The U.S. operations could be scheduled with none of the time pressure of World War II.

The war between India and Pakistan was frighteningly intense. The air forces were more evenly matched. The war was fought with Korean War-vintage aircraft. More to the point, it was fought with great vigor and skill by evenly matched pilots. The loss rate was disturbing.

In the 1973 Arab-Israeli War, the Israelis had a substantial edge. The Egyptians made the best of a bad situation and built up their ground-based air defenses. Unlike the Americans in Vietnam, the Israelis ignored this development. Once the war began, Israel did not have the time systematically to eliminate the formidable Egyptian flak defenses, and ran up large losses.

As matters stand now, Western air forces would most likely cause great carnage among Russian aircraft. Not unaware of this possibility, Russia has fielded enormous flak forces. In Central Europe, East Germany, western Czechoslovakia and Poland, Russia has 16,000 surface-to-air missile launchers and 2300 interceptors. Over two thirds of this force is concentrated in East Germany. The only bright spot is that at night or in bad weather only 2000 of the missiles and none of the aircraft are usable. The day has come when pilots pray for bad weather and the cloak of darkness to ensure their safe return from a mission.

Casualty Rates by Branch

These charts show the levels of casualties of the components of combat divisions: infantry, armor, artillery, support. Two division types are shown, U.S. infantry and Russian tank. The Russian figures are based on divisions stationed in East Germany. These nineteen divisions are about 20 percent more powerful than any other tank or rifle divisions in Russian service.

CASUALTY RATES BY BRANCH

A constant in combat attrition is that the infantry, the most numerous combat arm, possesses the least killing power and takes the most casualties.

Russian Tank Division

Overall Casualty Rate of % 30   Combat Strength Lost % 51

Branch	Full Strength Men	% of Men Strength	Full Strength Combat Cmbt Str	% of Str	Men Lost	% of All Losses	Casualty Rate %	Survi-vors	% of Division Strength	Surviving Combat Strength Cmbt Str	% of All Combat Str	Branch
Infantry	2204	20	18	5	1613	52	73	591	7	5	3	Infantry
Armor	3544	32	255	75	1056	34	30	2488	31	103	62	Armor
Artillery	2069	19	62	18	253	8	12	1816	23	54	33	Arty
All Other	3163	29	3	1	151	5	5	3012	38	3	2	Other
Totals	10980	100	338	100	3074	100	28	7906	100	165	100	

US Mechanized Infantry Division

Overall Casualty Rate of % 30   Combat Strength Lost % 46

Branch	Full Strength Men	% of Men Strength	Full Strength Combat Cmbt Str	% of Str	Men Lost	% of All Losses	Casualty Rate %	Survi-vors	% of Division Strength	Surviving Combat Strength Cmbt Str	% of All Combat Str	Branch
Infantry	6306	36	50	16	4614	78	73	1692	15	14	8	Infantry
Armor	2176	12	157	51	648	11	30	1528	13	63	38	Armor
Artillery	3205	18	96	31	392	7	12	2813	24	84	51	Arty
All Other	5866	33	6	2	281	5	5	5585	48	6	3	Other
Totals	17553	100	309	100	5936	100	34	11617	100	167	100	

Two casualty levels are shown (0 percent and 30 percent), along with the resulting personnel and combat power losses of each branch. These combat effects are averages based on historical experience. It is assumed that such elements as leadership, surprise, training, etc., are equal; they usually aren't.

OVERALL CASUALTY RATE OF % is the overall personnel casualty rate for the division in that chart.

COMBAT STRENGTH LOST % is the percentage of the division's combat power lost, given a certain level of personnel loss.

BRANCH is the four major segments of a division. Each has a different function, loss rate, amount of combat power per man and size.

INFANTRY is the branch with the most casualties. This category also includes reconnaissance troops. Most current divisions consist of less than 50 percent infantry, often closer to one third. As recently as World War II, infantry comprised nearly two thirds of many divisions' strength.

ARMOR is the branch with the most firepower per man. Armor combat strength declines rapidly because the heavy combat vehicles tend to break down easily.

ARTILLERY is the combat branch that inflicts the most casualties and receives the least. As the armor and infantry waste away in combat, artillery becomes the principal provider of combat power.

ALL OTHER includes the support troops that may come under fire but do not regularly confront the enemy in combat.

MEN is the number of men assigned to each branch. These are those whose primary jobs are as infantrymen, members of tank crews, artillery crews or others. For the most part, these men are found in battalions composed exclusively of troops of that branch. Where appropriate, the troops of a particular branch are counted as part of their branch no matter where they are. In the Russian Army, artillery is often assigned to infantry units. Most armies assign mortars to infantry units. Mortars, because of their short range and small ammunition supplies, are considered infantry weapons. % OF DIVISION (MEN) is the percentage of men in a division assigned to each branch category.

COMBAT STRENGTH is the amount of combat strength of each branch based on the average capabilities of that branch's normal weapons and equipment. % OF DIVISION (COMBAT STRENGTH) is the percentage of combat strength in a division provided by each branch category.

MEN LOST is the number of men lost given the overall casualty rate. % OF ALL LOSSES is the percentage of total casualties each branch has received. The infantry obviously takes the most losses. The greater the proportion of infantry in a division, the fewer the casualties for the other branches.

CASUALTY RATE is the percentage of each branch lost. The branch casualty rate often differs considerably from the overall division rate, as each branch has different exposure to enemy fire.

SURVIVORS is the number of men in each branch still available for service after casualty losses. % OF DIVISION is the number of survivors of each branch as a percentage of the division total. Compare this with the % OF DIVISION (MEN) to see how quickly some branches decline in combat.

SURVIVING COMBAT STRENGTH is each branch's surviving combat strength. For armor, combat strength declines much faster than personnel strength because the armored vehicles fail much more quickly than the troops are lost. % OF DIVISION is the amount of surviving combat strength as a percentage of the division total.

National Differences

DIVISIONAL ORGANIZATIONS

The Russians have developed two primary division types: the tank (armored) division and the motorized rifle (infantry) division. Both are smaller than their Western counterparts. This is a characteristic some Western armies are coming to appreciate.

Another Russian characteristic is the lower level of support troops. The Russian tank division has 40 support troops for each 100 combat troops. Their infantry division has 36. United States armored and infantry divisions have 52 and 50, respectively. The difference in support troop levels is due to differences in national policy. United States and Western divisions depend more on artillery and technological support in general. Greater technology requires greater support. The Russians also regard their divisions like a round of ammunition: Fire one and then load and fire another. The expended divisions are then rebuilt.

The next major difference is in the larger number of artillery troops in a division. The Russians are less generous when it comes to providing sufficient support, particularly ammunition. The Russian artillery with heavy support is organized into artillery brigades and divisions.

There are fewer differences between Western armored and infantry divisions. One type has somewhat more armor than the other.

COMBAT POWER

Russia puts most of its efforts into providing many tanks. Seventy-five percent of the combat power of a full-strength Russian tank division comes from its tank forces. Although this is a high percentage, the lowest value of any division is 50 percent—for the U.S. infantry division. More critical is the rate at which tank forces melt away in combat. After a division has taken 30 percent personnel casualties, tank forces suffer the most in lost combat power. At this point divisions have lost 39 percent to 55 percent of their combat strength. Armor forces have lost over 60 percent.

Infantry suffers even more, while artillery assumes a more important role as it retains most of its combat power. Depending on division organization, some strange things happen as combat takes its toll.

PATTERNS OF COMBAT LOSSES

One day of heavy combat or several days of light action usually result in 10 percent combat losses. At this point the heavier loss rates of the infantry and armor branches begin to show.

At 30 percent most divisions show serious signs of disintegration. This is the optimal time to take a division out of battle and rebuild its combat branches. Combat beyond this point will practically wipe out a division's combat power. It is anticipated that in a future war, divisions will routinely be pushed beyond the 30 percent level.

Once the 50 percent level is reached there is little left but support troops. In most cases, support troops are serving, without much enthusiasm, as infantry. Most tanks, although not all the crews, will be out of action. Only the artillery will be intact, and without infantry to protect them, soon the guns will be gone also. Continuing to fight beyond the 50 percent level quickly results in the complete destruction of the division. Because the support troops are less effective as combat troops, they are easily killed. Their skills as technicians are harder to replace. Without competent infantry up front, the artillery and support units will be overrun. Once their equipment is destroyed, the division ceases to exist.

The Russian army is an army of extremes. The infantry division is built to take enormous losses and still retain remnants of all its combat arms. The tank division, in a similar situation, is quickly reduced to a unit of artillery, with vehicle-less tank crewmen and support troops serving as infantry.

The Russians achieve this miracle with a motor rifle division consisting 46 percent of infantry. Even the U.S. infantry division is only 36 percent infantry. The Russian infantry division, even with the infantry's higher rate of loss, still has 10 percent of its men left when the division as a whole has lost 50 percent. This is in contrast to the tank division, which started with only 20 percent infantry; at the 50 percent loss level, infantry is but a memory. Tanks, as usual being lost more rapidly than their crews, are similarly gone.

25

VICTORY GOES TO THE BIGGER BATTALIONS: THE COST OF WAR

WARS ARE VERY expensive. People don't complain much when the shooting is going on. When the gunfire quiets down, taxpayers get louder. Supporting a military establishment in peacetime is more expensive in many ways, and certainly more painful, than wartime.

As much as people complain, victory in warfare almost always goes to the bigger battalions. To put it another way, victory is a property of the wealthy. Battles may be won by a David, but the Goliaths win the wars.

The wealthier nations tend to be conscious of their material advantages and are quick to arm themselves in self-defense. By 1945 the United States was maintaining 12 million troops at an annual cost of $33,000 per soldier. Nearly forty years later, 4 million are still directly involved in military affairs at a cost of $56,000 per individual. (Unless otherwise indicated, all prices and costs are given in 1982 dollars.)

Aside from the debatable cost effectiveness of this expensive situation, it is debatable whether or not the United States could afford another war of the magnitude of World War II. Since 1945 per capita income, after being adjusted for inflation, has increased four times. The wealth is there; what might not be available is the time to hammer all those goodies into weapons.

Can We Afford Peace?

Wars may be expensive, yet the high fiscal and human cost tend to limit their length. Peace lasts longer, and eventually costs more. During 1982 the nations of the world are spending over $700 billion on armed forces, more than was spent during the peak year of World War II (1944). Half is spent by just two nations, the United States and Russia. NATO represents nearly half and the Warsaw Pact countries about one quarter.

Defense spending accounts for 6 percent of global GNP (gross national product). Nationally, the range varies from 7 percent for the United States and 14 percent for Russia to 3 percent to 4 percent for the rest of NATO and a bit less for the rest of the Warsaw Pact nations. A major war will generally consume 30 percent to 50 percent of GNP for as long as the war lasts. One year of major war thus equals five to six years of peace. From a financial point of view World War II has been refought at least five times since 1945.

From an economic point of view this has lowered standards of living and decreased the rate of economic growth. Wealth can be spent for investment (to increase production capacity), maintenance (of workers and their tools) and consumption. Because weapons do not reproduce or maintain themselves, they fall into the consumption category. Because weapons use resources that would otherwise increase productive capacity, they have a debilitating effect on the economy.

Many nations get by with spending less than 2 percent of their GNP on defense. But what of the nations which have over the past five years spent an average of 10 percent or more on defense— Russia, Iraq, Egypt, Israel, Jordan, Saudi Arabia, Iran, China, Syria and North Korea? Other major nations have spent less than 2 percent: Japan, Canada, Brazil and Mexico. The high-spending nations have in common available wealth or a perception of many enemies. The low-spending nations are either neighbors of the United States or under U.S. protection (Japan). Think of it as another form of foreign aid, not to mention a great deal of trust and lack of paranoia, on the part of the low spenders.

The low and high spenders aside, most nations spend 2 percent to 5 percent of their GNP each year on their armed forces. What do they get for their money? Often one never really finds out. Wars are not that frequent and the results are not always conclusive. The German Army, for example, has lost every war it has fought for over a hundred years and is still highly regarded. The Russian Army has won most of its wars during the same period and is held in low regard. Quality will only prevail up to a point. In most cases the economically stronger nation, the "bigger battalions," will prevail. The remainder of this chapter will cover what money can buy when you want to go to war.

Why Does Everything Cost So Damn Much?

Weapons have always been expensive. The following major characteristics of military spending have produced the high prices and questionable performance with which we have become familiar.

Historically weapons have become more complex and less reliable. The first weapons were rocks. A rock is quite simple. It can also be very cheap. No doubt the first weapon-cost overrun occurred when the chiseling of specially formed war rocks took longer than anticipated. Military hardware tends to get more complex. Consider, for example, the radar in fighter aircraft. Basic radar with a 30-km range that can track ground and air targets costs $150,000. By raising the price to $350,000, we obtain longer range (40 km), better accuracy at long ranges, the ability to guide missiles to the target and resistance to countermeasures. Let's get a little more range so the enemy won't be able to shoot back. For $450,000 we get a range of 70 km. By making the ground tracking more accurate, the aircraft can fly on the deck in bad weather. Add to this a data link to a ground station and the price rises to $700,000. Finally, we get still longer range (180 km) with equal accuracy, more resistance to countermeasures and more bells and whistles in general. The price is now nearly $1 million.

As things get more expensive, they become less reliable. Studies have shown the following statistical relationships between electronic components cost and MTBF (mean time between failures). A component costing less than $1000 fails, on the average, once every 1500 hours. A $5000 component fails every 250 hours. A $10,000 component fails every 120 hours. A $100,000 component fails every 12 hours and a million-dollar component fails before it gets warmed up. This phenomenon is often overcome with duplicates of particularly failure-prone components.

This is why the space shuttle took five identical flight-control computers aloft. Sure, it's expensive, but did you ever try to land a shuttle while doing the flight calculations on a pocket calculator, which, because of its low cost, would be much less likely to fail? The first shuttle flight was held up because of a failure in one of the backup computers.

Each increase in weapon effectiveness was obtained with more complex, and usually but not always more lethal, weapons. As they have become more complex, they have also become less reliable. We went from rocks to spears (warped shafts, lost heads), swords (impure metal caused brittleness), bows (broken strings, warped arrows), muskets (wet powder, delicate trigger mechanisms), machine guns (delicate machinery), aircraft (delicate instruments), electronics, and guided missiles.

If complexity is held constant, the cost will decline. In the civilian market, similar products maintain their original design long enough to realize cost reductions. Simply consider TV sets, stereos and civilian firearms.

As a component matures it becomes more reliable. The military, however, rarely allows a component to remain unaltered for five or ten years.

The average weapons system (tank, aircraft, missile, etc.) becomes obsolete in ten years. The major problem with this obsolescence within ten years is that it takes an average of eight years to get the weapons to the troops. New ones must be constantly produced to keep up with the enemy's. Because these armaments often use new technology, there are many unexpected problems and delays. The problem is exacerbated in the West by a lack of continuity of project managers during the weapon's development. Thirty months is the average tenure for a military project manager in the United States.

Often new weapons are obsolete soon after, or even before, they are put into use. Never before has there been as much technological development as in the last forty years. The result is an unprecedented number of untried weapons. Worse yet, by the time many of the bugs have been worked out of them, they will be replaced with equally untried systems.

Often an original system is so extensively modified that, aside from external similarity, it is basically a new weapon. Take the M-60 tank. The initial M-60a1 was a refinement of the basic late World War II heavy tank. It cost $700,000, including $41,000 for the fire-control system. The M-60a3 model, which came out ten years later (1973), cost $940,000, including $190,000 for the fire-control system. The current M-1 tank costs $1,200,000, with a $450,000 fire-control system. These costs, incidentally, do not include development or operating costs, which triple the price.

The M-60a3 was a greater improvement over the M-60a1 than the "all-new" M-1 was over the M-60a3. Different versions of the same weapon often differ more from each other than from the next new model of that type.

While it may be technically possible to calculate accurately the cost of a new weapon system, political pressures and human nature conspire to prevent it. No one who is supposed to know the cost of the new system will admit to ignorance. An initial cost is conjured up and then modified by political considerations. This is all of little consequence, as the price will invariably rise. Indeed, any project will cost, on the average, twice the initial estimate.

The original estimated cost implies a threshold of intolerance, a price which, if exceeded, will result in project cancellation. The threshold is a natural reaction if a handful of new programs obliterates all other projects. The formula for this threshold is: $10^{10}/$(number of units to be produced)$^{1.2}$.

In plain English this means that if there is to be only one item, it can cost $10 billion. Two items can each cost no more than $4.3 billion, 10 items can each cost $631 million, 100 items, $40 million each, 1000 items $2.5 million each, 10,000 items $158,000 each, 100,000 $10,000 each, 1 million items $631 each. These averages were made from dozens of weapon systems and were based on 1980 dollars.

This phenomenon has another insidious aspect. Costs rise toward the limit even if there is no other reason. A more expensive tank tends to have more expensive common components (driver's seat, heater, paint job, etc.) than a less expensive tank. There is no other reason other than that a higher price attracts higher costs.

Also, as weapon unit costs increase, the number of weapon tests decrease dramatically. As we have previously seen, more expensive and complex weapons tend to be less reliable. Therefore, the costlier ones should be tested more than the cheaper ones. As this is not the case, the so-called more effective weapons tend to be less effective.

For example, antiaircraft cannon have been around for seventy years, but the missile has only been used in combat during the past three decades. Sufficient combat experience has demonstrated that cheaper cannon systems account for a higher proportion of aircraft damage than more expensive missile systems. So why continue to build missile systems? Partially because they are possible. Also, once a new type of weapon appears, it acquires a life of its own. It cannot be killed except by a newer and usually more expensive system. Missiles are currently in danger of being replaced by laser-charged, particle-beam "death ray" systems. Another $10 billion, please.

The longer a project takes, the more expensive it becomes. The more expensive a project becomes, the more complex it becomes. Greater complexity breeds still more complexity. It is often a violent process to get a weapon system away from the development people and into the hands of the troops.

Us vs. Them

Weapon development projects are usually begun in response to a perceived rather than an actual threat. Because it takes ten years to develop a new weapon, work must begin before the enemy version appears. Most of the really new weapons come from the West which, although more open, is not always sure what a new technology will be good for. The Russians then start working on a countermeasure to the Western weapon. This sometimes prompts the West to develop still another to counter the initial Russian response.

The result is a large number of weapon systems begun and precious few canceled. After its first year of existence, a system has a 4 percent chance of cancellation each year right up until it goes into production. This has been the experience during the past thirty years.

A good example is the U.S. B-70 bomber. This was an early 1960's project, a replacement for the B-52, which would be ten years old and "obsolete" in the late 1960's. ICBM's seemed a better investment, and the B-70, which showed every sign of being hideously expensive, was canceled. The Russians, in the meantime, began developing a weapon to counter the B-70, the MiG-25 Foxbat. The MiG-25 was not canceled and was available in 1970. Without a fast, high-flying bomber to intercept, the MiG-25 didn't have much to do. It could fly fast and high, but was not very maneuverable. It did not have long range or large carrying capacity.

It was specialized for one job and the job no longer existed. Initially the MiG-25 was turned into a camera-carrying reconnaissance plane. Operating alone, it could zip in and out of an area before it could be intercepted.

In the West the MiG-25 was viewed with alarm. Here was this huge, fast, ominous-looking plane. Something had to be done. No one mentioned that something had already been done when the B-70 was canceled. No one mentioned that the MiG-25 had only one useful mission left, and that was to frighten Western governments into countering the MiG-25's mythical capabilities. The result was a number of very capable, and expensive, Western aircraft and missile systems.

RUSSIA: COMPENSATING

Russia, not unmindful of its technological and financial disadvantages versus the West, has compensated in several ways.

First, Russia copies Western technology extensively. Although Russian industry is often incapable of duplicating Western technology, it comes close enough by producing a large number of less sophisticated weapons.

Second, Russia does not leap from one technological breakthrough to another. The Russians allow their systems to evolve slowly. Even totally new systems are usually progressive upgrades of the previous one. Their tanks and aircraft are famous for this.

Third, Russia often mass produces weapon designs that are not technologically advanced but simply are not made by Western armies in large quantities, such as the BMP, APC, many surface-to-air missile systems, etc.

Fourth, this approach tends more toward form than substance. In practice, less capable designs and operators often reduce effectiveness many times. This lower effectiveness is compounded by the Russian tendency to perfect weaponry in actual use. Many Russian weapons go through their entire life with serious defects that could not be put right. The Russians tolerate a much higher discrepancy between practical and theoretical performance than Western nations. This merely reflects their lower industrial standards.

Fifth, Russian doctrine calls for mass production. Quality is nice, but quantity comes first. The important thing is to have as many of a particular weapon as possible with as much commonality as possible.

All the above is changing as the Russians succumb to the pursuit of high technology. Thus, in addition to their usual shortcomings, they are beginning to suffer from the same problems found in the West.

THE U.S.: THE HIGH COST OF BEING FIRST

One of the worst side effects of putting a man on the moon was the feeling that, given sufficient funds, the United States could accomplish any technological feat. In the case of weapon development, where the budgets are smaller, the public support often nonexistent, the problems less well defined and the tasks often more difficult than staging a lunar press conference, the end result is a lot of money down the rathole. The resulting weapons often do not perform as expected and cost more than anticipated. They cost more than can be afforded, so fewer are produced or corners are cut in the design so as to bring the price down but at a huge decline in performance. They give new technology a bad name and make it more difficult for deserving new projects to get funds. Bigger lies are told than are needed, and in a vicious circle it becomes difficult to get anyone involved to speak honestly. Being first is very expensive.

A number of major changes occur when war breaks out. Because you now have a means of determining what works and what doesn't, much of the indecisiveness and overbuilding or weapons stops. Because new weapons are needed quickly, less time is available to spend money. Because more of each type is built, the unit cost comes down. Typically, if you were building a new tank and planned to produce 10,000 for peacetime use, producing three times as many tanks as for peacetime would reduce the unit cost by 20 percent. This is because the development cost represents 30 percent of the total cost. Further, during wartime it will take as little as a year or two before production, which would probably cut the developmental bill by more than half. All in all, the cost of a weapon developed and produced under wartime conditions would be less than half its peacetime cost. In addition, the wartime weapons would be less complex and more effective because of constant feedback from combat. You would be more certain of what you wanted the weapon to do and would be able to avoid designing it to face so many potential contingencies.

The cost of fighting a war today will be substantially higher than for peacetime operations. This is largely due to the high cost of ammunition. Currently a ton of conventional ammunition costs about $7000. A ton of missile munitions costs over half a million dollars. Some improved conventional munitions (ICM) cost ten times more than standard shells and bombs. The high cost of the more expensive munitions represents two things. One is the greater developmental cost. Second, their greater complexity requires much more labor during manufacturing. Under wartime conditions, economies of scale could reduce their cost by five or more times. Still, the price of an average ton of munitions could still be $22,000 or more.

With U.S. divisions consuming at least 1000 tons a day, the bill would be $22 million per division per day just for munitions. Fuel, at $500 a ton, would only be half a million dollars, although fuel would represent as much weight as the munitions. Replacing lost and damaged equipment, assuming 2 percent casualties on an average day plus food, spare parts and anything else, brings the daily bill to over $50 million a day per division. The eight U.S. divisions that are expected to be fighting in Europe would, in thirty days, use up over $10 billion of resources. The air force supporting the army, at over $400,000 per sortie, would cost another $40 billion. Thus, in one month of heavy combat, a quarter of 1982's defense budget would be expended.

Consider the overall situation. The U.S. economy, if mobilized, could raise $1500 billion a year for a major war. This would be half the peacetime GNP of $3000 billion. In wartime the overall GNP would increase somewhat, perhaps by 50 percent.

A force of 30 divisions, each in combat for 100 days a year, would cost $280 billion a year.

A force of 3600 combat aircraft, each lasting 50 sorties and averaging 350 sorties each a year, would cost $500 billion. Air operations have always been more expensive than ground operations. A hundred aircraft are generally considered to cost as much as a division. Looking at it this way, we devote twice as many resources to aircraft, including the fact that an air force unit of 100 aircraft requires one third the manpower of an army division. Air power is more expensive but less costly in human life.

A navy of over 300 ships and 1400 aircraft would consume over $300 billion in a year. Over 60 percent of this would be for land- and carrier-based aircraft operations.

If operations were carried out at the above levels, the cost would be over $1100 billion a year. Only about 10 percent of this is for replacement of destroyed equipment. The majority would be spent for munitions. Guided missiles and other so-called smart munitions would account for most of the cost. In addition, there is the expense of building additional ships, aircraft and weapons for new divisions. In all, we cannot afford to wage a major war using the munitions we currently plan to use.

Completely mobilizing for war would take several years. Procurement of weapons and munitions consumes less than 30 percent of the peacetime military budget. Munitions themselves cost $2 billion a year. Munition stocks on hand will sustain combat for thirty days, at best. Three quarters of this $2000 billion needed for the first year must be procured. This implies increasing procurement thirty times, but the most optimistic increase in one year would be by a factor of three to five. This, plus the thirty-day reserve, would provide no more than 25 percent of first-year requirements.

In 1914 the war ground to a halt after two months largely because both sides ran out of ammunition. During World War II, battles often wound down as the attacker ran out of ammunition and fuel. The same pattern appeared during the 1973 Arab-Israeli War, to many an expert's surprise.

Both Russia and its potential Western opponents perceive the solution to be a quick victory. This is another myth regularly worshiped by military planners. While possible, a short war is unlikely. More likely, the weapons of mass destruction will be avoided and a war of attrition will ensue. Given the lack of real emotional issues between the major potential antagonists, negotiation may be more possible than in the last two world wars. Having been burned badly twice in this century, should nations be politically inept enough to get caught up in another worldwide slaughter, they will have powerful incentives to settle. Heavy initial losses in a future war will further reinforce the lesson.

Gold-plated little wars are even more expensive than less-than-total wars. Here the natural temptation to use weapons that are safer for the users, and more expensive, will be difficult to resist. Vietnam was such a war, and the expensive munitions were just starting to get really expensive. Warfare has never been cheap, Vietnam showed that it will never get any cheaper.

The Price of Things to Come

Several cost trends have been inexorably progressing through this century. Aircraft costs, for example, have been increasing four times every ten years for the last seventy years. During the past twenty years, this disease has spread to land and naval weapons. Currently all aircraft cost $400 a pound to produce. Armored vehicles cost $15 a pound and ships $50 a pound. Many missiles cost three to five or more times as much as aircraft. The chief culprit is electronics. Promising much, and sometimes delivering it, electronics tend to become so complex that mere humans cannot easily ensure their reliability. During World War II, most combat aircraft had less than 100 pounds of electronic gear. Today, a ton, 2200 pounds, is the norm in Western aircraft, with the Russians catching up. In the last twenty years, tank fire-control systems have gone from 6 percent (M-60a1) of total vehicle cost to nearly 40 percent (M-1).

If the increases of the last seventy years were to persist, in some seventy-two years the entire defense budget of the United States would be spent on one combat aircraft.

To examine the relationship between cost and effectiveness, let us compare a modern carrier task force with one of World War II vintage. Task Force 58, in 1944, had 112 ships and nearly 1000 aircraft. These aircraft could deliver 400 tons of ordnance. They cost $520 million. A modern task force of 9 ships and 90 aircraft can deliver the same amount of ordnance. The aircraft cost twice as much. However, less than 10 percent as many aircrew are needed. In addition, there are 100 fewer ships. The modern aircraft can fly more often and have fewer accidents. The modern task force costs $9 billion and has nearly 9000 men. Task Force 58 cost five times as much after adjusting for inflation, and contained more than five times more sailors.

This is where the cost equation stands. Cost has increased, but *potential* capability has increased faster. More important, far fewer men are needed to man the weapons. When the cost of training and maintaining manpower is added, the modern task force is actually cheaper. Unfortunately, the costs of the weapons are reaching the limits of national budgets. Unit costs are increasing far faster than the ability of national economies to afford them.

The future is already here; fewer weapons are being purchased. Russia has crippled its economy with huge armament expenditures. The United States, to a lesser extent, has done the same. Japan, spending less than 1 percent of its GNP on arms, compared with more than 5 percent for the United States, has used the difference to build up industries producing a wide range of industrial and consumer goods. An arms race is a luxury few can afford. Those that can afford it pay a hidden price in lost economic growth. These lessons are catching up with the United States, Russia and Western Europe.

Warfare, and its peacetime buildup, only becomes overly expensive when the wealth exists to start the cycle. Historically these periods of wealth and arms building culminate in a series of wars. This usually impoverishes the survivors and takes the fight out of their descendants for a generation or so until the cycle begins anew.

Now we have nuclear weapons. We also have more people possessing more wealth and good living than ever before in human history. We also have a technological war of sorts for the first time. The high cost, and often dubious effectiveness, of many modern weapons has inflicted significant fiscal casualties on many nations. In addition to pouring billions into devices which themselves produce nothing, armament spending produces one third fewer jobs than does non-arms spending.

The result of all this spending is often of dubious value. As was pointed out in the chapter on leadership, there is a tendency to seek technological solutions to personnel and leadership problems. The primary objective of peacetime arms expenditures appears to be spending money. Those nations which can break out of this cycle will not only be better defended but much wealthier.

The ultimate cost of military spending may be political unrest leading to a war. This may happen in Russia first. High military expenditures have also caused political unrest in the United States. Other Western nations must constantly be wary of popular reactions to such spending. One way or another, the cost of war is felt even in times of peace.

Cost of War

This chart shows the cost of conducting a war for U.S. forces and Western armed forces in general. The costs cover a period of one year's fighting. Costs are expressed in dollars and in tonnage of supply that must be transported to the theater of operations. This chart is divided into five sections: Land operations covering all land combat not involving the air force; air operations in support of ground forces; air operations in support of naval forces; naval operations, not including transport of amphibious forces, or their combat on land; and a summary of all these operations.

LAND OPERATIONS

COMBAT DIVISIONS are the number of divisions involved in combat.

COMBAT DAYS are the number of days in the year that the division will be in combat. This is important because on these days the heavy munitions and fuel expenditures occur.

DAILY LOSS RATE % is the average percentage of personnel strength lost for each day of combat. Losses of weapons and equipment are tied to it.

COST OF DIVISION $ MILLION is the cost to arm and equip a division with new and repaired equipment. This is used to calculate replacement costs.

SUPPLIES are the various categories of goods and services the division requires in and out of combat. TONS are the tons of consumable supplies required per division per combat day. COST PER TON ($1000) is the average cost of each ton of supplies in thousands of dollars. TOTAL COST (MILLION) is the total cost of each category of supplies; % is the percentage of total cost each category of supplies represents. TOTAL TONS (THOUSANDS) is the total tonnage of each category of supplies.

AMMUNITION is all munitions used in combat. FUEL is all fuel used for vehicles, generators, aircraft, etc. MAINTENANCE is spare parts and other supplies required to maintain equipment. Also includes food, wages, clothing and other matériel. NONCOMBAT MAINTENANCE is the supplies required to maintain the equipment of noncombat support forces. NONCOMBAT SUPPORT is the cost of maintaining noncombat support troops.

TOTAL gives the total cost in billions of dollars and thousands of tons for all divisions for both combat and noncombat days. COST PER DIVISION COMBAT DAY is the cost (in millions of dollars) of one day's combat for one division. COMBAT CASUALTIES are the annual casualties for all combat and noncombat units. COMBAT UNIT CASUALTIES are the casualties just for the combat units. AS % OF DIVISION'S STRENGTH, total combat unit casualties as a percentage of division strength, indicates the severity of personnel losses (see the chapter on attrition for more details).

COST OF WAR

One year of a major war, as now planned, is more than the United States can support.

Land Operations

```
Combat Divisions                    30
Combat Days                        120
Daily Loss Rate %                    2
Cost of Division  $ Million       1700
```

Supplies	Tons	Cost per Ton ($1000)	Total Cost (million $)	%	Total Tons (thousands)
Ammunition	1200	22	95040	34	4320
Fuel	1000	1	1800	0.64	14400
Maintenance	40	10	1440	1	10
Replacement			122400	44	12240
Noncombat maintenance			24500	9	2450
Noncombat support			36000	13	
Total			281(billion)		33420

```
Cost Per Division Combat Day            67 $ Million
Combat Casualties                  1684800
Combat Unit Casualties             1296000
As % of Division's Strength             42
```

Air Operations in Support of Ground Forces
(Air Force & Marine Corps)

```
Combat Aircraft                   3600
Sorties per Aircraft               350
Sortie Loss Rate %                  .2
Cost of Aircraft                    15
```

Supplies	Tons	Cost per Ton ($1000)	Total Cost (million $)	%	Total Tons (thousands)
Ammunition	5	110	450450	84	6300
Fuel	10	1	4095	1	12600
Maintenance	1	15	12285	2	819
Replacement			37800	7	
Noncombat support			32400	6	
Total			537 (billion)		19719

```
Aircraft Lost                         2520

Aircrew Casualties                    1764
Groundcrew Casualties                19757
------------------------------------------
Total Casualties                     21521

Sorties per Division Combat Day        228

Cost per Sortie ($1000)                426              16   Tons
```

Air Operations in Support of Naval Forces

Combat Aircraft	1400	
Sorties per Aircraft	400	
Sortie Loss Rate %	0.05	
Cost of Aircraft	22	

Supplies	Tons	Cost per Ton ($1000)	Total Cost (million $)	%	Total Tons (thousands)
Ammunition	3	180	226800	89	1400
Fuel	12	1	3024	1	6720
Maintenance	1	15	7560	3	504
Replacement			6160	2	
Noncombat support			11200	4	
Total		$ Billion	255		8624

Aircraft Lost	280
Aircrew Casualties	952
Ground Crew Casualties	6950
Total Casualties	7902

Cost per Sortie ($1000)	455	15	tons

Naval Operations

Major Combat Ships	350
Fleet Overhead per Ship ($Million)	100
Sea Days per Ship	250
Average Ship Size (tons)	6500
Supply per Ship (1000 tons)	153
Cost per Ton of Supply ($1000)	1.76
Total Ship Supply (1000 tons)	13366
Supply Cost ($Million)	23534
Other Costs ($Million)	35000
Total Cost ($Billion)	59

Summary

	Billion $	% of Total
Total Land-Oriented Costs	818	72
Total Naval-Oriented Costs	313	28
Land Operations Only	281	25
Air Operations Only	792	70
Ship Operations Only	59	5
All Cost	1131	100

Combat Unit Consumption Only	Tons x 1000	%	Ships Required to Transport	
Fuel	44413	59	1255	250,000 Ton Tankers
Ammunition	13357	18	668	20,000 Ton Capacity Cargo Ships
Other	17359	23	868	20,000 Ton Capacity Cargo Ships
Total Supply Tonnage (x1000)	75128	100	2791	Total Shiploads

AIR OPERATIONS IN SUPPORT OF GROUND COMBAT

This is the cost of air operations in support of ground combat. Most of the missions represented here are ground attack, air superiority or reconnaissance.

COMBAT AIRCRAFT is the number of aircraft involved in operations. In this case strategic bombers are included, as they are quite effective with conventional munitions.

SORTIES PER AIRCRAFT is the average number of times each aircraft takes off in a year. In wartime the readiness level, percentage of aircraft not in repair or maintenance, will vary from 10 percent to 50 percent or more. Modern aircraft can fly an average of three sorties per day. This top performance is called surge. On many days aircraft will not fly at all. Most sorties will be flown on days the divisions are in combat.

SORTIE LOSS RATE % is the percentage of aircraft lost per 100 sorties. The figure .2 means 2 aircraft lost per 1000 sorties. Depending on pilot quality, level of aircraft maintenance and flying conditions, the *noncombat* loss rate can be 1 per 1000 sorties, or higher, particularly in Russian-style air forces. On the average, however, noncombat losses will be about 1 aircraft per 4000 to 10,000 sorties.

COST OF AIRCRAFT is the average cost of aircraft in millions of dollars. This is used to calculate replacement costs.

SUPPLIES, TONS, COST PER TON, %, TOTAL TONS are the same as for land operations above.

AMMUNITION includes missiles, cannon shells, bombs and any other disposable items, like chaff and flares to avoid enemy missiles. Although the majority of sorties will carry ground-attack munitions, many will be for reconnaissance and air superiority. These missions carry much lower weights. Therefore, the average weight carried is less than half the average maximum carrying capacity. FUEL is the average fuel carried on each sortie. MAINTENANCE is primarily spare parts, but also includes supplies necessary to support personnel (food, clothing, etc.). REPLACEMENT is the cost of replacement aircraft. There is no weight given, as these aircraft fly under their own power. NONCOMBAT SUPPORT is the cost of all other support operations, particularly those in the homeland.

AIRCRAFT LOST are the number of aircraft destroyed or damaged beyond repair. Such craft can sometimes serve as a source of cannibalized parts. As you can see, the tempo of operations would quickly destroy many of the air units. Given the present annual U.S. production of under 1000 aircraft, such losses would quickly reduce the level of air operations.

AIRCREW CASUALTIES are aircraft crew killed, seriously injured or taken prisoner. Not every lost aircraft results in total aircrew loss. The percentage of aircrew lost varies with pilot quality and the size, construction and mission of aircraft. Western aircraft are much more resistant to Russian weapons than the reverse. Western aircraft are generally larger and more robust. Russian anti-aircraft weapons are less devastating than Western weapons. A mission places the aircraft either high and over friendly territory (good) or low and over enemy territory (bad). Depending on all these factors, the percentage of aircrew emerging unhurt from a lost aircraft varies from 30 percent to more than 90 percent.

GROUND CREW CASUALTIES are losses of ground-crew and other airbase-support personnel. In all past wars, and even more in current ones, airbases are a priority target. Enemy aircraft are easier to destroy on the ground. The increasing complexity of modern aircraft has also increased their vulnerability to loss of ground-support facilities and technicians.

SORTIES PER DIVISION COMBAT DAY are the total number of sorties divided by the number of division combat days. Not all these sorties would be in direct support of the division. Many would be sent against enemy airbases, airborne aircraft and communications targets. Over half, however, would support the division. Air operations are substantial on noncombat days, too. Reconnaissance in particular proceeds at all times.

COST PER SORTIE is the average cost of a sortie in thousands of dollars. Also given is the average tonnage of supply needed. Even in peacetime, when minimal amounts of ammunition are used, the cost per sortie is about 30 percent of the wartime cost.

AIR OPERATIONS IN SUPPORT OF NAVAL FORCES

This section follows the same format as air operations in support of ground forces. There are some differences between naval support and ground-support air operations. At sea, aircraft spend much more of their time on patrol in search of enemy naval forces (surface or submarine) or in defense of friendly shipping. Although most naval aircraft are armed, they use these weapons far less frequently than land-oriented aircraft.

Naval aircraft suffer lower attrition. At sea, aircraft are less likely to encounter other aircraft. Most of the combat is against targets that can't shoot back—cruise missiles, submarines, merchant ships. Against armed ships, U.S. naval aircraft unleash a large array of electronic countermeasures.

Naval aircraft tend to be larger, with larger crews. They are more expensive and fly more sorties.

NAVAL OPERATIONS

Warfare at sea does not lend itself as easily as air or ground combat to terms like "combat days" and "sorties." Combat is irregular, if it occurs at all. Antisubmarine operations are the closest to attrition warfare.

The bulk of naval warfare is little more than keeping ships at sea and ready to fight. This is not an easy task. The supply requirements are enormous. The largest item, by weight, is fuel. More expensive weapons and equipment have made naval items the most expensive.

MAJOR COMBAT SHIPS are the combat ships (over 1000 tons displacement) that must "maintain station," stay at sea for long periods. This number does not include amphibious ships and other transports. The amphibious ships tend to be massed at a port and then sent off for a landing, from which they return as soon as possible. Other transports spend their time going to and fro delivering supplies to the fleet.

FLEET OVERHEAD PER COMBAT SHIP is the annual cost (in millions of dollars) to each ship of the fleet overhead, including the establishment of land bases and support ships.

SEA DAYS PER SHIP is the average number of days per year each ship stays at sea. A ship should spend 10 percent to 20 percent of its time in port being overhauled. The crew needs time ashore to maintain morale. Without such respite, both ship and crew performance declines.

AVERAGE SHIP SIZE (TONS) is the average tonnage (displacement) of each ship.

SUPPLY PER SHIP (1000 TONS) is the average tonnage of supply a ship requires per day at sea. COST PER TON OF SUPPLY ($1000) is the average cost per ton of supply in thousands of dollars. TOTAL SHIP SUPPLY (1000 TONS) is the total daily supply for all ships per day when at sea. SUPPLY COST ($MILLION) is the total cost of supplying all ships at sea for one day. OTHER COSTS ($MILLION) are all other costs, including payroll, shore-based support, repair and replacement.

SUMMARY

This section gives the total costs for land, naval and air operations in billions of dollars. Also given is a summary of the supply tonnage in thousands of tons and the number of ships required to carry it. The cost of replacement troops and the movement of seriously wounded back to the homeland is included in the various overheads and supply movements.

Several patterns emerge. The most expensive single item is ammunition, accounting for over two thirds of the cost. By weight, fuel accounts for over half. Aircraft account for over two thirds of the expense.

The Next War, 30 Days Long

Should a major war break out in Europe, U.S. forces could probably not continue heavy combat beyond thirty days. If anticipated rates of ammunition use and combat losses occurred, within thirty days the United States would have no ammunition left, and its combat divisions would be reduced to less than half their combat strength.

The United States maintains a stock of 500,000 tons of munitions in Europe. A force of 7 divisions could each participate in 25 days of combat during the first 30 days. About 1200 combat aircraft could be available to fly an average of 50 sorties each during that 30-day period. This level of combat would consume 468,000 tons of munitions.

Several factors can speed up the rate of consumption. One is enemy destruction of munitions stocks by either ground or air forces. Munition stockpiles, for obvious reasons, are prime targets. Losses of 30 percent to 50 percent can be expected from enemy action. Desperate battlefield situations will call for higher use of munitions. Some armies, such as West Germany's, normally plan on higher expenditure levels than the U.S. Army. The Germans may be right. Faulty munitions will reduce the percentage of available stocks. Inadequate storage can cause stored munitions to become defective. You never really know if the stuff is good until you use it. If you can't get the munitions from the dumps to the guns or aircraft, it is as good as lost. This has happened in the past, and truck drivers still get lost. Desperate allies, who did not stock sufficient supplies, may require munitions to prevent them from being overrun. Many U.S. allies fit this description. Because they use U.S.-built weapons, they can use U.S. munitions.

Other factors may slow down consumption rates. It is not likely that needs will be lower than expected. Even if needs were lower, it would be difficult to dissuade gunners and pilots from delivering the stuff. Blowing things up is a hard habit to break. It is also unlikely that the war will end before thirty days are over. If the Russians were stopped, they would be stopped inside friendly territory. There would be a great clamor to throw them out. This would take still more munitions. The munitions will last longer if the United States runs out of guns and aircraft to deliver the munitions at the planned rate. This is highly possible. Delivering new supplies is a strong possibility if the navy can keep Russian submarines at bay. Twenty thousand tons a day would keep things moving. Munition stockpiles in the United States could keep the fighting going full blast for another few months. They would then dry up, as U.S. production capacity currently produces in one day less than 10 percent of the munitions that would be consumed in one day's fighting by those seven divisions in Europe. Modern munitions are often more complex than the weapons of past wars. Setting up production facilities and training workers will take years, not months. Overall, it is more likely that munitions will not last as long as expected.

So what will happen? World War I opened with a roar as thousands of tons of munitions were expended. Within two months, most armies had run out of ammunition. Trench warfare then set in and lasted four years until the losing side ran out of munitions again. In World War II, many battles ended when one or both sides ran out of munitions.

There is no reason to believe that future operations won't stalemate when both sides run out of ammunition. What, however, is to prevent the Western forces from running out first? The Russians have plenty of munitions but will have considerable difficulty in moving the goods. Western air forces are equipped and trained to perform one mission extremely well: stop enemy supplies from moving. The Russians are more casual about moving supplies, as they are devoted to the "short war." Russian combat units are not burdened with a lot of munition-carrying transports. Combat vehicles would load up as much ammo as they could and head for the Rhine.

In wartime ammunition has a very short shelf life. People want to either use it or destroy it as quickly as possible.

Cost of Manpower: U.S. and RUSSIA

This chart shows the extent and distribution of armed-forces manpower and spending in the United States and Russia.

SERVICE is the branch of the armed forces. The United States and Russia have fundamentally different organizations.

ARMY represents the land combat forces of both nations.

NAVY represents seagoing forces. The naval strength of the Russian marines is included.

MARINES. The U.S. Navy has built up substantial land forces since World War II. These units, the U.S. Marine Corps, comprise a significant portion of U.S. land-combat capability. Although the USMC belongs to the U.S. Navy, it is recognized as a separate service.

STRATEGIC represents a combination of the Russian strategic rocket forces and air-defense units. In both nations submarine-launched missiles belong to the navy.

AIR FORCE in both nations shows air units. In neither nation does the air force represent all air units. The U.S. Navy, in particular, has substantial air units, as do the army and the marines. In Russia the strategic forces possess substantial air units (interceptors), while the Russian Navy has fewer, primarily land-based air units. The Russian Air Force is basically an army support force. The U.S. Air Force performs the same function, as well as having bombers and a few interceptors. As happened in Vietnam, all U.S. Air Force units, including the strategic bombers, can be applied to supporting ground combat. The U.S. armed forces have 25,000 aircraft, including helicopter, noncombat and reserve units. The U.S. Air Force has 31 percent, as does the navy, including the marines. Russia's 14,000 aircraft belong primarily to the air force (64 percent). Twenty-nine percent are under the control of the strategic forces, and 7 percent of the navy. All army aircraft come from air force units.

ARMED FORCES MANPOWER

Armed forces manpower (in thousands) remains remarkably stable from year to year. In the last ten years Russian military manpower has increased about 6 percent. During the same period, the Russian population increased over 12 percent. A major war and the elimination of conscription would change this, as Vietnam did to the United States. Manpower in the U.S. armed forces has fallen by nearly 40 percent since 1970. Further increases are not likely without conscription.

PERCENTAGE DISTRIBUTION OF ARMED FORCES MANPOWER shows the significant structural differences between the U.S. and Russian armed forces. A large proportion of Russian forces is devoted to essentially defensive missions

COST OF MANPOWER: U.S. AND RUSSIA

When all forms of military manpower are considered, the United States and Russia are roughly equal. Compared with Russia, the United States spends more on less in order to give each man more combat power. The money is real, the amount of combat power is difficult to prove short of a major war.

US Armed Forces Manpower

Service	Active	Civilian	Reserve	Total	Major Units	Men per unit
Army	776	361	742	1879	26	72269
Navy	534	296	174	1004	324	3099
Marines	185	20	88	293	4	73250
Air Force	564	263	195	1022	4700	217
Total	2059	940	1199	4198		
% of Tot.	49	22	29	100		

Percentage Distribution

Service	Active	Civilian	Reserve	Total
Army	18	9	18	45
Navy	13	7	4	24
Marines	4	0	2	7
Air Force	13	6	5	24
Total	49	22	29	100

Russian Armed Forces Manpower

Service	Active	Civilian	Para-military	Total	Major Units	Men per unit
Army	2440	310	400	3150	193	16321
Navy	433	140	75	648	686	945
Strategic	1085	130	0	1215	6400	190
Air Force	420	148	·50	618	8300	74
Total	4378	728	525	5631		
% of Tot.	78	13	9	100		

Percentage Distribution

Service	Active	Civilian	Para-military	Total
Army	43	6	7	56
Navy	8	2	1	12
Strategic	19	2	0	22
Air Force	7	3	1	11
Total	78	13	9	100

US Armed Forces
Annual Costs (x $1,000,000)

Service	Pay	Procure	Operate	Support	Total	% of Total	Major Units	Cost per Unit
Army	18400	12100	19900	9800	60200	27	26	2315
Navy	9720	23500	22450	19130	74800	34	324	231
Marines	2544	1970	2900	4786	12200	6	4	3050
Air Force	10900	21700	23100	18900	74600	34	4700	16
Total	41564	59270	68350	52616	221800	100		
% of Tot.	19	27	31	24	100			

Russian Armed Forces
Annual Costs (x $1,000,000)

Service	Pay	Procure	Operate	Support	Total	% of Total	Major Units	Cost per Unit
Army	8846	16872	14136	10397	50251	26	193	260
Navy	2006	17784	9394	7387	36571	19	686	53
Strategic	5050	21158	16872	14410	57490	30	6400	9
Air Force	2155	20064	17602	8299	48119	25	8300	6
Total	18058	75878	58003	40493	192432	100		
% of Tot.	9	39	30	21	100			

(strategic and paramilitary, 31 percent of the total). United States forces, on the other hand, are organized for offensive operations. Strategic forces comprise only 4 percent of total manpower.

ACTIVE is the full-time, uniformed troops.

CIVILIAN represents the full-time civilian employees who work for the armed forces. Because so few of the uniformed soldiers are in combat units, many perform jobs that could just as easily be held by civilians. It has long been traditional for the most technical jobs to be held by civilians. For example, when artillery was introduced to warfare 500 years ago, the guns were manned by civilians. This tradition persisted for hundreds of years. Even today, many of the more complex missile systems and aircraft are maintained, at least partially, by civilians. The reason is simple, people with sufficient technical skills cannot usually be kept in uniform.

RESERVE represents soldiers who spend *most* of their time as civilians. See Chapter 5 for more details.

PARAMILITARY is a category unique to Russia. These are combat troops used for border patrol and internal security. See Chapter 5 for more details.

MAJOR UNITS are the major combat units of each branch of the armed forces. These figures are used to calculate the cost of maintaining combat power. For the army and marines, major units include fully equipped combat divisions or groups of smaller units equivalent to divisions. For the United States they include reserve divisions. The Russian paramilitary ground forces become equivalent divisions. For the navy it includes major combat ships (over 500 tons). For the air force it means combat aircraft and ballistic-missile launchers. For strategic forces it means combat aircraft, ballistic-missile launchers and air-defense missile sites.

MEN PER UNIT is the average number of personnel for each major unit. Western units have higher manning levels. Russian Army units would almost double in manpower during a war, due to the call-up of reserves. For the navies Western manpower levels are higher, partially because of larger ships and higher overheads for research and support. Larger air force manpower levels in the West are due to more complex weapons and greater use.

ARMED FORCES ANNUAL COSTS

Annual costs (in millions of 1982 dollars) are the anticipated spending levels for the 1980's. During the 1970's, U.S. spending averaged $150 billion a year, Russian spending about $180 billion. These costs can be broken down in many ways. One alternative to this way is by functional categories, for example (for the United States), general-purpose forces (traditional army, navy, air force), 37 percent; strategic forces (ICBM's, etc.), 8 percent; airlift and sealift forces, 2 percent; reserve forces (add to general-purpose forces), 5 percent; intelligence and communications, 7 percent; research and development, 9 percent; central supply and maintenance, 10 percent; all other support, 22 percent.

The costs shown are estimates based on past trends and present pronouncements.

PAY is the wages and other allowances paid to the troops. There is a significant difference between the U.S. and Russian practice. About 50 percent of the Russians' uniformed manpower consists of conscripts, most of whom receive less than $100 a year. The regulars receive pay and benefits superior to most civilians. Still, the pay per man in the United States is many times the Russian rate.

PROCURE is procurement, the purchase of new weapons and equipment. In the United States this breaks down as follows: aircraft, 36 percent; missiles (primarily for aircraft), 16 percent; ships, 14 percent; combat vehicles (tanks, etc.), 6 percent; heavy weapons (artillery, torpedoes, etc.), 9 percent; electronics and communications, 8 percent; everything else (including munitions other than missiles and torpedoes), 11 percent.

Procurement in Russia follows a slightly different pattern. Less is spent for ships and aircraft and more for missiles (primarily ICBM's), combat vehicles and heavy weapons.

OPERATE is the cost of operations and maintenance. This covers everything from food and fuel to clothing and spare parts. Russia scrimps here by using its equipment much less and adopting an austere life-style.

SUPPORT includes research, development, testing, administration, pensions and construction. Research is a particularly high burden for Western armed forces. Construction costs also tend to be higher. Pensions are a particularly high burden for the United States, where they are closing in on $20 billion a year.

COST PER UNIT ($\times$ $1,000,000) is the annual cost of maintaining each combat unit. This is more than simply paying, feeding and maintaining the troops, and buying and maintaining equipment. It includes every cost incurred by the armed forces. As the two major armed powers, the United States and Russia, cannot buy better weapons from anyone else, they must develop them themselves so their relative combat effectiveness does not decline. Therefore, the high costs of developing and building new weapons become part of maintaining armed forces. Russia and the United States are the most extreme examples of their respective military systems. No other countries pay as much per unit for armed forces. This is because the two superpowers pay for weapon development as well as for maintaining the expensive strategic nuclear weapons systems. Additionally, few other nations maintain their armed forces at such a high state of readiness.

ARMED FORCES COST PER MAN

This section puts the different spending strategies into perspective.

Cost per man is the average cost per man for pay, procurement, operations and support.

The categories are the same as for armed forces annual costs, with the following additions.

US Armed Forces
Cost per Man (Dollars)

Service	Pay	Procure	Operate	Support	Total	Total with Russian Pay Levels	Total with Russian Pay Levels
Army	12121	6440	10591	5216	32038	30220	56784
Navy	13729	23406	22361	19054	74502	72443	72732
Marines	9319	6724	9898	16334	41638	40240	11790
Air Force	14361	21233	22603	18493	72994	70840	72398
Total	12758	14119	16282	12534	52835	50921	213705
% of Tot.	24	27	31	24	100	96	

Russian Armed Forces
Cost per Man (Dollars)

Service	Pay	Procure	Operate	Support	Total	Total with US Pay Levels	Total with US Pay Levels
Army	3115	5356	4488	3301	15953	26743	84239
Navy	3950	27444	14496	11400	56437	70118	45437
Strategic	4655	17414	13886	11860	47317	63440	77080
Air Force	4584	32466	28482	13429	77863	93743	57933
Total	3683	13475	10301	7191	34174	46931	264689
% of Tot.	11	39	30	21	100	137	

WITH RUSSIAN PAY LEVELS indicates what the per-man cost would be if U.S. forces adopted the same pay policies as Russia.

WITH U.S. PAY LEVELS indicates the cost to the Russians if they paid their troops on the same level as U.S. troops. This is the technique the CIA has used to produce Russian budgets comparable to U.S. budgets. One element the CIA left out of their analysis, or did not give as much weight to as I did, was the higher nonpayroll costs due to the priority of the military in every aspect of the economy. If you add equivalent U.S. pay standards to the nonpayroll expenses, the armed forces costs approach 30 percent of the Russian GNP. See the section titled "Apples and Cabbages" below.

TOTAL WITH RUSSIAN (U.S.) PAY LEVELS is the cost (in millions of dollars) of each service. Not much cheaper for the United States and much more expensive for Russia.

Apples and Cabbages

Comparing U.S. and Russian armed forces spending is much like trying to compare apples and cabbages with equivalent terms. The Russian economy is fundamentally different from that of the United States. First, in the United States you usually have to pay for higher quality. In Russia prices are set arbitrarily, often regardless of quality and utility. Better housing does not cost more, but it is more difficult to obtain. In effect, there is an official black market for scarce goods. This style of trade extends to the arms industry.

When pricing military goods within the Russian economy, one must keep in mind that the military go first class while everyone else goes last.

In the civilian sector Russian citizens simply put up with goods that would be unsalable in the West. On the battlefield inferiority threatens national survival. The Russian solution was to develop two separate economies. The military economy gets the best of everything: raw materials, labor and priority in general. The civilian economy gets what's left. Not only are massive quantities of national resources devoted to the armed forces, but the best of anything. One Western estimate is that 33 percent of the machine tool industry, 20 percent of the metallurgical industry, 18 percent of the chemical industry and 18 percent of all energy are devoted to the armed forces. Beyond those stark numbers, the armed forces get the best, which makes them the higher effective proportion of each economic sector.

Lower labor costs are obtained by conscription and a lower standard of living. Even allowing for lower productivity, production costs are nearly twice as high in the West. This is not a critical factor, as the key ingredient missing from Russian armaments is better technology. This affects quality and effectiveness for all but the simplest weapons. Increasingly electronic gear for fire control is vastly more effective. In this the Russians cannot keep up. Their labor costs are lower, but so is the effectiveness of their labor in developing, or duplicating, better technology. Components requiring quality finish or innovative design are difficult to produce within Russian industry.

Much of Western technology cannot be duplicated in Russia. Give the Russians an advanced Western weapon and they will, at best, produce a second-rate copy. An example is the U.S. Sidewinder missile, first used in the late 1950's. The Russians followed a year later with their AA-2 Atoll missile. Its appearance was a direct copy of the Sidewinder. Its performance was far less. This deficiency can be traced to a less effective heat sensor, which is a high technology component, and a less efficient control system, attributable to less exacting production standards. An even more precise example is shown by comparing two jet engines of equal age, takeoff thrust, weight and purpose: the Russian D-30 and the U.S. JT8D-11. The U.S. engine is 19 percent smaller and has 37 percent greater cruise power. In addition, the United States engine is much more reliable and requires less maintenance.

Wartime production levels give the Russian military economy a singular advantage. By producing weapons all the time in large quantities, in effect at wartime production rates, they reduce the cost. To give a U.S. example, producing the M-1 tank at the rate of 120 a month instead of 60 a month reduces the unit cost 30 percent.

Different concepts of effectiveness allow the Russians to produce weapons unacceptable in the West. Their tanks, for example, are extremely cramped. This allows for smaller, cheaper and relatively more cost-effective vehicles. To solve the interior space problem, the Russians recruit their tank crews from the shortest members of the population. Russian vehicles are so cramped that combat effectiveness is often reduced. In warm climates and during extended operations, crew fatigue is noticeable.

Russian weapons may look like their Western counterparts, but they have a profoundly different conception and design. Thus, it's impossible to compare exactly the cost of equivalent Russian and U.S. weapons. One can estimate what a Russian weapon would cost to build in the West, and this method has been used to

COST OF RAISING A DIVISION

As expensive as divisions are to raise, their annual maintenance expense exceeds their initial cost after a few years.

Cost of a Combat Division (US)

Item	Quantity	Avg Unit Cost	Total Cost	% of Total
APC's	1100	600	660000	29
Tanks	248	2000	496000	22
Munitions, Tons, 30 Days	15000	22	330000	14
Aircraft	37	5000	185000	8
Misc Equipment & Supplies			140000	6
Trucks	3000	35	105000	5
Communications and EW Equipment			120000	5
Air Defense Weapons	48	1500	72000	3
Field Artillery	66	900	59400	3
ATGM	380	110	41800	2
Personal Equipment	17000	2	34000	1
Infantry Weapons	24000	1	24000	1.05
Fuel Stocks (30 Days Supply)			7500	0.33
Food Stocks (30 Days Supply)			3500	0.15
Total (Millions of Dollars)			2278	100

Summaries

Combat Vehicles	1341000	59
Other Weapons	197200	9
Other Equipment	399000	18
Supplies	341000	15

Cost of a Combat Division (Russia)

Item	Quantity	Avg Unit Cost	Total Cost	% of Total
APC's	673	300	201900	18
Tanks	266	1200	319200	29
Munitions, Tons, 30 Days	10000	12	120000	11
Aircraft	3	3000	9000	1
Misc Equipment & Supplies			55000	5
Trucks	1500	30	45000	4
Communications and EW Equipment			85000	8
Air Defense Weapons	156	800	124800	11
Field Artillery	148	400	59200	5
ATGM	486	120	58320	5
Personal Equipment	14000	.7	9800	1
Infantry Weapons	22000	.9	19800	1.77
Fuel Stocks (30 Days Supply)			7500	0.67
Food Stocks (30 Days Supply)			2500	0.22
Total (Millions of Dollars)			1117	100

Summaries

Combat Vehicles	530100	47
Other Weapons	262120	23
Other Equipment	194800	17
Supplies	130000	12

an extent. Adjustments must still be made for the lower cost of labor, the economies of scale, and the greater cost to the Russian economy to produce items of "exceptional performance."

Although Russia has generally produced less capable weapons than the West, its technology is advancing. But as the Russians produce more technically advanced weapons, they fall victim to the same disease that afflicts the West. Just as the West has many technically superior weapons that work infrequently, so will Russia. As Russia's technicians are far less capable than the West's, and there are fewer of them, this disease will be even worse for them. If history is a guide, technically advanced weapons will be more of a curse than a benefit for the Russians.

Cost of Raising a Division

A combat division isn't cheap. The division costs shown here are for hardware only, weapons and equipment.

In addition to the costs shown on the chart, additional "support" equipment and personnel training will double or triple the final expenses. Each division will have nondivisional combat and noncombat support troops. The combat support will be in the form of additional artillery, missiles, air defense, electronic warfare. Noncombat support will include supply, transportation, medical and so on. The hardware cost for this combat and noncombat support will amount to a 30 percent to 80 percent surcharge on the division's hardware cost.

There is also the software, the personnel. The cost of properly trained personnel is enormous, a 50 percent to 100 percent surcharge on the expense of the division's equipment.

The training time to obtain qualified staff varies'from a few months for the simplest jobs to years for the more technical ones. Time spent in schools and other training programs can amount to several "division man-years" (all the division's personnel spend a year in school). These trainees must be paid and maintained. Their instructors are an additional expense. Even on-the-job training can be costly if you overstaff in order to increase the number of people trained. Training can add 50 percent to 100 percent of the equipment cost. Anything cheaper will result in inefficiently used equipment, which will defeat the purpose of the entire exercise, although it is often done.

Equipment without qualified operating personnel is useless. To operate a division requires 12,000 to 16,000 men. About half can be trained to an acceptable degree of competence in six months. The remaining cadre requires increasing amounts of technical training and experience. Consider the division's management, which is 20 percent of the unit's strength—the officers and NCO's, the generals, colonels, captains, lieutenants and sergeants.

If you have ever needed a repair technician to fix some complex piece of equipment, you know the manufacturer always attempts to send the youngest, and least qualified, technician first. The most qualified personnel are always in short supply and can usually be found dealing with the most difficult service problems. A more experienced technician can usually handle the same repair in a fraction of the time. This phenomenon applies to general management as well as technicians.

Armed forces attempt to solve this problem with a number of techniques, most of them very expensive. The most common is using up more equipment. Without sufficient qualified personnel, equipment becomes expendable anyway. An attempt is often made to mask this process by sprucing up appearance. Freshly painted equipment is often passed off as fully functional.

ITEM is the main equipment categories.

QUANTITY is the quantity of that item found in the division, for these charts, a U.S. mechanized infantry division and a Russian motor rifle division.

AVG UNIT COST is the average unit cost of each item of equipment in thousands of 1982 dollars. Only new equipment is considered. Where the range of item types within a category is too great to be meaningful, no number appears in this column.

TOTAL COST is the total cost of all items in that category.

% OF TOTAL is each item's percentage of the total cost of the division.

SUMMARIES show general categories of the division's equipment.

COMBAT VEHICLES comprise the largest single expense for most modern armies. Until recently, tanks represented the bulk of this category. With the development of the IFV (infantry fighting vehicle), the armored vehicles for the infantry are looming larger in the equation. IFV's are becoming tanks in their own right. Compared with many tanks of thirty to forty years ago, modern IFV's have equal or superior firepower. Aircraft assigned directly to the division are also becoming a major factor, particularly in Western armies. Drone aircraft may speed up this trend. Currently most armies devote about half their divisional equipment budget to combat vehicles. This ratio is not likely to increase.

OTHER WEAPONS consist primarily of artillery, particularly antitank and antiaircraft artillery. This includes missiles, whose cost is rapidly overwhelming conventional artillery.

OTHER EQUIPMENT is a rapidly growing category, as it includes many electronic items.

SUPPLIES are often overlooked items. Thirty-day stock for the early stage of a war is the accepted standard, although many armies skimp. Such frugality is fatal if a war breaks out. Another potentially serious problem is improper storage and maintenance of these supplies. All these items are perishable and must be rotated. Fuel is the most perishable of all and is usually the best cared for. Ammunition is fired off regularly in practice and replaced. This is becoming more expensive as the cost of missiles skyrockets. Storage is another problem, since there is a temptation not to disperse these supplies in protected areas in order to save money. Supply dumps are no secret from the enemy and make ideal targets for aircraft and artillery.

Part 7

MOVING THE GOODS

26

THE NAVY:
THE TONNAGE WAR

IN WORLD WAR I and II, and probably in World War III, landlocked nations have attempted to destroy the merchant shipping of the oceanic powers. Today, all major economic world powers save one are oceanic nations. Only Russia remains a self-contained continental nation. Guess who has the largest submarine fleet?

The navy is responsible for mobilizing and protecting the civilian merchant fleets in wartime. These ships are needed not only for moving and maintaining ground forces but for supplying the combat fleets.

The Tonnage Numbers

Because of the vastly increased need to move bulk raw materials, 70 percent of the world's merchant shipping tonnage is devoted to the transportation of bulk products; 44 percent is petroleum. The remaining one third can carry military cargo, men, supplies and weapons. This third equals 120 million gross register tons (GRT). To simplify a bit, one gross register ton equals one metric ton of dry military cargo.

Want to move a modern combat division? You'll need about 250,000 GRT. Keep it supplied for a month? Another 30,000 to 50,000 gross register tons, depending on how intense operations are. What about nondivisional troops? Ten tons per man to get them there. Figure 40,000 men per division. That's 400,000 tons. Supplying them? One ton per man per month for support, another 40,000 tons. Getting one new division over there with air force and other support will require 650,000 tons initially and up to 100,000 tons a month for support. Most of the latter will be tanker tonnage.

Want to support the combat fleets? Assume one gross register ton for every two or three displacement tons of combat shipping. This will vary, particularly with the carriers, according to the intensity of operations and the distance from bases.

It should be noted that the United States has taken the precaution of prepositioning in Germany the equipment for three of the reinforcing divisions. Only the troops have to be flown over. All personnel will be flown in; only the equipment will move by sea. But 500,000 tons of armored vehicles remain in the States. These must accompany the other equipment of the approximately nine heavy divisions that may be whipped into shape and shipped out within ninety days. Some of this equipment must also be moved quickly to replace combat losses.

Getting Organized

From the navy's point of view, the biggest problem is lack of control. Although the merchant shipping fleets are vast, they are not under any major navies' control. About 65 percent of the world's merchant shipping fleet is under the control of eight Western nations. Flags of convenience, Liberia and Panama, control 25 percent of all shipping. Two major naval powers, Britain and the United States, control only 12 percent of merchant shipping. All this means is that British and U.S. shipping companies control these ships and that, in most cases, the crews are British and U.S. subjects. Control does not do much for the ability to muster these vessels quickly into military service. Let us examine the problems of mobilizing merchant shipping fleets for a military emergency.

Assembling crews. Aside from possible problems with unions or noncitizen crews, there is the very basic problem of convincing merchant seamen to serve in a war zone. As losses to enemy action increase, it becomes more difficult to man the ships. It is especially difficult to do so in the winter, when a dunk in the frigid North

Atlantic is certain death. Each time a ship goes down, about 25 percent of the crew dies, varying with the severity of the weather and the volatility of the cargoes. The government can offer inducements, such as danger pay and conscription. The best inducement is to protect the merchant ships from enemy attack. All these measures will take a while. Meanwhile, the ships cannot move without crews. Modern merchant ship crews are mostly technicians, and substitutes will not work.

Mobilizing merchant ships. The bulk of the world's shipping travels a few routes. Two thirds the total shipping activity takes place between North America, Europe and the Persian Gulf. Adding Japan and Australia, more than 80 percent of the world's shipping has been accounted for. The 70 percent of merchant shipping that carries bulk products, mostly petroleum, is of less concern than the cargo carriers, which must be mobilized as quickly as possible without too much damage to military production. As it is, rapid commandeering of shipping will disrupt industrial schedules. Martial law or hastily drafted agreements with friendly merchant shipping nations will secure some vessels. Fewer ships, proportionately, than in previous wars are suitable for key military cargo like armored vehicles. Container ships predominate among the ranks of the most modern vessels. Container ships were not built to move armored vehicles. How successfully the initial muddle phase is passed will determine how quickly the reinforcing units arrive in the war zone.

Forming convoys. This procedure is practiced only during wartime. Forming convoys is like any other technical operation; make mistakes and you pay for it. The basic cost is a lot of ships waiting around for the convoy to sail. A considerable amount of coordination has to take place between the military planners, the war matériel producers and the convoy control staff. Based on World War II experience, a convoy schedule would be established. Every few days so many thousand gross register tons would go east or west. Coastal feeder convoys would leave ports up and down the coast for the transoceanic convoy assembly points. So that the valuable escorts' time would not be wasted, ships that don't make it to the assembly point would have to wait for the next convoy. All this discipline and coordination is not unknown to modern shipping operations. Since World War II, ships have become more conscious that time is money.

Moving convoys. The navy cannot afford the expense of

practicing with civilian merchant shipping. Some exercises are held to give the escorts practice. But the biggest problem will be getting civilian shipmasters to conform to the discipline of convoy operations. Steaming in formation and staying calm under the stress of combat are qualities not usually expected of civilian masters and crews. It is equally disorienting for the navy crews. They have trained for this, but nothing quite prepares one for carrying on while merchant and combat ships explode about one. Mistakes will be made, unnecessary losses will be taken, and eventually hard experience will be won.

On the plus side, a lot of enemy shipping will be seized. More enemy ships will be in our ports, or in our seas, than our shipping will be in similar situations.

Time and Space Variables

In order to organize merchant shipping on a war footing, reallocations will have to be made, quickly. The warship escorts are normally in or near their home ports. The cost and drain on morale does not permit frequent long-range cruising by escorts. Most of the merchant ships are on the high seas. There is no central command center for the 20,000 or more merchant ships the West must utilize. Improvising will cost more wasted time, lost ships and, initially, an inefficient use of merchant shipping.

The reallocation problem is compounded by the differing capabilities of the various ship types. Many current merchant vessels are very specialized. Much of the tanker tonnage consists of supertankers, 100,000 or more gross register tons each, built to steam between a handful of specialized loading and unloading terminals. They are not as flexible as smaller tankers. Container ships are preferred for modern cargo shipping. Some of these are even more specialized. The RO-RO types are built so cargo on vehicles can "roll on-roll off." Their ramps will often only fit two specific facilities—where they pick up and drop off cargo. Military planners in peacetime are only generally aware of all these details; in wartime these details will have to be well known. The research will probably have to be done under fire.

One major variable is distance. Steaming at 20 knots (36 km per hour) merchant shipping can move 800 km per day. In wartime there will be a certain amount of going to and fro to avoid suspected enemy submarines or even land-based air or surface ships. Forming into, and operating as, convoys will lose more travel time. Figure making some 500-km-per-day headway.

From the east coast of the United States to Europe, you must travel 6200 km. That's 13 days at sea. Depending on the condition of the ports on each end, a few days are needed for loading and unloading. One round trip a month. In the Pacific, it takes 8800 km to travel from the west coast to the mainland (Korea). That's 18 days at sea, 2 round trips every 3 months. Going to the Persian Gulf? Assuming a dangerous situation in the Mediterranean, you'll have to go around Africa. That's 21,600 km, a long 44-day haul, 100 days for each round trip.

If the seas are swept free of enemy ships and loading and unloading at each end is optimal, these transit times can be cut by as much as 30 percent.

One assumption is that 90 percent of the tonnage going to Europe in a war will have to go by sea and the rest by air. This will require 6000 ships, about 72 million GRT. It will also require more escort ships than are currently available. This has happened in previous wars and simply means that three things will happen. There will be fewer escorts, larger convoys, or even faster merchant ships traveling alone. New resources will be discovered, like more land-based aircraft, armed merchant ships, and putting ASW helicopters and other weapons aboard selected ships. At worst, merchant shipping losses will increase. Current NATO figures estimate that 3000 ships would be lost during the first 90 days of operations in the Atlantic. This is a somewhat pessimistic evaluation, as the discussion below will show.

Attacks on Shipping

One can only guess at the enemy's attack strategy and tactics. Chapters 9 through 11 explain the mechanics of naval combat. Will the Russians go after the Persian Gulf by land and air first? Or will they concentrate on the convoy routes? Will they first go after the major surface combat ships? Will they use nuclear weapons at sea? They can use chemical warfare at sea; will they?

Assume the Persian Gulf is put to the torch by enemy air, land and naval power. From a strategic point of view, this would appear more effective than sending submarines after tankers. Better to send the submarines after Europe's own oil- and gas-producing resources. We must protect them at all costs even though they are in the North Sea, so the enemy submarines would be going into a hornet's nest. Of course, these vulnerable targets could be taken out with air power. All of this just shows how simple the task of deploying forces to protect merchant shipping will *not* be. All indicators point to a major battle in the North Atlantic, and a smaller, but equally hard-fought, shipping battle in the Pacific.

The Russian Navy is divided into four fleets. The two largest are based at Murmansk, just east of northern Norway, and at Vladivostok, north of Korea in the Sea of Japan. The other two fleets are bottled up in the Baltic and Black Seas. Either of these last two fleets could break out into the Atlantic, but Western naval forces, mines and air power make this unlikely.

The northern fleet's main responsibility is the severing of Europe's merchant shipping lifeline to North America and points south. To do this the fleet normally has approximately 135 attack submarines, 16 major surface combat ships and 107 escort ships. In addition, 80 to 100 long-range aircraft are available.

Facing this force the Western navies muster approximately 140 attack submarines, 20 major surface combat ships and 260 escort ships. Aircraft available for naval operations will vary with the demands of the land campaign. They could be as many as 400 to 500.

There are only 135 Russian submarines in the northern fleet. About 60 of these are nuclear, the rest are older diesel boats modeled after the World War II German U-boats. To sink the 3000 Western merchant ships in 90 days, each would have to sink 22 merchant ships, or better than 7 ships per submarine per month, assuming no submarines would be sunk.

It is doubtful that Russian subs can be ten times as effective as U-boats. In their best months during World War II, the Germans barely destroyed one merchant ship per month for each submarine at sea. During their best sustained periods, they sank about .7 merchant ships per month per submarine at sea. During all of World War II, the Germans managed to sink about two merchant ships for each sub lost. The Russians, again with the world's largest submarine fleet, managed to sink *less* than one merchant ship for each submarine lost from 1941–1945. Also, one would hope the 100 or more attack submarines (30+ nuclear) of the Western navies will be sinking some Russian submarines. While submarine technology has improved vastly since 1945, so has antisubmarine-warfare technology. And the bulk of the Russian submarines are essentially still 1945-style boats. It is therefore unlikely that the Russians will sink anywhere near 3000 ships in the first 90 days of the war. The Russians can do substantial damage, particularly with nuclear weapons. But that would push us into another form of warfare, a war from which few will emerge.

Can the Russians make it to the high seas? Russian submarine commanders may have a field day for a while, with many targets and not much organized opposition. The largest variable here is how many Russian subs will make it to the high seas. In peacetime Russia has only 15 percent of its subs at sea at any one time. Given a running start, it can get 75 percent out there. Much depends on how quickly a war starts and how effectively the Western forces can blockade the Russian fleets. Western antisubmarine-warfare forces have a good worldwide submarine location system. One of the most likely first objectives of the Russians will be to cripple this system. The fate of thousands of merchant ships hangs in the balance of such variables.

The Russians also station a substantial naval air force with their northern fleet. These aircraft could do substantial damage to merchant shipping, but only if they were unopposed, which is unlikely. The Russians maintain the world's largest peacetime stock of naval mines. Most of their ships and naval aircraft are equipped to lay them. These little nasties sink anything that floats. The primary means of avoiding mines is to prevent the enemy from dropping them off near friendly ports or shallow-water shipping lanes. Submarines are the most likely means of delivering pressure mines undetected. Therefore, the enemy submarines must be destroyed early and often, before they can even get into friendly waters.

You can never fight the next war like you fought the last one. Barring a holocaust, the bulk of this chapter gives a reasonable description of the probable outcome. But there are other possibilities. There may be an undeclared war in which ships start to have a God-awful time getting to certain ports without disappearing. This might also lead to a steadily escalating conflict that eventually becomes a full-blown world war. Any one of many technical variables could alter the situation. The only constant is that the navy will have to deliver the goods. Should it fail, the war is lost.

WORLD SHIPPING CAPACITY AND PORT ACTIVITY

Most of the world's shipping is owned and used by the West to move raw materials. Close down the Persian Gulf, and over 20% of the world's ships are out of work.

Rank	Nation	Million GRT Shipping	% Cargo	% Tanker	% Bulk	Total % of World	Million Tons of Port Activity	% of World
1	Japan	37	25	43	33	9.61	760	10.83
2	Britain	31	28	47	25	8.05	250	3.56
3	Greece	30	30	33	37	7.79	41	0.58
4	Norway	30	11	54	36	7.79	61	0.87
5	US	14	47	40	13	3.64	880	12.54
6	France	12	17	67	17	3.12	250	3.56
7	Italy	11	18	45	36	2.86	265	3.77
8	West Ger.	10	40	40	20	2.60	155	2.21
9	Sweden	8	14	57	29	2.08	91	1.30
10	Spain	8	29	57	14	2.08	115	1.64
	Liberia	83	8	64	29	21.56	24	0.34
	Panama	20	53	37	11	5.19	8	0.11
	Russia	17	76	19	5	4.42	190	2.71
	Other WP	7	71	14	14	1.82	121	1.72
	All Other	67	45	27	28	17.40	3809	54.26
	Top 10+	294	22	50	27	76.36	2900	41.31
	USSR+	24	75	18	7	6.23	311	4.43

World Shipping Capacity and Post Activity

RANK is the ranking of nations by the size of their merchant marine fleets. Liberia and Panama are not included, since the ships flying their flags are not actually controlled by those countries.

NATION is the nation of the government or company headquarters controlling the shipping.

MILLION GRT SHIPPING is the amount of shipping controlled, expressed in millions of gross registered tons. There are, unfortunately, four widely used methods of computing a ship's size. Displacement, how many tons of water the ship displaces, is primarily used for warships. Deadweight tonnage is the ship's total carrying capacity in metric tons. A measurement ton is 40 cubic feet of cargo space. Gross register tons are 100 cubic feet of cargo space. One ton (2200 pounds) of average military cargo takes up 102 cubic feet (one register ton). The deadweight carrying capacity of most ships is closer to 1.6 deadweight tons per register ton, but much military cargo is bulky and not heavy (trucks, electronic equipment, etc.).

% CARGO is the percentage of that nation's shipping that consists of dry-cargo vessels, which can carry anything that is prepackaged in bales, boxes, containers. This is a typical cargo, or break-bulk, ship. More modern ships are container ships, designed to carry nothing but, say, 8-by-8-by-8-foot containers, each holding 5 register tons. Also becoming more popular are RO-RO ships (roll on-roll off). These are basically floating parking lots for truck trailers and other vehicles. They are ideal for military cargo, and the U.S. Navy is buying them directly from civilian sources. LASH ships carry their own landing craft for unloading at ports that cannot handle the mother ship itself. These come in either break-bulk or container versions.

The older "tramp" steamers weigh in at less than 10,000 deadweight tons, many weigh 2000 tons. These smaller ships are usually used for coastal and other short-range shipping. The more modern containers, RO-RO and LASH ships, vary from 10,000 to 20,000 tons.

% TANKER is the percentage of that nation's shipping which is capable of carrying only liquids, primarily petroleum, but also sulphur and liquid natural gas. These are very heavy ships, some weigh as much as 500,000 deadweight tons. The average is 100,000 tons. Most of these ships move from the Persian Gulf to Europe or Japan.

% BULK is the percentage of that nation's shipping that carries dry bulk cargo only. Typical loads are ores, coal and feed grains. These are not as large as the tankers but tend to be larger than break-bulk ships.

TOTAL % OF WORLD is the percentage of all the world's shipping controlled by that nation. Liberia and Panama are flags of convenience; shipping companies establish nominal headquarters in these countries in order to avoid taxes and regulations. Most of these ships are owned by companies of the top ten nations. In wartime they can be brought back under national control with a certain amount of arm twisting and bending of the law.

MILLION TONS OF PORT ACTIVITY is the annual activity of that nation's ports in terms of tons of cargo loaded or unloaded. This is a good indication of a nation's dependence on seaborne commerce.

% OF WORLD is the percentage of the world's port activity each nation represents. It is interesting to note that the United States and Japan account for nearly 25 percent of the total. The top ten shipping nations account for more than 40 percent. Including the loading activity in the Persian Gulf, which goes primarily to these top ten nations, over 50 percent of the world's merchant marine activity goes to just ten nations. Russia accounts for less than 3 percent.

OTHER WP is other Warsaw Pact (Russian-allied) nations whose shipping would likely come under Russian control in wartime. These countries include Poland, East Germany, Rumania and Bulgaria. One could make a case for including Cuba also.

ALL OTHER includes the shipping of all other nations not given on the chart. Note that this group controls only 17 percent of the world's shipping, yet handles over 50 percent of the world's port activity. These are primarily raw-material-producing nations, the Persian Gulf countries being prime examples.

TOP 10+ are the top ten countries shown above plus Liberia and Panama. This, in effect, is the merchant fleet of the industrialized Western countries. Seventy-six percent of the world's shipping and 41 percent of port activity is found in these nations.

USSR+ is Russia and the Warsaw Pact nations. Note that this group has only 8 percent as much merchant shipping as the top 10+ countries and one tenth of the port activity.

27

THE AIR FORCE: AIRFREIGHT

"GET THERE first with the most" is the motto of the air force transport units of the world. Technical advances in the last forty years have made air transportation only about twice as expensive per ton mile as land movement, not a very extravagant cost escalation by military standards. The primary limitation to air transport is the lack of aircraft and the weight and size of military equipment.

The Mechanics of Air Transport

The major armed forces of the world have fleets of specialized cargo aircraft. The United States and Russia together account for more than 80 percent of all military air-mobility capacity. Most nations have also made arrangements to militarize their civilian air-transport fleets in the event of a war. Russian- and U.S.-controlled aircraft account for over 80 percent of this capacity. How much capacity? The U.S. military fleet could lift over 40,000 tons 5000 km in one lift. The Russian fleet could lift less than half as much. The U.S.-Allied civilian fleet could lift over 50,000 tons; the Russian civilian fleet about half.

A U.S. mechanized infantry battalion weighs about 2500 tons. A Russian battalion weighs about 1500 tons. These weights include the APC's, which can only be carried by cargo aircraft, and normally transported ammunition, fuel and other supplies for two to three days' combat. Because an aircraft has restrictions on size ("cube," or cubic feet) as well as weight, the movement of large, light equipment wastes capacity. Therefore, it will take about 60 C-141 or C-5 aircraft to move a U.S. battalion's vehicles. Civilian aircraft could be used to move most of the remaining men and supplies. Only three wide-bodied passenger aircraft would be required to move the battalion's 900 men, including personal equipment, weapons and supplies in the aircraft's cargo containers. This still takes about 100 aircraft in all. You can't move more than five battalions at one time because only about 300 aircraft can carry heavy vehicles. Forget about tanks; only the C-5 can carry them, and only one at a time. A tank battalion has 54 tanks and the U.S. Air Force has only 70 C-5's.

The Russians are somewhat better off. Because of smaller battalions and fewer support vehicles, they require only about 50 aircraft to lift a battalion. They can lift 4 battalions 5000 km at any one time. Better yet, Russia's geographical position is closer to likely areas of conflict. This allows use of the shorter-range AN-12 aircraft, thus giving wings to another 8 battalions. Because there is no road connection between Moscow and Russia east of the Urals, and only one vulnerable railroad, air transport assumes critical strategic importance.

If you are content to carry only nonmotorized infantry, the carrying capacity increases quite a bit. War can be waged without tanks, particularly when defending. Antitank missiles weigh, at most, less than 50 pounds each; 107mm mortars can be put in a cargo container. Except in primitive areas, you can commandeer local trucks. A light infantry battalion of 900 men armed with 18 107mm mortars (and 90 tons of ammo), 60 ATGM launchers (and 1000 missiles), 50 tons of mines plus the usual armament of machine guns, rifles, grenades, sensors and other supplies, will require only 20 wide-bodied civilian aircraft.

Lift Capacity Restrictions

These theoretical lift capacities are misleading for several reasons. First, how much you can lift is dependent on how far you are going. With an average cruise speed of 500 to 800 km per hour, a 5000-km "hop" would take 7 hours. Landing, unloading, refueling (up to 100 tons) and reloading take another hour or two. Round-trip flight time: 14 hours. And that's surging. The following typical distances in hours of flying time (at 800 km per hour) do not allow for refueling stops every 6 to 10 hours for aircraft that cannot refuel in the air. From Washington, D.C. to Berlin—8.5 hours; to Cairo—12; to Istanbul—10.5; to London—7.5; to Madrid—7.5; to Tehran—13; to the Persian Gulf—12; to South Africa—16. From San Francisco to Hong Kong—14 hours; to Hawaii—5; to Melbourne—16; to New Delhi—15.5; to Tokyo—10.5; to Peking—12; to Singapore—17; to Saigon—16. From Moscow to Berlin—2 hours; to Tokyo—9.5; to Tehran—3; to Nairobi—8; to South Africa—12.5; to Peking—7.5. It takes 7 hours, 84 tons of fuel, to get across the Atlantic.

Lift capacity also depends on refueling opportunities. U.S. military aircraft can refuel in the air, Russian aircraft cannot. No civilian aircraft can refuel in the air. A B-747 jumbo jet burns 12 tons of fuel per hour of flight. There has to be fuel at both ends of the trip, as well as a stock of spares, technicians and maintenance equipment.

There aren't that many large airfields around. They make such good targets for enemy aircraft, missiles or ground forces. Europe has about fifty that can support long-range aircraft, but most of the support capacity is concentrated at less than thirty. Losing an airfield is bad enough. Losing the maintenance personnel is worse as they are harder to replace.

Russia has airfields at no less than 1000-km intervals along its entire border, except the arctic north. Soviet aircraft can often operate from unpaved fields and with less ground equipment. For example, many Russian aircraft can be refueled without fuel pumps. This takes longer but eliminates another piece of ground equipment. Russian aircraft often travel with a larger crew consisting of both flight and maintenance personnel. These aircraft have a lower readiness rate for sustained operations, but they can operate under more primitive conditions than Western aircraft. On the other hand, we can always use helicopters in such situations.

Although extremely agile, helicopters have very short ranges, usually under 500 km one way, and small load-carrying capacities, usually under 3 tons. The entire U.S. helicopter fleet, at over 8000 aircraft, the largest in the world, could lift about 6600 tons of weapons and equipment at one time. In function, helicopters have more in common with trucks and APC's than they do with aircraft.

Helicopters fly, and are based very close, to the fighting. Unlike transport aircraft, they are often armed. Most helicopters, in fact, perform primarily as combat systems or in direct support of combat units.

Not all aircraft are available at all times. As many as 20 percent will be out of service for maintenance. This figure will generally be higher for Russian aircraft. With sufficient crews an aircraft can theoretically be kept going 24 hours a day for a month. After a 12-hour maintenance check, it can go another month. Every 3000 to 4000 hours it must be pulled out a month or so for overhaul. As a practical matter, this tempo of activity would soon exhaust available maintenance crews. Ten or 12 hours a day on a sustained basis, with occasional surges of longer activity, could be maintained indefinitely, at least up to the 20,000- to 30,000-flight-hour life of the aircraft.

Allied civilian aircraft cannot be obtained with absolute certainty, due to potential political and labor problems.

Finally, in wartime there is enemy resistance. The U.S. Air Force estimates that in a major war its long-range transports will have an attrition rate of 2 losses per 1000 sorties. With 300 military and 600 civilian aircraft in operation, this would mean 300 aircraft lost in 6 months of operations. Let's hope not. And transport pilots don't usually get combat pay.

Strategic Military Airlift

This chart shows the strategic military transport aircraft available to the major military powers in the world. Only the United States has significant fleets of strategic transport aircraft. The Russian strategic transport aircraft fleet is closely integrated with the single national airline, Aeroflot. The U.S. Air Force fleet could be called the world's largest airline, at least in terms of lift capacity.

Not included are the twin-engine tactical transports that equip most other nation's air-transport fleets. Most lift five tons or less. Few nations have more than twenty or thirty. Transports for battlefield movement of men, equipment and supply, they were designed for short-range, 1000 km or less, movement and exist primarily to support defense.

OF AIRCRAFT simply adds up the total number of aircraft each nation has of the types shown on the chart. Many types are still in production except for the C-5, C-141, AN-12, C-123, KC-135, IL-18. The AN-72 and KC-10 are just entering production, while the CX is still in development and may not be produced. The AN-72 is just entering production and was designed for extremely short takeoff performance.

STRATEGIC MILITARY AIRLIFT

Most military airlift is owned by the United States. Because of the shorter distances to be covered, Russia possesses the largest effective capability.

Aircraft Lift

Aircraft Types (Long Range, Heavy Lift Models)

Nation	# of Air-Craft	Tons Cargo (x1000)	Pass.	C5	C141	C130	KC135	C123K	KC10	CX	AN72	IL76	IL18	An-22	AN-26	AN-12
Russia	760	20.46	83								0	150	20	50	40	500
US	1376	47.45	210	70	236	360	640	60	10	0						
Totals	2136	67.91	293													
US % of Tot.	64	70	72													

Aircraft Characteristics

	C5	C141B	C130	KC135	C123K	KC10	CX	AN72	IL76	IL18	An-22	AN-26	AN-12
Passengers	345	168	220	80	260	0	0	32	90	80	330	40	100
Cargo (tons)	120	41	34	22	38	77	59	7.5	40	13	80	5	20
Average Range (km)	4800	6400	4600	3700	2200	7000	5500	1000	6000	3700	5000	2400	4000
Empty Weight (tons)	151	66	33	45	14	109	80	16	100	31	110	15	28
Max Takeoff weight (tons)	349	156	79	135	27	268	181	30	170	64	250	24	61
Max Fuel Load (tons)	130	72	30	73	12	53	52	6	66	24	43	5.5	15
Minimum Airfield Length (m)	2600	1000	1100	2800	1100	3300	1000	500	850	3900	1300	800	700
In-flight Refueling?	Yes	Yes	No	No	No	No	Yes	No	No	No	No	No	No

TONS CARGO (X 1000). Total cargo-carrying capacity in 1000 tons.

PASSENGERS are given in thousands. Transporting troops is a secondary role for military-transport aircraft. Their primary function is freight. Truck drivers.

TOTAL IN USE is all aircraft in service.

U.S. % OF TOT. shows the United States' advantage over Russia. There are substantial qualifications to this seeming advantage. Nearly half of the U.S. aircraft are tankers (KC-135, KC-10). These can also carry cargo, but the bulk are assigned to the Strategic Air Command in support of the B-52 bomber fleet. As the war in Vietnam demonstrated, these tankers could be shifted quickly to the support of tactical operations. With more military-transport aircraft being equipped for in-flight refueling, this tanker fleet will play an increasing role in future nonnuclear conflicts.

Given that many of the potential battlefields are in areas adjacent to Russia, the practical lift capacity of the Russian military-transport aircraft fleet increases, while the longer flying distances work against U.S. aircraft.

AIRCRAFT CHARACTERISTICS give the operating characteristics of each type of aircraft.

PASSENGERS are the number of seats installed available for troops. These seats are temporary, easily removed to make room for cargo.

CARGO (tons) is the average cargo load that can be carried. Containers are often used, although military-transport aircraft are primarily designed to accommodate military vehicles. The largest military-transport aircraft can even move tanks. The C-5 can carry two, the Russian AN-22 can carry two tanks, the IL-76 one. It is far more efficient to carry lighter military vehicles, armored personnel carriers of twelve tons each, guided-missile vehicles, trucks, etc. The Russians employed just such an airlift when they went into Afghanistan. Pallets full of ammunition are another favorite cargo. Most of these aircraft have large doors in the rear that can be opened in flight. Their cargoes can land with parachutes.

AVERAGE RANGE (kilometers) is the average range with a full cargo load. An empty aircraft with a full load of fuel can go 20 percent to 60 percent beyond its full load range. Keep in mind that, unlike civilian commercial aircraft that travel essentially one way, military-transport aircraft must often be prepared to get out again without refueling. This is another role for the tankers.

EMPTY WEIGHT (tons) is aircraft weight without fuel or payload. A glance at the potential fuel capacity, possible cargo capacity and maximum takeoff weight shows that tradeoffs have to be made. Military transport must be able to move very heavy loads for short distances, or lighter loads over long hauls.

MAX TAKEOFF WEIGHT (tons) is the maximum takeoff weight of the aircraft. This suggests the aircraft's size. The C-5 (347 tons), for example, is 245 feet long,

65 feet high, with a wingspan of 223 feet. The C-141B (155 tons) is 168 feet long, 39 feet high and has a span of 160 feet. The AN-22 (250 tons) is 190 feet long, 41 feet high and has a span of 211 feet. The C-130 (79 tons) is 98 feet long, 38 feet high and has a span of 133 feet. In addition to being larger than comparable civilian transports, the military aircraft are more robust to handle the heavier equipment.

MAX FUEL LOAD (tons) demonstrates that the primary cargo of military-transport aircraft is fuel. Large quantities can be carried without sacrificing space by putting almost all the fuel in the wings. A major problem in wartime is providing sufficient fuel at both ends of the trip, particularly in the war zone, to keep the aircraft moving.

MINIMUM AIRFIELD LENGTH (meters) is the minimum-size airstrip needed for takeoff. Takeoff always requires a longer airfield than landing. Taking off with maximum load requires 30 percent to 40 percent more space than that in the chart. To get off the ground in the shortest possible space, the aircraft will have to go half loaded.

IN-FLIGHT REFUELING? This indicates whether or not the aircraft can refuel in flight. The Russians only use in-flight refueling with their naval reconnaissance aircraft. Modified Tu-16 bombers refuel other Tu-16's.

The tankers rely more on their own fuel capacity than on the additional fuel they carry in their cargo spaces, generally the space beneath the cargo deck. The KC-135, a militarized 707, normally carries 73 tons of fuel plus another 15 tons as cargo. The aircraft can draw upon all this fuel for its own engines. Depending on how far the tanker has to travel, it can transfer up to 90 percent of its fuel load. The larger KC-10, a militarized DC-10, normally carries 108 tons of fuel plus 53 tons as cargo. Of this 161 tons, about 90 tons are available for transfer.

The most efficient use of the tankers is to allow aircraft to take off with low fuel but a higher cargo, or weapons, load. The fuel tanks can then be filled up and the aircraft weight increased to a level that would prohibit taking off but not flying. Tankers can also meet aircraft returning from a mission and, particularly in the case of fighters, which can be real fuel hogs, give them enough to get home. Cargo aircraft can be refueled during long flights instead of landing to save time.

The quantity of fuel that large aircraft carry is shown on this chart and on the similar one for civilian transports. By far the worst offender is the B-52, which carries 141 tons of fuel. For this reason, two tankers are usually assigned to each B-52.

Smaller combat aircraft are at the other extreme. Fuel load for the F-4 is 6 tons; for the F-16, 3 tons; for the F-15, 5 tons; for the F-18, 5 tons; for the A-6, 7 tons; and for the F-14, 7 tons. To accommodate the larger number of combat aircraft that can be refueled by one tanker, the KC-10 may be equipped to refuel three aircraft at once.

Wartime Use of Civilian Aircraft, Long-Range Heavy Lift Models

This chart shows the heavy lift aircraft available to the major military powers. The air fleets selected are those that can be easily and quickly militarized, primarily major airlines of nations likely to be involved in hostilities. Not included are minor military powers, major military powers that are neutral and many small charter airlines.

It is assumed that local air service will be maintained with the shorter-range aircraft not shown on this chart—the B-727, DC-9, B-737, etc. Under wartime conditions all nonessential travel will be curtailed, and aircraft will fly with nearly all seats occupied, as against 50 percent in peacetime.

The national air fleets are ranked in order of the number of aircraft available. This, as you can see, is not the primary determinant of airlift capability. Cargo and passenger capacity vary considerably.

OF AIRCRAFT simply adds up the total number of aircraft each nation has of the types shown on the chart. Of these types, most are still in production except for the DC-8, 707 and AN-12. Although the 757 and 767 will not enter service until 1983, they will replace, in numbers if not in capacity, the retiring DC-8's and 707's. Longer-range versions of the A300 and 767 will replace some of the DC-8-707 aircraft. For the most part, the 757, 767, and A300 are intended to replace shorter-range aircraft.

TONS CARGO is given in thousands of tons. Most civilian aircraft are designed to carry passengers plus cargo containers in those spaces not suitable for passengers. About 20 percent are equipped to carry cargo only. Even these aircraft normally carry only freight containers that are preloaded to improve efficiency. Civilian aircraft operate on a fast turnaround. Anyone who has flown regularly has witnessed how quickly cargo is unloaded and loaded, crews changed and the aircraft put into the air again. This container system limits the military cargo that a civilian aircraft can carry. Most military vehicles and heavy weapons will have to go by sea or military aircraft, but many critical military cargoes can be carried: spare parts for weapons and equipment; ammunition, especially missiles; and lighter and usually more valuable electronic equipment.

PASS. is passengers given in thousands. The transportation of troops makes civilian aircraft valuable. The United States plans to raise antitank-missile battalions with American National Guard and reserve troops. Their equipment would be light and transportable by civilian aircraft. The vehicles could be requisitioned from civilian sources at their destination, which would generally be possible, since large mechanized formations usually operate where there are many roads and vehicles. Western Europe is a prime candidate. Otherwise, light infantry weapons like mortars and small arms can also be moved by civilian aircraft.

Each passenger represents about 300 pounds of load. You can carry about 500 pounds more cargo for each passenger you don't transport in a convertible aircraft.

WARTIME USE OF CIVILIAN AIRCRAFT, LONG-RANGE, HEAVY LIFT MODELS

Civil air transport gives the West a decisive edge in air transport

| Nation | Aircraft Lift | | | Aircraft Types | | | | | | | | | | | | | |
	# of Aircraft	Tons Cargo	Pass. (x1000)	Boeing 757	Boeing 767	Airbus A300	747	L-1011	DC-10	DC-8	707	IL-86	IL-76	IL-62M	TU-154	AN-26	AN-12
Russia	1450	22.07	143									20	100	130	380	600	220
US	720	29.28	176				155	105	160	150	150						
Japan	117	5.99	32				45	21	18	33							
Britain	103	4.53	26			6	30	20	22		25						
Warsaw Pact	77	1.59	12									2		38	32	3	2
France	64	2.83	16			25	24	0	0		15						
Benelux	63	2.45	15			15	17			16	15						
Canada	59	2.43	14				11	12	4	32							
West Germany	59	2.20	14			15	12		12	0	20						
Italy	38	2.01	11			6	16	0	8	8	0						
Korea	31	1.44	8			8	12	0	4	0	7						
Israel	22	1.21	6				15				7						
China	18	0.46	3				3				10			5			
Totals	2821	78.47	477	0	0	75	340	158	228	239	249	22	100	173	412	603	222
Non US NATO	386	16.44	96			67	110	32	46	56	75						
Other Allies	170	8.64	47	0	0	8	72	21	22	33	14						
Total "West"	1276	54.36	318			75	337	158	228	239	239						
Total "East"	1527	23.66	155									22	100	168	412	603	222

	757	767	A300	747	L-1011	DC-10	DC-8	707	IL-86	IL-76	IL-62M	TU-154	AN-26	AN-12	Total
Total in Use	0	0	175	480	210	340	330	450	22	250	180	430	220	240	3327 109.6 643
Percent Militarized	0	0	43	71	75	67	72	55	100	40	96	96	274	93	85 72 74

Aircraft Characteristics

	757	767	A300	747	L-1011	DC-10	DC-8	707	IL-86	IL-76	IL-62M	TU-154	AN-26	AN-12
Passengers	180	211	220	350	260	260	200	150	240	90	170	160	40	100
Cargo (tons)	20	34	34	75	38	46	30	12	42	40	23	18	5	20
Average Range (km)	4400	5200	4600	9000	9000	6000	7000	7200	4200	6000	8000	3200	2400	4000
Empty Weight (tons)	59	74	78	177	110	120	65	65	120	100	68	51	15	28
Max Fuel Load (tons)	33	48	44	150	73	112	71	73	65	66	85	40	5.5	15
Max Takeoff Weight (tons)	108	136	150	370	210	252	152	153	195	170	162	94	24	61
Max Load Capacity (tons)	80	114	111	278	150	197	131	108	143	120	134	82	17	50
Practical Load (tons)	49	62	72	193	100	132	87	88	75	70	94	43	9	33
With Full Fuel (tons)	16	14	28	43	27	20	16	15	10	4	9	3	4	18

Flight Distances (km)

NY to Paris—5798	Montreal to Ireland—4600	London to Rome—1420	Rome to Cairo—2120
Cairo to Tehran—2000	Moscow to Berlin—1650	Moscow to Baghdad—2600	
Montreal to Iceland—3800	Iceland to London—2000	Moscow to London—2000	Moscow to Peiping—5900
California to Hawaii—3900	Hawaii to Tokyo—6200	Tokyo to Peiping—2100	
NY to Cairo—9100			

On the return trip, disabled wounded troops and even prisoners of war can be carried. On the other hand, considering the amount of fuel required (see the aircraft characteristics portion of the chart) it might be preferable to send the aircraft back as lightly loaded as possible if petroleum is in short supply in the combat zone.

TOTALS show the considerable airlift advantage of Western forces over Russia— over twice the cargo and passenger capacity. NON-U.S. NATO includes Britain, West Germany, Benelux (Belgium, Netherlands and Luxembourg), France and Italy. OTHER ALLIES includes Israel, Japan and Korea. TOTAL "WEST" includes everyone except Russia and China. TOTAL "EAST" includes Russia and the Warsaw Pact countries (Poland, East Germany, Hungary, Czechoslovakia, Bulgaria, Rumania).

TOTAL IN USE is all aircraft in service.

PERCENT MILITARIZED is the percentage of the TOTAL IN USE that shows up on the TOTALS line, that is, the percentage of the world's aircraft usable by the military.

AIRCRAFT CHARACTERISTICS gives the average characteristics of each type of aircraft. More so than military aircraft, civilian craft are built with many variations. Averages are perfectly suitable if you are dealing with large numbers of aircraft.

PASSENGERS are the number of seats installed for peacetime operations. Given enough time, more seats could be put in with a less luxurious standard. Yes, fellow air travelers, it can get worse.

CARGO (tons) is the average cargo load that can be carried in containers. Modern aircraft are weight, not space, limited.

AVERAGE RANGE (kilometers) is the average range with a full passenger load. An empty aircraft with a full load of fuel can go 20 percent to 60 percent farther. Keep in mind that civilian commercial aircraft must maintain reserves to allow for rerouting to another airport in case of bad weather or heavy traffic. These reserves can add as much as 1000 km onto an aircraft's range.

EMPTY WEIGHT (tons) is aircraft weight without fuel or payload. A glance at the potential passenger load (at 300 pounds each), fuel capacity, possible cargo capacity and maximum takeoff weight reveals that everything won't go into the air at once. Tradeoffs have to be made.

MAX FUEL LOAD (tons) demonstrates that the primary cargo of commercial aircraft is fuel. Usually large quantities can be carried without sacrificing space by putting almost all the fuel in the wings. A major problem in wartime is providing sufficient fuel at both ends of the trip, particularly in the war zone, to keep the aircraft moving.

MAX TAKEOFF WEIGHT (tons) is the maximum takeoff weight of the aircraft. It indicates the aircraft's size. The 747 (370 tons), for example, is 232 feet long, 63 feet high with a wingspan of 196 feet. The 707 (151 tons) is 153 feet long, 42 feet high and has a span of 146 feet. The Tu-154 (94 tons) is 157 feet long, 37 feet high and has a span of 123 feet. The AN-26 (24 tons) is 78 feet long, 28 feet high and has a span of 96 feet. We are dealing with very large machines.

MAX LOAD CAPACITY (tons) is the weight of everything you can get onto the aircraft. It includes passengers, cargo and fuel. Because of takeoff weight limits, you can't take it all. This number gives a good indication of the aircraft's capacity and flexibility.

PRACTICAL LOAD (tons) is what you can get off the ground. This is the maximum takeoff weight less the weight of the empty craft. This is an indicator of the aircraft's actual lift capacity.

WITH FULL FUEL is the practical load when carrying a full fuel load. It indicates the long-distance carrying capacity of the aircraft.

FLIGHT DISTANCES give point-to-point flying distances in kilometers. What this points out is that with sufficient airfields along the way, it is possible to cover long distances with short-range aircraft or with long-range aircraft carrying heavier loads. For this reason, Iceland is very important to the United States and Western Europe. Hawaii is important to the United States. Russia's extensive network of airfields is a considerable advantage.

National Differences

There are two major manufacturers of aircraft in the world: the United States and Russia. Most of the technological advances in design have come from the United States. Russian aircraft are thus less reliable and capable than their opposite numbers in the West.

Russian aircraft are less efficient load carriers. Ton for ton of aircraft weight, Western aircraft can carry greater loads longer distances. This results largely from Russian inferiority in the design and manufacture of large engines.

Russian aircraft require more ground support to sustain the same level of operations. Most Western aircraft require three to four man-hours of maintenance per flight hour. Russian aircraft require twice this level of maintenance. And at that they are more prone to accident and breakdown. The Russians attempt to surmount this problem by adding maintenance personnel to their heavier craft (AN-22, IL-76) so that some inspection and repair can be done during layovers. This practice ties up already scarce aircraft mechanics.

Russian maintenance is not evenly spread over all flight operations. Every 50 to 60 flight hours, a few man-hours are spent checking for any unusual deficiencies. Every 300 to 600 hours, an overnight check is performed, consuming perhaps 100 man-hours. During this check, parts of the aircraft will be disassembled and any worn parts replaced to avoid failure during operations. Failure often happens anyway, and is mainly responsible for those unexpected delays at airports. All aircraft have extensive sensor systems that give preflight warning when a system is ready to fail.

Every 3000 to 4000 hours, a complete overhaul is performed. Engines are often replaced, as well as any other major component that shows signs of fatigue. It is during these month-long overhauls that other modifications are made (new seats, instruments, paint jobs, etc.). The average civilian aircraft flies 3000 to 4000 hours a year and has a useful life of 10 to 20 years. Russia obtains more efficiency by using the same transport aircraft in its military and civilian fleets. The two fleets are merged for maintenance purposes. In wartime all transport aircraft, like the railroads, come under military control.

Military-transport aircraft do not fly nearly as many hours as civilian craft (1000 to 2000 at most). This saves their capacity for the rigorous demands of wartime operations. It also saves the cost of large maintenance staff. The average civilian aircraft requires the services of seven full-time mechanics year round. Military aircraft can get by with less than half that.

In general, Russian aircraft are required to operate under more primitive and rigorous conditions than their Western counterparts. Russia has no extensive road or rail network. Many parts of the country are served primarily by air. For this reason the shorter-range AN-26 aircraft have been added to the list. As the Russians do not have to fly over large expanses of water, the short-range AN-26 can hop from one airfield to another and be just as effective as a heavier, nonstop aircraft.

These airfields are often gravel and the maintenance facilities scanty or nonexistent. The weather is often severe, especially during the winters. Their aircraft are built for ruggedness. These operational factors, added to their lower level of technology, account for the performance advantage of Western aircraft.

Part 8

TOOLS OF THE TRADE

28

THE WEAPONS
OF THE WORLD

MOST PEOPLE can do nothing with the traditional recitation of weapon statistics. Analytical and subjective evaluations of weapons show what the weapon could do. This chapter will briefly discuss the general effectiveness of weapons. More importantly, one must discuss why they often do not work. Of even greater interest is the frequency with which they do not function as the users *think* they are functioning.

Untried Technology

Weapons are often conceived, designed, manufactured and used in a triumph of hope over experience. This was less true in the past, when weapon designs persisted for hundreds of years. Frequent use brought constant refinements. In the past hundred years, the development of new weapons has increased as never before. So rapid has this development process become that many weapons are now produced and replaced without ever having been used in combat. Untried weapons are often incapable of doing what they were designed to do. Even in the past this happened. It is happening more frequently today. Examples abound. After gunpowder arms had become widely used, edged weapons—swords and bayonets—remained in evidence. For 300 years, until the late nineteenth century, substantial effort was devoted to the purchase, training and use of edged weapons. These quickly became a minuscule force on the battlefield. Yet it took hundreds of years for this lesson to sink in Indeed, most armies still issue their troops bayonets and train soldiers to use them.

Dreadnaught battleships were the culmination of hundreds of years of warship development. During a period of less than 50 years, over 170 of these vessels were built at a cost of over $150 billion (in 1982 dollars). Over half never saw combat against another battleship. One reason was that they were too valuable to risk in action. Fifty-five were sunk, which leads to some interesting statistics. Of those sunk, 17 percent were by accident, usually an explosion while the ship was in port. Aircraft sank 44 percent; torpedo boats, submarine and surface ships got 10 percent. Indeed, torpedoes accounted for 38 percent of the sinkings. Shellfire from other battleships accounted for only five sinkings (9 percent). Battleships were well protected against each other but not against cheaper weapons. Originally designed to secure control of the oceans, battleships spent most of their careers fearfully lurking in safe ports. Cheaper weapons—aircraft, submarines and mines— made the high seas too dangerous for dreadnaughts. The first of these battleships was built in 1906, and it was not until after World War II that navies gave them up. Never before had so much been spent on a weapon system for such little return in combat effectiveness.

The battleship fiasco was bad, but the miracle of modern electronics may have truly monumental high costs and low benefits. Missiles often cost millions of dollars. Because of their expense and complexity, they are tested less thoroughly than cheaper weapons, thus giving them less intrinsic reliability. Also, the rapid improvements in electronics makes the weapons obsolete more quickly. Some missiles are rebuilt to the new standard. More often, they are simply discarded, usually without ever having been tested in combat. The missiles' complexity makes them more vulnerable to enemy electronic countermeasures. Never before has so much technology been available. Never before has so much been attempted with unknown and untried devices.

Aircraft engines are improved by increasing thrust without increasing weight, size or fuel consumption. Tradeoffs must be made—usually in greater fuel consumption and shorter component life. The latter is a result of running the engine at a higher rate than it can safely handle for long periods. Often this foray into unknown territory results in components that fail even after short use. Failures are often unpredictable. It frequently takes thousands of flight hours to debug the system. This makes pilots nervous. On the positive side, modern aircraft are much safer than any of their World War II predecessors. This safety factor is often achieved at a much greater maintenance cost. Modern maintenance test equipment is often good at spotting which expensive part has to be replaced. At worst, the test gear keeps a plane on the ground until the suspected failure is run down. This makes pilots feel better, although somewhat upset at the amount of time they are grounded.

World War II vs. Today

The following list covers all the weapon classes in this book. With each class the essential differences between common World War II era perception and actual current capabilities are given.

INFANTRY WEAPONS

These have changed the least since World War II. There are more automatic weapons. Nearly every infantryman today has an automatic rifle. Mortars, grenades and machine guns are basically the same. What *has* changed is readily obvious. For over twenty years, the infantry has used radar and other electronic detection devices. In the hands of well-trained troops, these weapons are quite effective.

Another area of obvious change has been in antitank weapons Most soldiers still have a "stovepipe" they put on their shoulders, aim at a tank and hope to achieve a hit with before being run over. Rocket-launcher weapons have become more refined, but against their intended targets they have become less effective. In most cases it is impossible to achieve a kill against the frontal armor of a modern tank. In some cases, even side and rear hits are chancy. Fortunately, modern tanks are spread out more than they were in World War II. Separations of 100 meters are commonly used. To give the infantry a fighting chance against armor, the ATGM (antitank guided missile) was developed over twenty years ago. Despite constant improvement and refinement, this weapon has consistently fallen short of its promise. The ATGM is not effective at the closer ranges (under 200 meters), where the infantry are most likely to encounter tanks.

The most significant change since World War II has been the mechanization of infantry. The half-track APC's (armored personnel carriers) of World War II vintage are still around. For the most part they have been superseded by fully tracked vehicles, which are now evolving into IFV's (infantry fighting vehicles). The IFV can fight by itself in addition to transporting infantry. This is a significant change, for the infantry is expected to fight from an IFV in addition to operating dismounted. IFV's have turrets with small-caliber cannon as well as ATGM. Experience thus far shows fighting from inside the IFV to be less effective than troops out in the open using the IFV as a support vehicle. Some think of the IFV as a lighter tank. Although not designed or with crews trained for that role, the IFV may indeed find itself operating more like a tank than an APC. Whatever the outcome, the infantry is no longer alone. It now has its own armored vehicles to look after and protect during combat. In this respect, the infantry becomes much like a tank crew. Its combat routine revolves around its IFV's.

TANKS

Today's tanks may look like their World War II cousins, but they are far more deadly and vulnerable. They are also larger, 50 percent heavier than in World War II. Unlike in World War II, the following little jingle is true: "What you can see you can hit and what you can hit you can kill."

ARTILLERY

There has been little visible change. The guns are more efficient and more frequently self-propelled. The ammunition has changed considerably. Current ammunition will do far more damage than World War II versions. Because of the cost and complexity, the less affluent, more numerous and more-likely-to-be-at-war armies will still be fighting with old-fashioned shells. Don't let that fool you; the big guys have some very deadly stuff.

THE AIR FORCE

High-speed, jet-propelled aircraft don't change air warfare as much as one would think. While the maximum speed of combat aircraft has more than tripled since World War II, the average speed during combat has increased less than 100 percent. Because the aircraft and their pilots cannot physically withstand the stresses of high-speed maneuvers, the average combat speeds are low (700 to 1100 km per hour).

Although highly capable airborne radar and weapon-control systems have proliferated, their inability efficiently to tell friend from foe has limited their use. Most post-World War II combat has been visual contact and engage. Air force suppliers keep promising that the next batch of electronic gadgets will finally change all that. Pilots, being pragmatic and somewhat cynical by now, prefer to rely on what they can see. For this reason cannon is still a preferred weapon for shooting down aircraft.

Bombing has changed radically. Much heavier bombloads are possible and can often be delivered in all weather conditions. In World War II, the B-17 was the typical heavy bomber, weighing 25 tons and capable of carrying 7 tons of bombs. The U.S. F-4 (a 1960's vintage *fighter*-bomber) weighs 28 tons and carries up to 8 tons of bombs. The F-4 normally travels three times as fast as the B-17 over longer ranges. The F-4 has a crew of two versus eleven for the B-17. The modern replacement for the B-17, the B-52, can carry over 25 tons of bombs. Not only have the bombers' capacities increased enormously, the effectiveness of the bombs have also. Leaving aside chemical and nuclear weapons, conventional bombs are five times more destructive than their World War II predecessors.

THE NAVY

Ships still float. That much hasn't changed. The navies' bad experience with the battleship and good with aircraft has radically changed fleets. Most ships are now designed to deliver and deal with airborne threats. Those navies that can afford them have aircraft carriers. All navies ship out with guided missiles (think of them as pilotless aircraft). There has also been a revolutionary change in submarine warfare. Indeed, without aircraft to assist them, most surface ships are at a grave disadvantage when fighting a nuclear attack submarine.

The only combat cliché to survive from World War II is that of the naval aviator flying off to sink the enemy fleet. Even that is not quite the same. There are no more torpedo or dive bombers roaring through maelstroms of enemy flak to deliver their loads. Nowadays the guy in the back seat stares at a TV screen and does what the computer tells him. Everything is very efficient and automatic, including the failures.

The skills of the computer programmer and the electronics engineer now count for nearly as much as the aircraft and naval crew. Without effectively programmed and maintained electronic devices, they are shooting blanks. The trend toward more technology and less reliance on seamanship and weapon-handling expertise has been ongoing for hundreds of years.

For each benefit received from technology, a price of dependence is extracted. The time is past when the commander of a ship understood how mot of his equipment worked.

Yes, the ships still float. That's about the only familiar aspect of naval warfare that remains from World War II.

AMPHIBIOUS OPERATIONS

It's comforting to see that some things don't change. Aside from improved amphibious shipping and landing craft, the only new development has been helicopters. The amphibious troops, being infantry, have these new weapons.

AIR DEFENSE

Today's air defense is an odd mixture of the familiar World War II and a form of science fiction known as the SAM (surface-to-air missile). The World War II cannon-type of air defense is still used, in some cases with original World War II weapons. More often, the small-caliber (under 75mm) cannon is controlled by radar and computerized fire-control systems. Manual override keeps them honest, and minimally effective.

SAM's have great potential. At least one in 50 will hit its target. Radar screens blink, obscure jargon is muttered, at times sounding like prayers, and, far from the darkened room, missiles leave launchers by remote control. So far the electronic-warfare defenses against SAM's have held the upper hand. This is small comfort to pilots, who must entrust their lives to a lot of black boxes. Aircraft are still hit. The SAM's keep getting better. More frequently, pilotless aircraft are sent into areas heavily defended by SAM's. We let the machines kill each other off.

Compared with much subsequent electronic technology, nuclear weapons are straightforward and effective. The chief means of delivering nuclear weapons—missiles—are another story. Incredibly, and increasingly, complex, missiles are the weak link in the use of nuclear weapons.

Items as simple as rifles have had lurking defects. The M-16 was adopted by the U.S. Army in response to conditions in Vietnam during the 1960's. This rifle had problems then. It still has problems, as a 1980 exercise in the Egyptian desert revealed. The M-16's had to be cleaned up to three times a day to prevent fouling by the ever-present sand.

Users of high-technology weapons cannot seem to resist asking for ever more capable equipment. Their suppliers are equally smitten, and the result has been an unending stream of almost-ready weapons. Whoever gets a new weapon invariably comes in last in the weapons-available-for-use category.

Most nations have been infected by the technology bug. Each nation, because of its historical experience, handles the resulting problems differently.

The United States, one of the most avid fans of the new and wonderful, usually attempts to solve problems arising from shaky technology by applying more technology. Having the largest and most adventurous arms industry in the West, the United States is the source of most new weapons. It resists ideas and weapons designs from nations with more military experience. In the past there have been many opportunities to purchase better, and cheaper, weapons from other nations: tanks from West Germany, or the FN rifle from Belgium, for example. The "we can do it better" attitude has prevailed, with predictable results like the abortive MBT tank and the unlamented M-14 rifle. This attitude is changing now, but much damage has already been done.

The U.S. arms industry also gets a lot of black eyes for being first so often. Whoever goes into the dark room first will stumble over the uncharted obstacles; those who follow will have an easier time. Being first is expensive and often leads to untrustworthy equipment.

The West European nations have a more pragmatic weapons tradition. They have more successfully introduced high-technology weapons into their armed forces.

Russia, despite its reputation for producing rugged, dependable, soldier-proof weapons, is also much taken by new gadgets. Russian weapons often appear the equal of those found in the West. Russian arms are not only crude and unreliable, but also often much less capable. They would argue that if only a fraction of their weapons work, they will still outnumber potential opponents. Perhaps, but the skill of the men behind these weapons will be more decisive than their number.

With many basic weapons like small arms, artillery and tanks, the Russians do end up with highly effective and reliable equipment after they have had a number of years to work out the bugs.

At the Movies

We gather most of our impressions of weapon effectiveness from our daily exposure to the media. The worst offender is the motion picture.

First, movies quite naturally depict World War II much more than potential contemporary conflict. Because the appearance of World War II weapons does not differ radically from current ones, we tend to equate their performance with the latter.

To further muddle our perceptions, movies enhance weapon effects in order to increase their visual impact on the screen. Weapon effects in the movies are more like fireworks displays than reality. For example, most real shell explosions are much smaller and lack flames. When bullets pass they make a pronounced crack. This is rarely heard in a movie sound track, probably because it would be distracting and confuse the dialogue.

Finally, soldiers always appear more in control of the situation in movies than they are in real battles. This is a particularly dangerous misconception for the young soldier going into battle for the first time.

29

THE ARMED FORCES
OF THE WORLD

THE OVERRIDING rationale for armed forces is self-defense. Armed forces serve as one more bargaining chip in a state's international diplomacy. If war comes, the armed forces have failed in their primary purpose. To avoid war the armed forces must appear as powerful as possible. If substance is sacrificed to increase form, why not? An apparently stronger armed force is more valuable in diplomacy than a less capable-appearing one.

Actual combat capability is difficult to measure. One can more easily count the number of tanks, ships, aircraft and men in uniform. Numbers make the loudest noise when you must rattle the saber. Should bluff fail, and you are forced to wage war, well, that's another set of problems.

Doctrine vs. Reality

Doctrine is the plan, reality is what actually happens. Most nations' military planning rests on their appraisals of their own military ability. This appraisal reaches a low point just before arms budgets are voted on, and rises swiftly during international crises and reelection campaigns.

When actual warfare approaches, the military becomes more realistic. It is always a touchy matter when the generals must confront the national leader, inflamed by patriotic optimism, with a sober appraisal. One of the more poignant examples was in 1914. The Kaiser, after declaring war, began to realize the enormity of his action. He asked the generals to stop the mobilization. They informed him that this would put Germany at a grave disadvantage as it would totally disorganize the armed forces. The war went on, and millions of lives were lost.

The peacetime gap between doctrine and reality is recognized by most national leaders. Unfortunately, people sometimes forget or are overtaken by events.

Most nations have traditional armed forces, capable of routine warfare. Countries look at their neighbors' past history and finances, and build up their armed forces accordingly. The idea is to have armed forces that can successfully resist the one or more potentially unfriendly neighbor.

Border disputes and excess wealth are the most common causes of building up armed forces above the levels needed for self-defense. The Middle East is a good example. Israel's existence is a border dispute in the eyes of the Arab nations. The size of the armed forces in the region has grown accordingly. Excess wealth in the region has also led to excessive armed forces. The more wealth one has, the more concerned one becomes about keeping it. Like good health, no price is too high for security, particularly if you have deep pockets. As doctors are concentrated in wealthy neighborhoods, so arms dealers flock to the oil-rich Middle East.

As a nation's apparent military capabilities grow, doctrine tends to follow. One defensive plan is to have armed forces mobilize near the borders to repel invaders. Increasing strength leads one to contemplate taking the war to the aggressor's homeland. An invasion has three attractions. The fighting is shifted to the enemy's territory. The aggressor is forced to choose between continuing his invasion or pulling back to defend his own lands. Retribution is made. Never underestimate the power of revenge in world affairs.

Most armed forces are capable of mustering a defense. An attack, particularly an invasion into hostile territory, is considerably more difficult. In the defense you dig trenches and wait. An attack requires moving large numbers of troops. Eventually defenders start shooting at you. Troops, difficult to control under any circumstances, are more so while moving and being shot at. Keeping large numbers of troops fed and healthy becomes more difficult in unfriendly territory. More supplies must be moved farther. Enemy attacks on these supplies create still more problems. All that movement uses and wastes much more supply than remaining within your own borders. Attacking often uses more supply than defending. Commanding, controlling and communicating with moving forces in enemy territory is enormously more difficult. Gathering information on enemy forces is obviously easier for the defender in his homeland. It is common for a defender to defeat an invader soundly and then suffer an equally disastrous defeat during a pursuit into the invader's homeland.

Attrition vs. Maneuver

There are two ways to fight a war: plain (attrition) and fancy (maneuver). The stronger military power has the option. If the stronger power has little military experience, the only choice is attrition. The United States used this approach successfully through most of its wars, including the Civil War. It is easier to be proficient at attrition warfare, which requires the simplest military skills and enormous quantities of arms and munitions.

The maneuver approach is being more efficient than your opponent. Instead of engaging in a mutual slaughter, you run circles around him. You destroy his will to fight, by everything from stunning him into surrender with your fancy footwork to the mundane destruction of his headquarters and sources of supply. Maneuver warfare is waged against leadership and the troops' confidence and sense of security.

The United States has engaged in maneuver warfare in several small wars. In the war with Mexico in the 1840's, small U.S. forces invaded and outmaneuvered the opposition. The Confederacy, during the Civil War, kept the war going for so long through superior ability. They fought a war of maneuver, but were ultimately defeated in a war of attrition.

Russia, despite all the talk of the sheer size of its armed forces, is a firm believer in maneuver warfare. Russian doctrine stresses destroying the enemy's will and means to fight. The means includes the armed forces, but the first targets are headquarters, transportation and logistics. The Russians learned this from the Germans in World War II. It's interesting to consider that much of current Russian doctrine is derived from the nation Russia soundly defeated in that war. Russia was victorious in World War II through its successful application of attrition.

Attrition should not lull the stronger industrial powers into a false sense of security, however. A massive disparity in military skills can defeat the larger powers. This does not happen often and when it does, the smaller, stronger power is usually incapable of invading and defeating the larger nation's homeland. Such small, skilled and determined countries usually adopt such a posture to dissuade larger neighbors from rearranging their borders. Examples are Israel, Finland, Switzerland and Sweden.

The Difference Between Wars and Disorder

Much of what currently passes for war is not war. It is simply insurrection, guerrilla warfare or general disorder involving the armed forces. This is an important distinction, as a great deal of military skill is not needed to create armed disorder. You don't need trained troops to create a proper insurrection or civil war. All you need is an armed citizenry at odds with itself.

A war, as it is meant in this book, is more than slaughter, mayhem and senseless destruction. A certain amount of skill is implied, perhaps even a reasonable excuse for the exercise. Not all the armed forces described in this chapter possess skill. Military skill is more than uniforms, display and awesome-looking equipment. Much of the military violence in the world is nothing more than large-scale disorder, banditry or worse. Uganda, Lebanon, El Salvador and Afghanistan are examples of disorder. In such conflicts combat values take on a different meaning. For example, during a disorder in which one side is clearly stronger than the other, the weaker side fights when and where it has a chance of success. When faced with overwhelming military power, the weaker force will turn into civilians, or otherwise seek sanctuary.

Afghanistan again comes to mind. If Russia put a million troops into the country, ten times the initial number sent in, the Afghans would simply wait them out. Sufficient outrages would occur to keep the populace in a properly hateful frame of mind.

A war is fought to a conclusion. Disorders may go on for hundreds of years. Wars are fought by powerful, and expensive, armed forces. Disorders are fought with whatever deadly force is handy, plus the legendary hearts and minds.

Making disorder is much simpler than making war, which is why it is so much more common.

ARMED FORCES OF THE WORLD

The NATO and Warsaw Pact alliances comprise over half the world's military might.

Nation	Combat Value Total	Man-power Total	Value per Man Average	Land Value Total	Naval Value Total	Divi-sions Total	AFV Total	Combat Planes Total	Land Border Total	Force Mult Average	Arms Cost Total	Cost per Man Average	Popu-lation Total	Per Cap GNP Average
Europe	2199	9307	24	1814	391	379	167910	14006	50160	0.22	399611	42937	666	66
Asia	1663	12645	13	1429	239	429	66760	12563	90340	0.08	164463	13006	2543	8
Americas	473	4624	10	378	95	76	31190	4952	78300	0.06	147960	31998	577	51
Middle East	650	3254	20	632	7	124	49400	3780	70836	0.08	75823	23302	273	15
Africa	94	1167	8	92	2	67	8900	650	49700	0.06	8435	7228	247	5
All Other	30	375	8	27	3	23	1702	371	NA	0.05	2831	7541	193	3
World Totals	5109	31372	16	4372	737	1098	325862	36322	339336	0.14	799123	25472	4499	22
Russia	1008	4200	24	900	110	193	112000	8500	20619	0.10	192432	45817	270	37
% of Total	20	13	147	21	15	18	34	23	6	72	24	180	6	10
United States	812	2800	29	438	385	30	34500	5300	12000	0.14	221800	79214	231	100
% of Total	16	9	178	10	52	3	11	15	4	106	28	311	5	23
NATO Europe	635	1723	37	527	108	59	31760	2344	12560	0.12	78740	45699	219	75
% of Total	12	5	227	12	15	5	10	6	4	87	10	179	5	17
Warsaw Pact	152	1181	13	150	2	58	35800	2220	16100	0.07	18900	16003	112	39
% of Total	3	4	80	3	0	5	11	6	5	51	2	63	2	4
Russia/WP	1160	5381	22	1050	112	251	147800	10720	36719	0.09	211332	39274	382	38
% of Total	23	17	132	24	15	23	45	30	11	66	26	154	8	14
US/NATO	1447	4523	32	965	493	89	66260	7644	24560	0.13	300540	66447	450	87
% of Total	28	14	196	22	67	8	20	21	7	95	38	261	10	40

ARMED FORCES OF THE WORLD: NATIONS BY CONTINENT

Each region has one nation that militarily dominates the others.

Nation	Combat Value	Man-power	Value per Man	Land Value	Naval Value	Divi-sions	AFV	Combat Planes	Land Border	Force Mult	Arms Cost	Cost per Man	Popu-lation	Per Cap GNP
Asia Total	1663	12645	13	1429	239	429	66760	12563	90340	0.08	164463	13006	2543	8
China	540	4700	11	535	5	129	14000	6100	4200	0.08	55000	11702	1020	5
Russia Far East	252	1050	24	225	33	48	28000	2125	24000	0.10	48108	45817	38	34
US Pacific	233	208	112	79	154	3	2200	400	0	0.14	19920	95769	.2	100
India	192	1450	13	185	7	32	3000	640	12700	0.12	5200	3586	680	1
Japan	70	250	28	53	17	14	1200	470	0	0.15	12000	48000	118	76
South Korea	61	602	10	59	2	23	1400	380	241	0.11	4500	7475	40	12
Vietnam	49	980	5	48	2	39	4100	400	4600	0.09	1200	1224	60	1
North Korea	48	820	6	47	1	44	3600	650	1675	0.07	1500	1829	20	6
Taiwan	48	550	9	41	7	22	1900	380	0	0.13	2000	3636	18	14
Australia	40	73	55	32	8	4	800	140	0	0.20	4000	54795	15	76
Pakistan	32	560	6	31	1	20	1600	225	5900	0.08	1600	2857	90	3
Thailand	29	275	11	29	0	7	600	190	4900	0.09	1300	4727	49	5
Indonesia	11	285	4	11	0	5	1600	50	0	0.05	2500	8772	155	2
Burma	9	210	4	9	0	12	110	15	5900	0.04	220	1048	35	1
Malaysia	8	121	7	8	0	4	300	40	2300	0.06	2400	19835	14	12
Philippines	8	156	5	8	0	5	100	120	0	0.06	1020	6538	50	6
Singapore	7	50	15	7	0	1	1100	95	0	0.11	650	13000	2	38
Bangladesh	6	107	6	6	0	5	50	15	2500	0.06	170	1589	94	1
New Zealand	4	13	34	4	1	1	80	34	0	0.08	450	34615	3	51
Laos	3	50	5	3	0	5	100	22	5100	0.04	50	1000	4	1
Mongolia	3	69	4	3	0	1	220	12	8000	0.05	140	2029	2	12
Britain in Asia	2	3	67	2	0	.4	60	0	24	0.25	160	53333	5	25
France in Asia	2	3	60	2	0	.4	20	0	0	0.18	150	50000	.8	10
Nepal	2	40	6	2	0	1	20	0	2800	0.06	25	625	15	1
Afghanistan	1	20	7	1	0	3	600	60	5500	0.04	200	10000	15	1

Nation	Combat Value	Man-power	Value per Man	Land Value	Naval Value	Divi-sions	AFV	Combat Planes	Land Border	Force Mult	Arms Cost per Man	Cost per Man	Popu-lation	Per Cap GNP
Americas Total	473	4624	10	378	95	76	31190	4952	78300	0.06	147960	31998	577	51
United States	322	2742	12	245	77	21	24710	3734	12000	0.14	133080	48534	230	100
Brazil	36	455	8	33	3	9	1300	170	13100	0.08	1700	3736	125	14
Canada	33	63	53	25	8	2	1040	220	9000	0.22	4200	66667	24	86
Cuba	20	210	9	19	1	6	1100	160	0	0.06	1100	5238	10	11
Argentina	16	228	7	15	1	4	900	240	9400	0.05	3500	15351	29	16
Chile	12	140	8	10	2	6	520	90	6300	0.05	800	5714	11	14
Mexico	10	372	3	10	0	7	300	30	4200	0.05	1200	3226	71	14
Peru	9	155	6	8	1	4	700	110	6100	0.04	450	2903	18	6
Venezuela	4	61	7	3	1	2	220	104	4200	0.05	1100	18033	17	23
Colombia	4	120	3	3	1	3	120	24	6000	0.05	270	2250	27	7
Ecuador	3	44	6	2	0	6	200	50	1900	0.05	200	4545	8	9
Bolivia	2	31	6	2	0	6	60	20	6100	0.04	210	6774	6	5
France in Amer.	2	3	60	2	0	.4	20	0	0	0.18	150	50000	.7	11
Africa Total	94	1167	8	92	2	67	8900	650	49700	0.06	8435	7228	247	5
South Africa	42	182	23	41	1	4	3900	240	2100	0.17	2800	15385	30	15
Ethiopia	10	400	2	10	0	29	1200	110	5200	0.06	410	1025	31	0
Nigeria	10	156	6	10	0	4	200	22	4000	0.09	1800	11538	80	5
France in Africa	8	18	44	8	0	1	150	0	0	0.18	900	50000	0	0
Somalia	4	92	4	4	0	7	300	30	2300	0.05	110	1196	6	1
Zimbabwe	4	40	10	4	0	1	120	35	3000	0.07	450	11250	8	3
Sudan	3	75	5	3	0	4	600	35	7800	0.05	260	3467	19	2
Tanzania	3	48	7	3	0	3	140	22	3900	0.05	220	4583	19	2
Cuba in Africa	3	35	8	3	0	3	400	0	0	0.06	200	5714	0	0
Angola	1	33	4	1	0	6	600	40	5100	0.06	260	7879	7	3
Ghana	1	23	5	1	0	1	100	12	2300	0.06	150	6522	13	7
Kenya	1	17	7	1	0	2	140	25	3400	0.06	220	12941	17	3
Mozambique	1	28	4	1	0	2	800	30	4600	0.06	210	7500	11	1
East Germany/Af.	1	3	33	1	0	.3	150	0	0	0.14	25	8333	1	0
Zambia	1	17	7	1	0	1	100	49	6000	0.04	420	24706	6	4
Europe Total	2199	9307	24	1814	391	379	167910	14006	50160	0.22	399611	42937	666	66
Russia Europe	534	2226	24	477	58	102	59300	4505	3000	0.10	101989	45817	144	37
West Germany	312	516	60	298	14	19	12700	660	4232	0.20	27000	52326	62	102
US Europe	268	300	89	114	154	6	8970	1378	0	0.14	133080	79214	0.50	100
France	186	580	32	144	42	29	6900	606	2888	0.18	26000	44828	54	94
Britain	181	344	53	104	77	9	6300	720	360	0.27	29000	45817	56	68
Russia Balkans	121	504	24	108	19	23	13440	1020	1000	0.10	23092	45817	32	37
Sweden	76	720	11	74	2	22	1400	430	2196	0.08	3900	5417	8	111
Yugoslavia	63	440	14	62	1	17	2200	340	3000	0.11	3800	8636	23	24

	Total	Total Average	Average	Total	Total	Total	Total	Total	Total Average	Average	Total Average	Average	Total Average	Average
East Germany	62	167	37	61	1	6	5000	360	2300	0.14	7000	41916	17	55
Italy	57	366	15	50	6	8	5600	310	1700	0.09	8900	24317	58	53
Switzerland	48	625	8	48	0	19	2000	370	1900	0.09	1900	3040	6	157
Spain	39	440	9	35	4	12	1500	202	1900	0.09	4100	9318	38	43
Finland	34	130	26	34	0	3	700	50	2500	0.15	750	5769	5	80
Rumania	28	185	15	28	0	11	4000	320	3000	0.09	1400	7568	23	36
Greece	27	194	14	25	2	15	2900	400	1200	0.08	1900	9794	10	36
Belgium	25	105	24	24	0	3	1800	140	1400	0.09	3700	35238	10	97
Netherlands	25	116	21	20	5	3	3600	190	1000	0.06	5000	43103	14	86
Bulgaria	20	164	12	20	0	10	3700	240	1900	0.07	1400	8537	9	35
Czechoslovakia	17	231	7	17	0	10	7700	470	3500	0.06	3700	16017	16	49
Poland	16	318	5	16	0	15	11900	700	3100	0.02	4200	13208	36	31
Portugal	13	71	18	12	1	5	200	80	1200	0.09	960	13521	10	19
Austria	12	160	8	12	0	9	700	35	1675	0.07	950	5938	8	74
Hungary	9	116	7	9	0	6	3500	130	2300	0.05	1200	10345	11	35
Denmark	8	33	25	7	1	2	1100	110	68	0.08	1600	48485	5	94
Canada in Europe	8	12	69	6	2	0	260	24	0	0.22	880	73333	.02	0
Norway	7	160	4	6	1	10	200	110	2600	0.07	1700	10625	4	114
Albania	3	69	4	3	0	3	200	90	0	0.06	200	2899	3	3
Ireland	1	15	8	1	0	1	140	16	241	0.06	310	20667	3	44
Middle East Tot.	650	3254	20	632	7	124	49400	3780	70836	0.08	75823	23302	273	15
Israel	330	410	80	327	3	16	7100	600	1036	0.30	7500	18293	4	46
Russia Asia	101	420	24	90	0	19	11200	850	9000	0.10	19243	45817	56	37
Egypt	55	405	14	54	1	13	4400	280	2500	0.16	2200	5432	44	3
Turkey	51	688	7	49	2	22	5500	320	2600	0.08	3300	4797	47	8
Iran	26	220	12	26	1	7	2100	100	5300	0.09	4800	21818	41	13
Syria	22	230	9	22	0	8	4400	380	2200	0.09	2500	10870	9	13
Iraq	16	280	6	16	0	13	3800	280	3700	0.06	3200	11429	14	9
Jordan	15	78	20	15	0	4	1600	80	1800	0.12	450	5769	3	7
Morocco	10	150	7	10	0	6	1500	70	5900	0.06	1300	8667	22	6
Saudi Arabia	9	87	10	9	0	2	1000	140	4900	0.07	28000	321839	11	77
North Yemen	3	52	6	3	0	3	900	60	2700	0.05	220	4231	5	6
South Yemen	3	42	8	3	0	4	600	110	0	0.05	140	3333	2	2
Libya	3	60	6	3	0	3	4300	400	0	0.03	600	10000	3	57
United Arab Emir	2	42	6	2	0	1	200	50	9400	0.07	800	19048	1	178
Tunisia	1	36	3	1	0	1	200	10	13100	0.05	110	3056	7	8
Kuwait	1	30	4	1	0	1	500	50	4600	0.04	1200	40000	1	185
Lebanon	1	24	4	1	0	1	100	0	2100	0.05	260	10833	3	8
Russia	1008	4200	24	900	110	193	112000	8500	20619	0.10	192432	45817	270	37
United States	812	2800	29	438	385	30	34500	5300	12000	0.14	221800	79214	231	100
NATO Europe	635	1723	37	527	108	59	31760	2344	12560	0.12	78740	45699	219	75
Warsaw Pact	152	1181	13	150	2	58	35800	2220	16100	0.07	18900	16003	112	39

Armed Forces of the World

This chart gives evaluations of the quantity and quality of each nation's armed forces. The quantity of each combat unit has been derived from various open sources. Quality has been determined by evaluating historical performance. All armed forces are not equal, and this inequality has been expressed numerically.

In calculating the numerical value of total strength it is important to differentiate between apples and cabbages. Aircraft carriers and tank divisions are very different instruments of destruction. Both cost about the same, but a carrier cannot march on Moscow, nor can a division fight a battle in the middle of the Atlantic. Destructive effect was the main consideration in assigning values. This was modified by the mobility and flexibility of the system. For example, a tank division is good at one thing—ground combat. An aircraft carrier can support ground combat as well as destroy enemy naval and air forces.

All these values are subjective. The reader is free to impose his own interpretations. The evaluations were felt to be the most reasonable in the light of current conditions and historical experience.

Ground and naval forces are grouped on the armed forces chart, while naval forces alone are considered on the naval forces chart. On the armed forces chart, naval forces are evaluated only in terms of their ability to directly support ground operations. On the naval forces chart, these forces are considered purely for their abilities at sea.

The first part of the chart gives continental summaries, world totals, and recaps of the two superpowers and the major alliances in Europe (NATO and Warsaw Pact). Following that are breakdowns by nation.

NATION lists every nation with a combat value of 1 or more. Nations with a combat value of less than 1 have little more than national police. Many smaller countries, especially those that lack a threatening neighbor, use their forces primarily for internal security. Any nation with a combat value of less than 1 may launch or repel an invasion, or fend one off, most effectively simply by arming the population.

Nations are grouped into four regions: Europe; Middle East and North Africa; the rest of Africa, Asia and the Pacific; the Americas. The United States and Russia have a number of geographically and functionally separate armed forces. Russia possesses no fewer than four distinct groupings. The largest force is in Eastern Europe and Western Russia. Another is oriented toward the Balkans and European Turkey. Deployed along a line from the Black Sea to the Hindu Kush mountains are Russia's Asian and Middle Eastern armies. Facing China are the Far Eastern armies. Supporting each of these armed forces is a local population. Showing these groups as separate entities is a more accurate reflection of Russia's strategic military situation.

While Russia is forced into such a situation by being surrounded by far-apart hostile neighbors, the United States has voluntarily deployed substantial forces in far-flung parts of the world. The United States has substantial ground forces in Europe and equally massive naval forces in the Pacific. Because these foreign-based forces consist primarily of combat troops, their combat value per man is higher than the national average. The only other nations to station substantial forces so far from the homeland are France, Cuba, East Germany, Britain and Canada.

COMBAT VALUE is the total combat ability of the nation's armed forces. Certain countries (Israel, Sweden and Switzerland) have a rapid mobilization capability which achieves the combat value shown within three days of mobilization. Their normal, unmobilized, combat value is about 30 percent of the value shown. As explained above, combat value is modified by geographical, climatic and political factors. The value given here is a combination of the quantity and quality of manpower, equipment and weapons. The raw combat value is then multiplied by the force multiplier (see below).

MANPOWER (in thousands) is the total uniformed, paid manpower organized into combat and support units. Because of the widely varying systems of organizing military manpower, this figure is at best a good indicator of the personnel devoted to the military. The industrialized nations hire many civilians to perform support duties, while other nations flesh out skeleton units with ill-trained reserves. The use of reserves varies considerably; see Chapter 6 for more details.

VALUE PER MAN is the combat value per 100,000 men. This value is obtained by dividing combat value by manpower and multiplying by 100,000. This number clearly shows the qualitative differences in equipment, weapons, manpower, national economy and, most important, leadership.

LAND VALUE is the combat value of land combat forces. This includes all air power devoted to the support of land operations—interceptors as well as ground attack aircraft.

NAVAL VALUE is the value of national naval forces' support of ground operations. It includes naval air power and amphibious shipping. See the naval forces chart for a naval power's value when it is devoted purely to naval operations.

DIVISIONS (or equivalent) represent the number of self-supporting ground combat units maintained by each nation. This figure includes marines. These are fully equipped, although not always full-strength, units. Most nations rely on reserves to fill out these units in wartime. Equivalent divisions are combinations of lesser units that could be used in combination at approximately division strength. A division normally has 10,000 to 20,000 men. See Chapter 2 for more details.

AFV (armored fighting vehicles). These include tanks, armored personnel carriers, and all other armored combat and support vehicles. AFV are the primary components of a ground offensive, and greatly enhance chances of success.

COMBAT PLANES TOTAL are the number of combat aircraft devoted to land operations. This, like AFV, is a good indicator of raw power. The quality of the aircraft, their pilots, ground crew and leadership is the most important factor in the air power's overall value.

LAND BORDER (in kilometers) suggests the nation's defense problems, particularly if it is bordered by not so friendly neighbors.

FORCE MULT (force multiplier) is a fraction by which raw (theoretical) combat power should be multiplied to account for imperfect leadership, support, training and other "soft" factors. Think of it as an efficiency rating, with 1 being perfect and .12 being 12 percent efficient.

ARMS COST is the current annual armed forces spending of that nation, in millions of dollars. All nations use somewhat different accounting systems for defense spending. Efforts were made to eliminate some of the more gross attempts to hide arms expenditures. Some of the figures, particularly for smaller nations, may be off by 10 percent either way.

COST PER MAN is the annual cost per man for armed forces in dollars. This is an excellent indicator of the quantity and, to a lesser extent, the quality of weapons and equipment. Some adjustments should be made for different levels of personnel costs, research and development, strategic programs and waste. Russia and the United States, in particular, are prone to all four afflictions. The precise adjustments are highly debatable. I would cut the U.S. cost per man by at least one third and the Russian figure by at least 20 percent. Other nations with strategic programs (Britain, France, China, etc.) should be adjusted less than 10 percent. Britain would also take a 10 percent cut for its highly paid volunteer forces. At the other extreme, there are many countries that get by on far less. Using conscripts and good leadership, many of them overcome most of the ill effects of stripped-down arms budgets. Most nations simply end up with what they're paying for.

POPULATION (in millions) indicates the nation's relative military manpower resources. Population is also a more meaningful indicator of a nation's size than territory.

PER CAP GNP (per capita gross national product) is given as a percentage of the United States' GNP to indicate the nation's economic power. This does not translate immediately into military power because of the time needed to convert industry from civilian to military production. Mobilization of some types of military equipment takes years. Other types, especially those using electronics, can be brought to bear in months.

Current Potentials for War

Each region varies in its potential for war and the potential destructiveness of such conflicts.

EUROPE has little potential for war, but the highest likelihood that a conflict would turn into a holocaust. Although Greece and Turkey have long-standing differences, most other conflicts would spread from internal disorders. Eastern

Europe is the principal hot spot. Russia's heavy-handed maintenance of an alliance has brought many of these nations to the boiling point. The 1981 Polish fracas was the third in twenty-five years (1956, 1970). Yugoslavia's internal trouble could also prove tempting to Russia. The ultimate horror might be a reunification movement in both Germanies.

MIDDLE EAST is the most volatile region. Arab animosity toward Israel runs a close second to their disputes with one another. While there have been four Arab-Israeli wars, there have been six conflicts between other nations since 1969 (Saudi Arabia-South Yemen, Syria-Jordan, Libya-Egypt, Syria-Lebanon, North Yemen-South Yemen, Iran-Iraq). These conflicts lacked the intensity and decisiveness of the Arab-Israeli conflicts. As the Iran-Iraq War has demonstrated, these little disputes can get out of hand. With the West still dependent on Arab oil, any local war threatening this resource will bring intervention. Russia's best move in such a situation would be to do nothing and let the West get mired in Middle East politics.

ASIA is another area where things can get out of hand. The biggest danger is on the border one hears little about: China and Russia's. Russia fears China retaking its Far Eastern provinces. China has a very good case, and need, to reclaim the Far Eastern areas Russia took by force in the last hundred years. Russians have long held the Asiatic "barbarians from the East" in contempt. The Chinese have a superiority complex going back thousands of years. For such round-eyed, big-nosed, unwashed Western barbarians to have the audacity to station armies in Chinese territory is unforgivable. The Chinese have patience, not just from their pragmatism and superior attitude, but because they have the numbers, and time, on their side.

The Chinese superiority complex also produces potential conflict with Vietnam, India and even Japan. Korean reunification would most likely bring in the Chinese, who for centuries have played the role of big brother in Korea. China has tried to assume a measure of control over the Japanese, an aspect of Asian politics of which the Japanese are ever mindful.

The central Asian tribes have been waging war with everyone within reach for thousands of years. Only in the last hundred years have the Russians actually invaded Central Asia. Their hold is tenuous.

INDIA itself is a collection of many different peoples. Despite this, they have long-standing grievances against Pakistan, which consists primarily of those same tribes mentioned before. What Attila the Hun started is not yet over.

Other conflicts in Asia are internal. The self-destruction of Cambodia (with an assist from the United States, Vietnam and China) creates little but local misery and speeches in the UN. Vietnam's centuries-old push to control all of Indochina has the potential, but not much likelihood, of spreading farther, unless, of course, the major powers run out of entanglements elsewhere.

AFRICA, south of the Sahara, is a mess dominated militarily by South Africa. The colonial period created long-term anarchy by setting political boundaries that ignored ethnic realities. Internal disasters in most African states have bred misery and the inability to develop significant military power. South Africa is immune to any external military threat from other African nations. Internally, it's another story.

AMERICAS are insulated by the United States and two oceans. North America prospers while South America slips in and out of anarchy. There are few serious disputes between South American nations. The biggest danger remains intervention by the United States. If Cuba is any example, the successful establishment of additional Communist governments would result in further internal revolution and increased strains on the Russian budget.

How to Determine the Losers of Future Wars

Imagine a major league sport in which the teams practiced, developed players and exotic training methods, but never played a game. How do you predict which team would win the championship? This is the problem with predicting military success. The solution is the same as in preseason sports: Look at the statistics of the teams and individual players. This speculation and handicapping can become a very popular indoor sport during international tension. If approached with a degree of precision, some common sense and a systematic method, such predictions can even be accurate. The following techniques have been used successfully in the past by intelligence analysts and wargame designers to forecast the performance of armies past and present. The major caveat is not to become mesmerized by the numbers. Counting resources and computing odds will take you only so far. The following procedure will advance you a little farther.

1. Choose your nations and determine what is being fought over. The victory conditions of either side are not always obvious. Sorting out all these motives will appear as more of a political than a military exercise. But then, military activity is alleged to be an extension of politics.

2. Choose your mode of combat. Will it be naval or land? Which side will be attacking?

3. Look at the combat values on the charts for each nation. The objective of the attacker is to obtain a greater than 1-to-1 ratio of his strength to the defender's strength. A ratio of 6+ to 1 assures an almost instant victory. Anything below 1 to 1 means almost certain failure of the attack. But doing that simple calculation, the armchair strategist must figure the probable effects of geography, surprise and human factors.

4. Climate and geography. Some terrains favor the defense. Severe terrain conditions will as much as double the value of the defenders. Add severe climatic conditions and the defender's value will be tripled. The attacker would be handicapped by invading Switzerland or Afghanistan in winter, the urban sprawl that covers most of West Germany or a jungle or other thick forest.

5. Surprise will benefit the attacker. The highest degree of surprise will multiply the attacker's strength five times. This is rarely attained; the Japanese came close in 1941. A more likely degree of surprise will multiply the attacker's strength two or three times. At the start of a war, the side that opens hostilities will usually obtain some surprise advantage, at least 10 percent to 50 percent

Basically surprise means attacking defending forces before they are prepared to resist. Examples are air attacks that destroy enemy aircraft on their airfields. Ground forces require at least a few hours to organize for combat. After that, the various smaller units—battalions, brigades—must march off to join similar-sized units. By interrupting this mobilization with air attacks, the already mobilized attacking forces will be able to destroy the defender piecemeal.

Surprise is also possible through unexpected weapons or tactics. This is technological surprise. An example is the German *blitzkrieg* during 1939–1941. During the 1973 Arab-Israeli War, the Israeli Air Force was surprised by the effectiveness of the Arab air defenses. In a future NATO-Warsaw Pact war in Europe, NATO forces are liable to be "surprised" by massive Russian chemical weapons. On the other hand, the Russians may be surprised by the technological superiority of NATO electronic weapons. Actually, both sides may be surprised by the effects of chemical and electronic weapons. It could be World War I all over again. In 1914 all armies were surprised by the lethality of machine guns and rapid-fire artillery. A four-year, blood-drenched stalemate resulted. But then, neither side had nuclear weapons, the ultimate technological surprise.

6. Nations at war do not, and often cannot, commit the whole of their armed strength against the enemy. There are often other threats, internal as well as external. It is also prudent to retain some strength as a reserve to exploit unexpected success or to recover from an unforeseen disaster. Also, armed forces are normally stationed throughout a nation. Thus, no more than 40 percent to 70 percent of a nation's armed forces will be available all at once for an attack. For defending a nation, a higher proportion of the armed forces can be gathered, up to 90 percent. An additional burden on the attacker is the need to police the overrun populations. Paramilitary forces or disloyal locals are often used. At best, 200 combat troops are needed per million occupied enemy population. At worst, with a resisting population, 2000 or more troops are needed per million.

7. The military value of small armies, particularly those of Third World countries, can drastically increase or decrease with an exceptionally effective or inept leader. Any national armed force of less than 100,000 men is prone to this syndrome. In larger armies one man cannot exert as decisive an influence simply because the organization is too large. The force multiplier on the armed forces chart reflects this leadership value. In small armed forces one man (or a small, cohesive group) can raise this value to .9 or lower it to .1. The result can be devastating.

8. Time is the defender's strongest ally. If the attacker doesn't win quickly, the defender has time to learn from his mistakes. The attacker will learn also, but the defender has more incentives. Most wars without a quick win bog down in attrition. World Wars I and II, Korea, Algeria, Vietnam followed this pattern. Victory went to the bigger, and usually more persistent, battalions. In some cases time simply works against a nation that mobilizes too much of its economic manpower and resources into the combat forces. The economies of Israel, Sweden and Switzerland could not support a major war for more than a few months without external aid.

EXAMPLES

The Arab-Israeli War of 1973 found Israel weaker and Egypt stronger than they are today. Still, Israel had a value of 200, Egypt about 75 and Syria at 20.

With the advantage of surprise attack, Israel should have been able to repeat its lightning six-day victory of 1967. A preemptive surprise attack would have raised Israel's effective strength to 500 to 600. But the Arabs attacked first. Not only did they achieve the advantage of surprise, but they caught the Israeli armed forces unmobilized. Like the Swiss and Swedish forces, the peacetime Israeli strength is only 30 percent of its mobilized combat value. This gave the Arabs a 320 to 150, or more than a two-to-one advantage, not enough for an immediate victory, but enough to get across the Canal and deep into the Golan Heights. Because Israeli mobilization was not disrupted, the force ratio changed quickly. After a week of heavy fighting, it was three to one in Israel's favor. With no further Arab reserves to call on, except Russian paratroopers, the war soon ended.

If replaying the past is interesting, looking into the future is compelling. Take the current situation in Europe. This military standoff currently consumes over $300 billion a year and ties up nearly 6 million troops. When the military speaks nervously of the Next War, this is it, the Big One.

The West tends to think only in terms of Russia attacking Europe. The Russians think in terms of themselves being attacked. As will be demonstrated, there is a case for both viewpoints.

A NATO worst case would be a surprise attack by Russia's European forces and Warsaw Pact allies on Germany. The combat value of Russia's land forces is 676. Let's assume 500 could be quickly committed to the attack and they obtained a 400 percent surprise bonus, producing an effective combat value of 2000. Defending would be NATO forces with a value of 600. This gives Russia a 3.3-to-1 ratio. If French forces are included, the ratio declines to 2.6 to 1. These ratios make a quick victory possible but not certain. The enormous naval advantage held by NATO would serve primarily to redress the imbalance of forces in Norway. In the Balkans, Russian forces would obtain a nearly 6-to-1 ratio over Turkey while opposition in Iran, Iraq and the Persian Gulf would be largely insignificant.

A large number of variables are rattling around inside this scenario. The chief unknown is the political and military reliability of the smaller allies on both sides. The situation is worst in Eastern Europe, where Russia has had to threaten or invade its erstwhile allies in order to maintain the alliance. In the West the smaller allies are more reliable because their compliance is voluntary, but as each goes its own way on key defense policies, there will be less effective military cooperation during combat.

From the Russian point of view, the worst case is a collapse of their Eastern European hegemony, followed by a NATO invasion. This leaves Russian forces without their allies as well as the portion of their own forces to cope with the ex-allies. The net Russian forces available for defense might have a combat value of only 300. NATO could muster a value of 500 for an attack. With surprise, this could be at least doubled, for a forces ratio of 3.3 to 1. This is a worrisome prospect, especially exacerbated by the political tensions within Russia. The Soviet Union is, after all, an involuntary confederation in which the Russians are a minority. From the Russian point of view, a NATO invasion is not absurd at all. Particularly threatening is the specter of East and West Germany swept by reunification fever. This, in addition to Polish labor unions and liberal Communist governments, keeps the Russian arms factories going far into the night.

The bright spot in all this is the unlikelihood of the worst case occurring for either side. Without a worst-case situation, neither side has a great enough prospect of success to attack.

Armed Forces of the World: Naval Forces

This chart shows the world's 40 most powerful fleets. As you can see, fleet numbers 41 and beyond are decidedly small change. The top 40 represent 98 percent of the world's naval power. The fleets are ranked in order of combat value.

NATION is the nation of the ships displayed. The figures include coast guard ships if they have a war time combat capability. Amphibious shipping is covered in the chapter on amphibious operations.

COMBAT VALUE is the numerical combat value of the nation's fleet. This value reflects the overall quantity and quality of ships and crew when used only for naval combat. Included is the effectiveness of support and the fleet's system of bases. Aside from the known quantities of ship numbers, tonnages and manpower, less firm data on quality have been taken in consideration. To put it more crudely, it comes down to who is more capable of doing what they say they can do. The quality factor was derived from historical experience, a less-than-perfect guide.

% OF TOTAL is the percentage of the world's total combat value each fleet represents.

1000 TONS is the weight of the nation's fleet in thousands of tons' full-load displacement.

SHIPS is the total number of ships.

MANPOWER is the number of men in the navy, in thousands. This includes naval air power and support.

AVG WEIGHT is the average weight of the fleet's ships, in thousands of tons. This indicates whether the navy goes in for large or small ships.

SHIP TYPES are the number of each ship type the nation possesses. CV are carriers using fixed-wing aircraft. SSBN are ballistic-missile submarines (including the few still powered by nonnuclear fuel). SSN are nuclear attack subs, including those equipped with cruise missiles. SS are nonnuclear-powered attack subs. CA/CG are cruisers, here defined as surface warships of over 5000 tons' displacement. DD/DDG are destroyer-class ships (3000 to 5000 tons). FF/FFG are frigate-class ships (1000 to 3000 tons). COMBAT PATROL are combat ships under 1000 tons armed with missiles or torpedoes. MINE OPS are mine-warfare operations ships. AIRCRAFT are naval, fixed-wing combat aircraft.

CLASS TONNAGE AS A PERCENT OF TOTAL TONNAGE is the percentage of total weight of each class of ships.

THEATER BALANCES are the totals for sea areas where naval combat would take place in wartime.

ARMED FORCES OF THE WORLD: NAVAL FORCES

The United States alone possesses nearly half the world's effective naval power. The "Western" nations together possess over 70%.

Nation	Combat Value	% of Total	1000 Tons	Ships	Man-power	Avg Weight	Ship Types CV	SSBN	SSN	SS	CA CG	DD DDG	FF FFG	Combat Patrol	Mine Ops.	Air-craft
United States	2677	50.59	2974	397	530	7.49	14	33	93	5	70	153	25	4	3	2200
Russia	1120	21.17	2489	743	440	3.35	3	85	119	94	47	99	166	130	375	1050
Britain	287	5.43	359	91	67	3.95	2	4	12	16	6	51			38	44
France	242	4.58	356	106	70	3.36	2	6	1	21	7	29	26	14	22	146
Japan	112	2.12	139	68	45	2.04				15	2	7	39	5	46	130
Netherlands	75	1.43	105	75	16	1.40				6	2	16	12	39		22
China	71	1.34	185	562	322	0.33		1	2	85		7	22	445	370	750
Italy	70	1.32	116	56	40	2.07				12	3	10	22	9	42	36
West Germany	70	1.32	98	100	37	0.98				26		13	11	50	58	160
India	57	1.07	121	50	47	2.42	1			8	1	4	28	8	13	40
Taiwan	54	1.02	106	45	35	2.36				2		22	13	8	17	27
Spain	46	0.87	111	41	36	2.71	1			10		12	20	6	15	13
Canada	46	0.86	80	26	6	3.08				3		12	11			42
Australia	42	0.80	71	19	17	3.74	1			6		6	6		3	38
Brazil	28	0.54	79	28	32	2.82	1			8		19			6	28
Turkey	27	0.51	75	53	40	1.42				16		12	2	23	40	22
Sweden	27	0.51	42	57	10	0.74				14		2	6	35	80	
Peru	26	0.49	73	27	19	2.70				10	3	4	4	6		11
Greece	22	0.41	53	51	19	1.04				11		8	4	28	16	12
South Korea	20	0.39	53	29	25	1.83						10	10	9	8	19
Argentina	20	0.37	60	18	26	3.33	1			4		10	3		6	55
Pakistan	16	0.29	35	18	13	1.92				6	1	3	4	4	6	5
Norway	16	0.29	23	70	9	0.33				15			8	47	13	7
Denmark	15	0.29	21	34	6	0.63				6			12	16	15	
Portugal	14	0.26	31	20	11	1.55				3			17		4	

North Korea	13	0.24	48	137	30	0.35				18			4	115	50	
East Germany	11	0.20	21	94	17	0.22							4	90		5
Israel	10	0.19	12	30	9	0.40				3			1	26	10	25
South Africa	9	0.17	14	14	6	1.00				3			3	8		18
Philippines	8	0.16	28	21	19	1.33						4	21	52	12	
Egypt	7	0.13	31	68	19	0.46				8		3	4	6	5	
Iran	6	0.11	21	17	9	1.23							8	35	46	2
Yugoslavia	6	0.11	15	44	15	0.35				8		1	1	29	46	52
Poland	5	0.09	12	34	23	2.25				4					26	
Belgium	5	0.09	9	4	5	0.91							4	6	7	
Thailand	4	0.07	11	12	18	1.75							6			12
Mexico	3	0.07	21	12	16	0.66						2	10			14
Bulgaria	3	0.06	15	23	10	0.28				4			5	14	18	
Indonesia	2	0.03	6	23	40					4				9	4	24
Rumania	1	0.01	2	38	10	0.05							10	38	30	
Totals	5291	100	8122	3355	2164	2.42	26	129	227	454	142	519	552	1314	1450	5009
Class Tonnage as Percent of Total Tonnage							17	11	12	8	16	23	10	2		

Theater Balances

Total West	3862	73	4831	1348	1060	3.58	20	43	106	176	90	361	270	290	373	2953
Total East	990	19	2587	1069	530	2.42	3	85	119	120	47	100	179	416	519	1102
All Other	439	8	704	938	574	0.75	3	1	2	158	5	58	103	608	558	954
Atlantic West	1916	36	2307	565	436	4.08	11	24	59	65	47	173	84	108	113	1306
Atlantic East	520	10	1352	291	239	4.65	2	57	87	-7	26	47	48	31	70	275
Pacific West	1281	24	1578	358	363	4.41	6	13	35	26	29	109	110	30	82	1101
Pacific East	268	5	711	351	147	2.02	1	28	31	54	7	28	45	157	95	300
Med. West	594	11	847	311	225	2.72	3	7	12	58	15	69	55	96	117	419
Med. East	124	2	330	186	75	1.77	3	0	0	30	13	17	44	82	118	245
Baltic West	71	1	100	114	36	0.88	0	0	0	27	0	10	21	56	61	128
Baltic East	78	1	195	241	69	0.81	0	0	1	43	1	8	42	146	236	282

TOTAL WEST are all the navies allied with the United States: all the NATO nations (including Spain, Israel and France) as well as the Pacific alliances (Australia, New Zealand, the Philippines, Japan, Thailand, South Korea). In a war not all of these countries could be depended on. If the war involved attacks on merchant shipping, all the above nations could be expected to join the United States in the defense of their mutual trade routes. Alliance of necessity could also bring in many neutral nations.

TOTAL EAST are all the navies allied with Russia: the Warsaw Pact nations plus North Korea. In a war not all these allies could be depended on. It is also possible that other allies, particularly currently nonaligned nations, could be gained, depending on the situation.

ALL OTHER is the totals for the fleets not falling under the above two categories.

An immediately obvious pattern is the tremendous superiority of the Western navies. In combat strength and aircraft it is over three to one. In ship tonnage it is almost two to one. In manpower it is over two to one. Only in numbers of ships does the East approach parity. If SSBN's and coastal patrol craft are subtracted from both sides, the West again has a two to one margin. The situation appears more difficult for the East when one considers the individual theaters.

ATLANTIC includes primarily the Britain-Iceland-Greenland (BIG) gap through which all Russian ships must travel to reach the high seas, the mid-Atlantic shipping lanes, the North Sea, and the South African shipping lanes. The Russian fleet is the only eastern force in this theater, and it is heavily outnumbered. The main Russian base in the Arctic is easily defended and just as easily sealed off from the Atlantic. Should the Russians capture Norway and Denmark, they would still have to run a gauntlet of Western naval and air power to get their larger ships down south. Smaller ships like destroyers can be moved through the inland waterway system, although not in winter.
Western nations are primarily concerned with the security of their seaborne trade. They would be quite content if the Russians adopted a defensive attitude. Should the Russians go after Western shipping, Russian naval power will be reduced by half or more because of the long and dangerous voyages. Western naval power would be less affected, as their bases in Britain and the Continent are much closer to the shipping lanes. Western forces in the Atlantic consist of half the U.S., French and Spanish fleets, plus all the British, Belgian, Dutch, Portuguese and Norwegian forces as well as 20 percent of West Germany's ships. The United States' home water deployments are included in the Atlantic and Pacific figures, as these areas are where the trade routes start.

PACIFIC is a much larger area than the Atlantic. It consists of the Indian Ocean to the west, the Australian-Indonesian-Malaysian (AIM) gap, the western Pacific trade routes and the eastern Pacific trade routes.
The most critical area is the AIM gap, as this blocks passage to the Persian Gulf oil fields. This gap and Japan are the key to Western control of the Pacific. Russia is vastly outgunned here. Local navies are more than a match for the Russian Far East fleet.

MED. is the Mediterranean and the Black Sea. Because so many major Western nations border this area, Western naval superiority is massive. Assuming half the French and Spanish fleets, plus the whole of the Greek, Italian and Israeli forces, the Russians are already outnumbered two to one. United States strength alone outmatches the Russian by two to one. The United States deploys 12 percent of its strength in this area, as opposed to 50 percent in the Atlantic and the remainder in the Pacific.

Russia gets the Black Sea by default, with the help of Rumania and Bulgaria. Some Russian strength must be quickly thrown against Turkey, for as long as the Turks hold the Dardanelles, the bulk of the Russian forces cannot enter the Mediterranean. Russia's forces in the Med will probably be destroyed before reinforcements can be moved from the Black Sea.

BALTIC is one area where Russia has parity, helped by significant East German and Polish forces. West Germany and Denmark have their hands full. Even so, Germany devotes 20 percent of its forces to the North Sea. The Russian situation is not very attractive. They must run a gauntlet of Western ships, aircraft and mines as they attempt to open the Danish straits. The roles are clearly defined. The Russians must attack an equal force possessing excellent defensive positions. Most likely, Western naval forces would lose the Baltic through the loss of its coast, and friendly bases, to enemy ground forces.

Even if one assumes that the Russian fleet is as powerful as that of the United States, it is not in a position to sweep the seas. The historical record of its naval exploits does not support an assumption that it will perform well. In contrast to Germany in World War II, when small numbers of efficient submarines sank over 2000 ships, the Russians do not have allies similar to the powerful World War II Italian and Japanese fleets. Unless the Russians can start a major war in stages as the Germans did and pick off significant naval forces like the French, Dutch and Belgians, they will be outnumbered and outclassed from the beginning. The only thing that will defeat the West at sea is complacency or bankruptcy from building too many ships.

SOURCES AND SUGGESTED READINGS

Periodicals

Periodicals are a good source of current events and documents. The world defense industry is a $700-billion-a-year business with a wide range of "consumer"-oriented periodicals. Security considerations notwithstanding, it is difficult to prevent salesmen from touting their wares and capabilities.

The easiest periodicals to obtain are *Army* (1529 Pennsylvania Avenue, N.W., Washington, D.C. 20036), *Air Force Magazine* (1750 Pennsylvania Avenue, N.W., Washington, D.C. 20006) and *Naval Institute Proceedings* (Annapolis, Md. 21402). These cover current and historical experience within their respective branches. For more specialization there is, *The Marine Corps Gazette* (Box 1775, Quantico, Va. 22134), *Military Review* (Fort Leavenworth, Kans. 66027), *Infantry* (Fort Benning, Ga. 31905), *Armor* (Fort Knox, Ky. 40021), *Field Artillery Journal* (Fort Sill, Okla.).

Similar British journals are available. Information can be obtained through the British Embassy. Other major nations also have such periodicals; contact the appropriate embassy or write directly to that country's ministry of defense.

Some periodicals have well-deserved reputations for regularly divulging classified information. These are largely trade journals available to the general public. The ads are also interesting. ("Might is right, power is also a question of possessing highly operational military avionics," reads one ad in *Aviation Week.)*

The following publications often have very technical descriptions of weapons and equipment. Also reported are user evaluations, obtained either through testing or combat experience. Many large libraries subscribe to these fairly expensive publications. In addition to the periodicals mentioned below, there are hundreds of others. These are often mentioned or advertised in the following magazines.

Aviation Week & Space Technology also covers land and naval weapons developments (Box 503, Hightstown, N.J. 08520).

International Defense Review really gets down to details (8 6th Avenue, Louis Casai, Box 1621216, Cointrin, Geneva, Switzerland).

Ground Defense, Aviation & Marine and *Military Electronics* (Skybooks, 48 East 50th Street, N.Y., N.Y. 10022).

Military Technology is an English-language European publication. Much like *International Defense Review* (German Language Publications, 560 Sylvan Avenue, Englewood Cliffs, N.J. 07632).

Books

One primary source is military training manuals. Increasingly these are a fertile source because of their detailed descriptions of "enemy" weapons. Russian official publications are a great source for information on Western weapons and capabilities. The same has been true of Western manuals on Russian arms. These publications can be obtained from the U.S. Government Printing Office. If resistance is encountered, invoke the Freedom of Information Act. Russian manuals can be obtained from Russian book publishers. It's not impossible, just difficult. Translations are sometimes published by agencies of the United States government.

Another primary source of information in the government is the numerous studies done for the Defense Department by hundreds of think tanks and consultants. These studies are often deposited in the Library of Congress or in military libraries open to the public. The best is the U.S. Army Historical Collection at Carlisle Barracks, Pa. Finding these reports is often an adventure in itself; they are usually *most* interesting.

Handbooks are often long on illustrations and short on hard data. The authors of these books are usually quite knowledgeable and frequently slip in copious quantities of technical data, not quite comprehensible but interesting to look at. Britain's Salamander Books has produced a notable collection, published in the United States by Crescent, Simon & Schuster, St. Martins Press and others. Another notable British publisher is Arms & Armor Press, which publishes new material every year, including English translations obtained from Continental publishers. In the States, the Naval Institute Press puts out an impressive list of naval titles. Arco and Squadron/Signal publish technical titles on aircraft. Because this is contemporary material, get your hands on the latest offerings. Your local bookstore should be able to help.

Annuals have been around for most of this century and began with *Janes Fighting Ships.* The annuals are handbooks that are updated each year. They use everything they can dig up for sources. They are expensive. The various *Janes* volumes are $125 each. In addition to the *Janes* books on ships, aircraft and ground weapons, the British International Institute for Strategic Studies *(IISS)* publishes *The Military Balance,* also often consulted and quoted. These volumes are not the most accurate. They depend, for the most part, on open sources. The general press, and any data released by the military, comprise the vast majority of their input. Anyone who plows the same ground will notice that both *Janes* and *IISS* often miss many minor, and occasionally major, details. Keep this in mind if you come across something which contradicts the two bibles. In fact, if you write to them with your discovery, you will quite likely see it incorporated into the next year's edition. Both sources encourage contributions from readers.

An up-and-coming annual is the Stockholm International Peace Research Institute's (SIPRI) *World Armaments and Disarmament Yearbook.* Investing somewhat more diligence in their work than *Janes* or *IISS,* SIPRI covers some details better and fills in many gaps. Unfortunately, they have not been as regular as the earlier two annuals.

In addition to their annuals, all three of the above organizations publish separate studies.

Other annuals are updated every two, three or more years. For example, *Weyer's Warships of the World,* Trevor Dupuy's *Almanac of World Military Power,* and John Keegan's *World Armies.*

Governments publish extensive documents on their military activities; budgets and legislative proceedings are particularly helpful. Bibliographies are also published.

Books on modern military affairs tend to be either handbooks of specific weapons and equipment or books on grand strategy or defense policy. No one seems interested in how to make war. That's either a good sign or an indication of moral cowardice in avoiding the consequences of one's actions.

Two recent books on "World War III" are one extreme of this genre, being fictional in approach with the addition of large quantities of technical information. At the other extreme are scholarly works that go out of print rather quickly. Go through *Books In Print* to see what is currently available.

One work that received wide distribution was John M. Collins's *U.S.–Soviet Military Balance 1960–1980*. This was a U.S. Congress report published by McGraw-Hill. It reads like a government report because that is what it is. There is a gold mine of accurate data in there.

Another excellent book is Trevor Dupuy's *Numbers, Predictions and War*. This volume sets forth a historical analysis of combat. Having spent many pleasant hours discussing the subject with the author, I can vouch for the thoroughness and scholarship of this work. As you can see from sections of this book, and much of my previous work, I concur with Dupuy's analysis and methodology.

People in the Business

While most of these folk are properly security conscious, once you get to know them and talk shop, the line between what is *secret* and what is not is crossed unknowingly and frequently. Having had a security clearance and access to state secrets years ago, I've witnessed this on both sides of the fence. I remember one evening of idle conversation with another teenage soldier when I was made privy to the inner workings of America's most recent tactical nuclear warhead. The primary topic of the conversation was a hypothetical science fiction story on how a terrorist could destroy the Empire State Building.

INDEX

About the Author

JAMES F. DUNNIGAN is the author of *The Complete Wargames Handbook* (Morrow, 1980) and the founder of Simulations Publications, the world's most prolific publisher of historical simulations. Defense establishments worldwide use the more than one hundred simulations he has designed. He lectures and is a consultant for the Department of Defense, the Institute for Defense Analysis, the Central Intelligence Agency, West Point, the Naval Post Graduate School, the U.S. Marine Corps, and the Army War College. Dunnigan lives in New York City.